Information Processing in the Visual Systems of Arthropods

Symposium Held at the Department of Zoology
University of Zurich, March 6–9, 1972

Edited by
Rüdiger Wehner

With 263 Figures

Springer-Verlag
Berlin Heidelberg New York 1972

The cover shows a scanning electron micrograph of the compound eye of a desert ant, *Cataglyphis bicolor* (University of Zurich).

ISBN 3-540-06020-0 Springer-Verlag Berlin Heidelberg New York
ISBN 0-387-06020-0 Springer-Verlag New York Heidelberg Berlin

Preface

It is now generally accepted for a variety of reasons – morphological as well as physiological – that the visual systems of arthropods provide a suitable model for the study of information processing in neuronal networks. Unlike the neurophysiology of the visual pathway in the frog and the cat which is more than adequately documented, recent work on the compound eye and optical ganglia of spiders, crustaceans, and insects has scarcely been summarized. In order to fill this void so that others, especially vertebrate neurophysiologists may become familiar with the advantages of these systems, our group at Zurich University organized here in March 1972, a European meeting to discuss the anatomical, neurophysiological and behavioral knowledge on the compound eye and the visual pathway of arthropods. Systems analysis was regarded as the main theme of the conference, but systems analysis of a network of neurons cannot be done as a mere "black-box" maneuver. The conference therefore tried to reconcile neurophysiology and behavioral analysis in order to make predictions about a necessary and sufficient neural structure. The "wiring diagrams" of such a structure might then be confirmed histologically. Hence the aim of the conference was not to deal only with the structure and function of the compound eye – i.e. data processing in the dioptric apparatus and in the layer of the retinular cells – but to apply this knowledge to the performance of the visual system in color discrimination, pattern recognition, and other central nervous processes.

Although the official language of the symposium was German, the contributors wrote their manuscripts in English, also reflecting the extensive discussions that arose both during and after the formal sessions. We thus hope to make any insights derived from the discussions available to those unable to attend.

Besides the authors, we would like to thank all the 77 participants of the symposium for their lively contributions both inside the lecture hall and outside. The discussions allowed the contributors to cross-check recent knowledge and made it clear that the visual systems of arthropods have much to contribute to neurophysiology. Much enthusiasm was generated for planning the further work along these lines. I wish most of all to express our appreciation to the Swiss National Foundation for Scientific Research, which provided generous financial support for the meeting, and to Springer-Verlag for the very rapid publication of the proceedings.

Zurich, June 1972   Rüdiger Wehner

Contents

VI.

PATTERN RECOGNITION

List of Participants

ANDERSON, A.M., University of Edinburgh, Dept. of Zoology, West Mains Rd., GB-Edinburgh,
    EH9 3JT
BAESSLER, U., Universität Trier-Kaiserslautern, Naturwiss.-Techn. Fakultät, D-675 Kaisers-
    lautern, Postfach 1019
BAUMANN, F., Université de Genève, Dépt. de Physiologie, CH-1211 Genève 4
BOSCHEK, C.B., Max-Planck-Institut für biologische Kybernetik, D-74 Tübingen, Spemannstr.38
BRAITENBERG, V., Max-Planck-Institut für biologische Kybernetik, D-74 Tübingen, Spemann-
    strasse 38
BRUNNERT, R., Zoologisches Institut der Universität Zürich, CH-8006 Zürich, Künstlergasse 16
BURKHALTER, R., Zoologisches Institut der Universität Zürich, CH-8006 Zürich, Künstler-
    gasse 16
BURKHARDT, D., Universität Regensburg, Fachbereich Biologie, D-8400 Regensburg, Universitäts-
    strasse 31
BUTENANDT, E., Max-Planck-Institut für Verhaltensphysiologie, D-8131 Seewiesen/Obb.
CAMPOS-ORTEGA, J.A., Max-Planck-Institut für biologische Kybernetik, D-74 Tübingen,
    Spemannstr. 38
CRUSE, H., Max-Planck-Institut für biologische Kybernetik, D-74 Tübingen, Spemannstr. 38
DORNFELD, K., II. Zoologisches Institut der Freien Universität, D-1 Berlin-41, Grunewaldstr.34
DUELLI, P., Zoologisches Institut der Universität Zürich, CH-8006 Zürich, Künstlergasse 16
EHEIM, W.P., Zoologisches Institut der Universität Zürich, CH-8006 Zürich, Künstlergasse 16
ERBER, J., Zoologisches Institut der Technischen Hochschule, D-61 Darmstadt, Schnittspahnstr.3
FLATT, E., Zoologisches Institut der Universität Zürich, CH-8006 Zürich, Künstlergasse 16
FLEISSNER, G., Zoologisches Institut der Universität, D-6 Frankfurt a.M., Siesmayerstr. 70
FRANCESCHINI, N., Max-Planck-Institut für biologische Kybernetik, D-74 Tübingen, Spemann-
    strasse 38
FRISCHKNECHT, M., Zoologisches Institut der Universität Zürich, CH-8006 Zürich, Künstler-
    gasse 16
GEIGER, G., Max-Planck-Institut für biologische Kybernetik, D-74 Tübingen, Spemannstr. 38
GOERNER, P., II. Zoologisches Institut der Freien Universität, D-1 Berlin 41, Grunewaldstr. 34
GOETZ, K.G., Max-Planck-Institut für biologische Kybernetik, D-74 Tübingen, Spemannstr. 38
HAMDORF, K., Ruhr-Universität Bochum, Lehrstuhl für Tierphysiologie, Gebäude ND, 5.O.G.,
    D-463 Bochum
HAUSEN, H., Max-Planck-Institut für biologische Kybernetik, D-74 Tübingen, Spemannstr. 38
HEISENBERG, M., Max-Planck-Institut für biologische Kybernetik, D-74 Tübingen, Spemann-
    strasse 38
HELVERSEN, O.v., Biologisches Institut I (Zoologie) der Albert-Ludwigs-Universität, D-78 Frei-
    burg i.Br., Katharinenstr. 20
HENGSTENBERG, R., Max-Planck-Institut für biologische Kybernetik, D-74 Tübingen, Spemann-
    strasse 38
HERRLING, P.L., Zoologisches Institut der Universität Zürich, CH-8006 Zürich, Künstlergasse 16
HUBER, M., Zoologisches Institut der Universität Zürich, CH-8006 Zürich, Künstlergasse 16
JAERVILEHTO, M., Zoologisches Institut der Universität München, D-8 München 2, Luisenstr. 14
KAISER, W., Zoologisches Institut der Technischen Hochschule, D-61 Darmstädt, Schnittspahn-
    strasse 3

KIEPENHEUER, J., Universität Frankfurt, Zoologisches Institut, D-6 Frankfurt a.M., Siesmayer-
strasse 70

KIRSCHFELD, K., Max-Planck-Institut für biologische Kybernetik, D-74 Tübingen, Spemannstr. 38

KLEIN, R., Zoologisches Institut der Universität Zürich, CH-8006 Zürich, Künstlergasse 16

KRETZ, R., Zoologisches Institut der Universität Zürich, CH-8006 Zürich, Künstlergasse 16

KUNZE, P., Max-Planck-Institut für biologische Kybernetik, D-74 Tübingen, Spemannstr. 38

KURTH, B., II. Zoologisches Institut der Freien Universität, D-1 Berlin-41, Grunewaldstr. 34

LABHART, T., Zoologisches Institut der Universität Zürich, CH-8006 Zürich, Künstlergasse 16

LAND, M., University of Sussex, Biology Bldg., Falmer, Brighton, Suss. BN1 9QG, Great Britain

LANGER, H., Ruhr-Universität Bochum, Institut für Tierphysiologie, Gebäude JC-4/43,
D-463 Bochum, Postfach 2148

LISKE, E., Zoologisches Institut der Technischen Hochschule, D-61 Darmstadt, Schnittspahnstr. 3

MASUHR, T., Zoologisches Institut der Technischen Hochschule, D-61 Darmstadt, Schnittspahnstr. 3

MENZEL, R., Zoologisches Institut der Technischen Hochschule, D-61 Darmstadt, Schnittspahnstr. 3

MEYER, E., Zoologisches Institut der Universität Zürich, CH-8006 Zürich, Künstlergasse 16

MEYER, H., Zoologisches Institut der Universität, D-6 Frankfurt a.M., Siesmayerstr. 70

MITTELSTAEDT, H., Max-Planck-Institut für Verhaltensphysiologie, D-8131 Seewiesen/Obb.

MITTELSTAEDT, M., Max-Planck-Institut für Verhaltensphysiologie, D-8131 Seewiesen/Obb.

MONDADORI, C., Zoologisches Institut der Universität Zürich, CH-8006 Zürich, Künstlergasse 16

MOTTE de la, J., Universität Regensburg, Fachbereich Biologie, D-8400 Regensburg, Universitätsstr. 31

NOWAK, H., Zoologisches Institut der Universität Zürich, CH-8006 Zürich, Künstlergasse 16

PAULUS, H., I. Zoologisches Institut der Universität Wien, A-1010 Wien-1, Dr. Karl Lueger-
Ring 1

PERRELET, A., Université de Genève, Dépt. Histologie, CH-1211-Genève 4, 20, rue de
l'Ecole de Médécine

PICK, H., Max-Planck-Institut für biologische Kybernetik, D-74 Tübingen, Spemannstr. 38

REICHARDT, W., Max-Planck-Institut für biologische Kybernetik, D-74 Tübingen, Spemannstr.38

REISSLAND, A., II. Zoologisches Institut der Freien Universität, D-1 Berlin-41, Grunewaldstr.34

RIBI, W., Zoologisches Institut der Universität Zürich, CH-8006 Zürich, Künstlergasse 16

ROTH, H., Zoologisches Institut der Technischen Hochschule, D-61 Darmstadt, Schnittspahnstr.3

ROTHENBACH, D., Zoologisches Institut der Universität Zürich, CH-8006 Zürich, Künstlergasse 16

SCHALLER, F., I. Zoologisches Institut der Universität Wien, A-1010 Wien-1, Dr. Karl Lueger-
Ring 1

SCHIFF-SERTORIO, H., Laboratoire d'Anatomie physique comparé, Université de Genève,
CH-1200 Genève, 4, rue Candolle

SCHNEIDER, V., Zoologisches Institut der Universität Zürich, CH-8006 Zürich, Künstlergasse 16

SCHNETTER, B., Zoologisches Institut der Universität, D-87 Würzburg, Röntgenring 10

SCHOENENBERGER, N., Laboratoire d'Anatomie physique comparé, Université de Genève,
CH-1200 Genève, 4, rue Candolle

SCHUEMPERLI, R., Zoologisches Institut der Universität Zürich, CH-8006 Zürich, Künstlergasse 16

SEGESSER v. BRUNEGG, S., Zoologisches Institut der Universität Zürich, CH-8006 Zürich,
Künstlergasse 16

SPATZ, Ch., Institut für Biologie III (Genetik und Molekularbiologie) Universität Freiburg,
D-78 Freiburg i.Br., Schänzlestr. 9-11

STRAUSFELD, N.J., Max-Planck-Institut für biologische Kybernetik, D-74 Tübingen, Spemann-
strasse 38

STRECK, P., Universität Regensburg, LS Biologie II, D-8400 Regensburg, Universitätsstr. 31

TOGGWEILER, R., Zoologisches Institut der Universität Zürich, CH-8006 Zürich, Künstlergasse 16

VIRSIK-STRAUSFELD, R., Max-Planck-Institut für biologische Kybernetik, D-74 Tübingen,
    Spemannstr. 38
VOGT, K., Max-Planck-Institut für biologische Kybernetik, D-74 Tübingen, Spemannstr. 38
WEHNER, R., Zoologisches Institut der Universität Zürich, CH-8006 Zürich, Künstlergasse 16
WEHRHAHN, Ch., Max-Planck-Institut für biologische Kybernetik, D-74 Tübingen, Spemannstr.38
WEILER, R., Zoologisches Institut der Universität Zürich, CH-8006 Zürich, Künstlergasse 16
WIEBKING, H., Springer-Verlag, D-69 Heidelberg-1, Neuenheimer Landstrasse 28-30
ZETTLER, F., Zoologisches Institut der Universität München, D-8 München-2, Luisenstr. 14
ZIMMERMANN, G., Max-Planck-Institut für biologische Kybernetik, D-74 Tübingen, Spe-
    mannstr. 38

Opening Remarks

It would be sending owls to Athens or – more appropriately – flies to Zurich if I were to begin by dilating upon the advantage we have in being able to use insects with their compound eyes as objects for neurophysiology. But if besides the wish and pleasure of welcoming you all here and of expressing my thanks to the Swiss National Fund and to all who have helped me in the preparation – if, besides these pleasant duties, I should also like to address a few introductory words to you, it is really for one reason only: to explain why we wish to consider the work of the whole visual system, and yet according to the program we shall be hearing papers on very specialized topics in optics, electrophysiology and neuroanatomy. It would almost seem as though the propagated system analysis were merely words, a kind of wishful thinking not to be realized until the far distant future, serving merely to justify, and to form a superstructure for, our specialized work in a great variety of smaller fields.

I believe that after these four days you will be convinced that this is not the case – if you do not already know it. For this we have to thank in the first place our experimental objects, the insects, in particular bees, flies and ants. For (1) the electrophysiologically relevant size of their visual cells, (2) the spatial separation of the different levels of information processing along the optical path, was (3) the clearly repetitive arrangement of the various neuronal elements in the retina and in the peripheral optical ganglia and finally (4) the comparatively small number of their interneurons, which are about 4 power of ten less than in vertebrates – all these characteristics give insects definite advantages over vertebrates from the neurophysiological point of view. Moreover, the stereotype repertoire of behavioural elements, the advantages of which reach into learning physiology, allows quantifiable measurements in behavioural work with a precision scarcely to be achieved with vertebrates. This behaviour repertoire also distinguishes the arthropods in comparison with such neurophysiologically admirable objects as Aplysia, where the behaviour spectrum is so narrow that there is scarcely anything to be correlated with neurophysiological findings.

These favourable starting conditions which insects thus offer for a joint procedure to both neurophysiologists and behaviour physiologists can, however, only come into effect when the questions posed by behavioural sciences are strictly correlated to those posed by neurophysiology. For what we need today in the framework of neurophysiology, and particularly in brain research, are not merely systems of conceptions arising from the observation of behaviour. To these belong such historical – because merely descriptive – conceptions as that of taxis, with all the classification scheme built up on it, – or, on closer inspection, also that of innate releasing mechanisms. What we need is a physiology of behaviour which – if it is to be a physiology at all – must be related to neurophysiology both in conception and method, not just at the discussion stage, but much earlier, in the program for procedure. Thus the formal cybernetic observations of the fifties, which aimed at purely functional models and were less interested in the analysis of the rule system which has been carried out biologically, may be said to have fulfilled their purpose. But above all, the times in which people investigating behaviour were more concerned with attempting to open the black boxes of their own world of ideas than those of the system in question seem to us to be definitely over.

It has been mainly with insects that the most impressive success so far has been achieved in making quantitative behavioural measurements within the framework of a field of questions marked out neurophysiologically. To take one example: not until there was a means of measuring the wing-buzzing of the male silkworm-moth was it possible to apply the electrophysiological and autoradiographical findings on olfaction in Bombyx mori to making statements on the quantitative side and thus also on the mechanism of the primary reactions on the sensory cell membrane. This particular work showed impressively that the sensitivity thresholds of the whole senso-neuro-motor system are only able to be examined by means of behavioural work, whereas electrophysiological measurements only present the thresholds of the various transmission levels. In this case the threshold concentration in Bombyx as determined on the basis of the buzzing reaction lies about two powers of ten lower than that obtained by the electroantennogram.

To mention another example, which leads directly into the program of these days: functions showing direction sensitivity measured electrophysiologically on directionally selective units of the lobula of Diptera always show a much greater half-width than comparative functions obtained by behavioural tests. Mere electrophysiological deductions cannot do more than interpret single events on the various levels of data processing. Therefore neuroanatomical analyses - preferably linked simultaneously with electrophysiological measurements - are as indispensible as behavioural input-output analyses of the whole system.

With this in mind, in making up the program, we have chosen various achievements of the visual system (intensity-evaluating systems, colour-selecting systems, pattern-recognition systems) and we have tried to muster working groups acting with different methods on different working levels of the corresponding "subsystems". Looked at objectively, this should be a kind of inventory of results gained at different levels of the optical pathway, an inventory which should show up the empty spaces - i.e. the transmission links which are still missing - and which should make rational connections between the working groups helping to fill them up; that would be the finest and most tangible success of this symposium. Perhaps the discussions in these coming days will play their part in re-formulating the questions to be asked and in applying the work of part-disciplines, in which field we are all active, effectively to the analysis of the whole system. To understand a computer, for example, may mean to understand its input-output functions, i.e. to be able to set its program; to understand the way a subsystem (such as an adder) works; or it may mean to understand the mechanism of the elements such as a transistor. If we could at least come to an agreement as to the level on which we wish to "understand" in order to analyse the system in question, the symposium will have fulfilled that purpose which we should so much like to give it.

Zurich, March 6th 1972   Rüdiger Wehner

I. Anatomy of the Visual System

1. Periodic Structures and Structural Gradients in the Visual Ganglia of the Fly

V. Braitenberg
Max-Planck-Institute of Biological Cybernetics, Tübingen, Germany

Abstract.   The orderly projection of the visual space onto the visual ganglia and of the planes of the four visual ganglia onto each other is briefly reviewed. The variation of the size of some of the elements in the eye and of the ganglia is then described. This variation follows different gradients for different elements. The gradient of the size of the lenses in the cornea and that of the thickness of the L3 fiber in the lamina have a similar shape, with a maximum near the anterior border of the eye. The thickness of L1 and L2 follow a different rule. A correlation of these gradients with some variations of the efficiency of visual stimuli in different regions of the visual field leads to a tentative proposal for the role of the laminar neurons L1, L2 and L3 in perception.

This paper is organized in two sections. In the first, I shall collect information about the structures of the eye and the visual ganglia of the fly, insofar as they fit into the periodic scheme which makes the mapping of the various levels onto each other possible. In the second part, on the background of this periodic structure we shall discuss some striking quantitative variations which follow various gradients throughout the visual system. This second part is largely drawn from a paper written together with H. HAUSER, which is now in press.

1. Terminology

In gross outline the following are the subdivisions of the visual system which we shall discuss. The cornea (Co) with its 3'000 lenses on each side, together with the associated crystalline cones represents the camera optics of the compound eye. The rhabdomeres (Re) act as light guides from the focal plane of each lens on and absorb light in the visual pigments which they contain. Each rhabdomere belongs to one retinula cell. The retinula is the pattern formed by the tips of the rhabdomeres in the focal plane of each lens. The complex lens + crystalline cone + associated retinula cells + associated supporting and pigment cells is called an ommatidium. The visual ganglia are 3 on each side, the third one subdivided in two portions. The lamina (or lamina ganglionaris, La) is the first visual ganglion. The periodic units of its structure are called cartridges or neuroommatidia, 3'000 on each side. The medulla (Me) is the second ganglion. The lobuli form the third ganglion, with the two portions lobula (the anterior one, Lba) and lobulus (or lobula plate, Lbs). The external chiasma (Che) is the mass of fibers between lamina and medulla. The internal chiasma (Chi) connects medulla, lobulus and lobula.

A summary diagram may help to identify these structures (fig. 1).

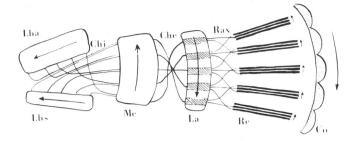

Fig. 1. Diagrammatic view of the right visual system of the fly, with 5, instead of 36 ommatidia in a row

2. The basic arrangement of elements in the lamina and in the medulla

All three ganglia of the visual system can be understood as neuropils of the "cortex" type. A certain pattern of connections along the thickness of the ganglion is repeated almost identically over the whole surface. A surface view (in reality: a reconstruction from tangential sections) reveals the repetitive arrangement of the input fiber bundles and of most of the internal elements of the ganglia. Horizontal rows of identical elements are easily recognized in the lamina and in the medulla, as well as in the external and internal chiasma. Two adjacent rows are shifted by a half period and their distance is such that neighbouring elements of two rows form fairly regular equilateral triangles (fig. 2a). The number of elements in each row varies between 36 and 37 in most instances in a large part of the eye, only in the upper and lower region the ganglia are rounded off and the rows are shorter. The number of rows varies between 100 and 110. Thus the surface of the ganglia is elongated, the long vertical axis being almost 3 times the short axis (lamina: 1000 μ by 360 μ, medulla: 700 μ by 270 μ).

3. Orientation of the optic axes of ommatidia

The visual field of each eye in front slightly overlaps (5° to 10° in common) that of the other eye. Whether they also overlap in the back is not known, but together they seem to cover almost the entire spherical visual field, only perhaps with a blind region corresponding to the body of the animal. How is the visual field of one eye, measuring about 180° by 180°, represented in the elongated lamina (and medulla)? The projection can obviously not be an isometric one, being strongly compressed along the horizontal coordinate of the ganglia. This can be understood by comparing fig. 2 a and b. The two patterns of dots, both "hexagonal", both having the same

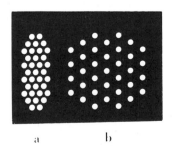

a   b

Fig. 2. Geometrical relation between the array of elements in the ganglia (a) and the array of the visual axes (b) in the visual field. The number of elements in reality is about 3'000

number of dots and the same number of horizontal rows, are related to each other by a horizontal stretching by a factor 3. The three axes of the hexagonal pattern do not have the same orientation in fig. 2 a and b. This is a fact which must be kept in mind when comparing the pattern of the elements in the ganglia to the pattern formed by the axes of the ommatidia intersecting a sphere around the animal. The two patterns are related to each other like fig. 2 a and fig. 2 b. There is a one to one correspondence between the ommatidial axes and the elements of the ganglia which is like the obvious correspondence of dots in fig. 2 a and b.

A hexagonal pattern cannot continuously cover a sphere. How are the rows of elements in the ganglia projected onto the visual sphere through the corresponding ommatidial axes?

The mapping of ommatidial axes over the entire visual sphere has yet to be completed (the endurance of the fly, rather than of the investigator, being the limiting factor). The following facts are fairly certain (BURKHARDT and BRAITENBERG, unpublished observations): (1) In an anterior sector of the visual field, rows of axes corresponding to horizontal rows (h in fig. 3) in the

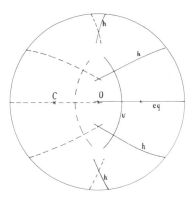

Fig. 3. Front view of a sphere onto which the projection of rows of ommatidial axes have been drawn

ganglion converge toward a center C located in the opposite half of the visual field; (2) horizontal rows of axes corresponding to the uppermost and lowermost parts of the ganglia are arranged in almost sagittal planes (almost vertically); (3) only at a level corresponding to the middle of the eye (probably corresponding to the equator in the ganglia, eq in fig. 3) is a horizontal row continuous between the right and left visual field; (4) vertical rows of elements in the ganglion correspond to rows of optic axes (v) which in the anterior part of the eye are arranged in fairly precise circles, with the centers at the point on the opposite side where the horizontal rows converge.

Fig. 3 illustrates these facts in a qualitative rather than precise way. It is to be understood as the front view of a sphere in whose center the fly is suspended and on whose surface the projections of some rows of ommatidial axes are drawn as lines. 0 is the intersection of the equatorial plane (see section on symmetry) with the medial sagittal plane. Whether 0 corresponds to the habitual direction of flight is not certain.

4. The visual field of individual ommatidia

The array of rhabdomeres in the focal plane of the optics of each ommatidium, the so-called retinula, is indicated in fig. 4 a for the four quadrants of the (right and left) eye. Thus, considering the inverting optics, an ommatidium in the upper half of the right eye has a visual field consisting of 7 points arranged as in fig. 4b. The set of points seen by one ommatidium partially overlaps the sets of points (the visual fields) of 20 neighbouring ommatidia. Each individual point of the discrete visual field is seen by 7 different ommatidia. A necessary condition for this is that both the period and the orientation of the pattern of points in the visual field of one ommatidium be the same as the period and the orientation of the pattern formed by the axes of the ommatidia, which is indeed the case (KIRSCHFELD, 1967).

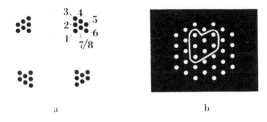

Fig. 4. a: Orientation of the retinulas in the four quadrants of the eye. b: The visual field of one ommatidium

5. Two systems of light-sensitive cells

The central rhabdomere, labelled 7/8 is different from the others in many respects. (a) It has a noticeably smaller diameter (BOSCHEK, 1971) at least at its peripheral end; (b) it is cylindrical, while the others are conical in shape (truncated cones with base pointing outward (BOSCHEK, 1971); (c) it is composed of parts of two separate cells, aligned to form a common light guide (TRUJILLO-CENOZ and MELAMED, 1966); (d) it is connected with axons which do not make synaptic contacts in the lamina, like the others, but rather in the medulla (TRUJILLO-CENOZ and MELAMED, 1966); (e) it is different in function, in the sense that it works under photopic conditions and is sensitive to the plane of polarization of the light, while the other six together form a skotopic but polarization insensitive system (KIRSCHFELD, 1969).

6. Neural projection of the rhabdomeres onto the lamina

The term "visual field of one ommatidium", used in a preceding section, is in a way misleading, since no operation or computation is performed by the ommatidium on the set of 7 points which constitute its visual field. On the contrary, the ommatidium (the lens and the associated rhabdomeres) just bundles the information derived from the seven points of the visual field in order to immediately redistribute it, at the neural level, onto the appropriate points of the lamina, in the sense of the correspondence of elements of the lamina and points in visual space schematically indicated by the correspondence of dots in fig. 2 a and b. This is done, for rhabdomeres 1 to 6, by means of a twisted bundle of fibers emerging from each ommatidium and distributing itself over the surface of the lamina (fig. 5). The region of the lamina which receives the fibers of one

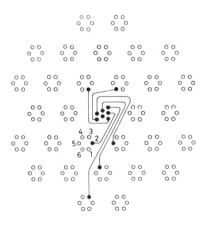

Fig. 5. Projection of one ommatidium onto the lamina (from BRAITENBERG, 1970)

ommatidium has the same shape and orientation as the visual field of that ommatidium, except compressed laterally for reasons connected with the geometrical relation of fig. 2 a and b. Each compartment of the lamina surface receives all the fibers derived from rhabdomeres of different ommatidia that look at the same point of visual space. These are 6 fibers derived from rhabdomeres 1 to 6 of neighbouring ommatidia, making synaptic contacts in that compartment, plus the two fibers derived from rhabdomere 7 (and 8) of the ommatidium corresponding to the position of the compartment. These have no synapses in the lamina.

7. Precision of wiring

The net formed by the fibers of the projection between the retina and the lamina is remarkable because every fiber in it can be explained on the basis of the simple principle described in the preceding paragraph. It is also remarkable for the faultless precision with which the principle is carried through. HORRIDGE and MEINERTZHAGEN (1970) followed 650 axons on serial sections and found none that did not enter the compartment predicted according to the wiring scheme that will make the projection of points of visual space onto the lamina isomorphic. In most instances also the order in which the fibers are arranged around the second order fibers in the lamina is kept constant, although this order is probably functionally quite irrelevant.

8. Mirror symmetry

The orientation of elements in the retinula as shown, for the four quadrants of the eye, in fig. 4 a is not the only example of a pattern mirror symmetrical with respect to a horizontal as well as the median-sagittal plane. The plane of mirror-reflection, which we may call equatorial plane, intersects the retina, the ganglia and the chiasmata, dividing each of these structures in two about equal parts. We have found mirror symmetry with respect to the equator, as well as between right and left, for (a) the pattern of the retinula; (b) for the pattern of the distribution of fibers of one ommatidium onto the lamina; (c) for the right- or left handedness of the 180° twist of the bundles of these fibers; (d) for the arrangement of all elements in the cartridges of the lamina; (e) for the right or left handedness of the twist of each layer of the external chiasma, which is the same as the handedness of the twist of the ommatidial bundles in the same quadrant, and (f) for the arrangement of the input fiber bundles in the periodic subunits of the medulla. In general, all structures in the whole visual system which obey the basic periodicity but do not have radial symmetry, obey the scheme of the two mirror symmetries with respect to the two perpendicular (equatorial and median-sagittal) planes.

9. Cartridges of the lamina

The most impressive structure present in each compartment of the lamina is the so-called cartridge. This is a bundle of afferent fibers from the lamina (the six retinula cell axons already mentioned), afferent fibers from the medulla (so called recurrent fibers) and of neurons originating in the lamina itself, with cell bodies located above the neuropil of the lamina and with terminations in the medulla. All synaptic traffic in the lamina presumably occurs within the cartridge. The cartridges are embedded in a matrix of cells, often called epithelial cells, which have some of the characteristics of glia cells but also have peculiar contacts with the axons of the retinula cells (the "capitate projections" of TRUJILLO-CENOZ), not seen anywhere else in the nervous system. Another type of cell, which may be glia, delimits the lamina downward, with processes extending half way up between the cartridges.

Fig. 6 shows schematically the cross-section of a cartridge.

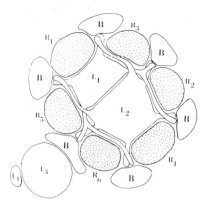

Fig. 6. Cross-section of a cartridge

The orientation is that of a cartridge of the upper half of the right eye seen from the outside. The numbering of the retinula fibers, R1 to R6, is the same as on figs. 4 and 5. The two fibers forming the core of the cartridge, L1 and L2, have processes extending into the spaces between the retinula fibers. Each of the fibers has a row or comb of about 20 such processes in each of the clefts between the R1 to R6 fibers. The synapses between the R- and the L-fibers are mainly located on these processes, the L-fibers being post-synaptic (TRUJILLO-CENOZ, 1969). Another laminar neuron, L3, shown outside the crown of the R-fibers in a constant position between R5 and R6, has processes above the level shown which also come in close contact with the six R-fibers, also by means of combs of processes, even if limited to at most the upper half of the cartridge.

Two more L-fibers, originating from cell-bodies above the lamina, L4 in close proximity to L3 and L5 (described by STRAUSFELD (1971) not shown on the diagram) have as yet obscure synaptic relations. L4 has collaterals reaching around the cartridge in its upper segment, and others forming a very regular network at the base of the lamina, with endfeet containing vesicles and therefore presumably communicating with neighbouring L1 and L2 fibers. L5 has very scanty processes within the lamina which are hardly comparable with the impressive dendritic apparatus of the first 4 L-fibers.

Also shown on the diagram, labelled B, are the cross-sections of fibers wedged in the interstices between the R-fibers from the outside. Each of the regions B is composed of at least two different components, already recognized by TRUJILLO-CENOZ (1965), with very impressive membrane thickenings between them, invaginated into one of them. The six branches of a recurrent fiber from the medulla which is seen in Golgi preparations to form a basket around the cartridge (CAJAL and SANCHEZ, 1915) are certainly part of the B complex.

10. Shape of L1, L2 and L3

I shall postpone to a later section a discussion of the variation of these fibers in different regions of the ganglion. At this point, we shall only consider the difference between them as they appear in Golgi preparations. Fig. 7 is to be understood as a caricature of the most salient features which can be recognized to a greater or lesser degree, in all cartridges independently of their position. L1 and L2 have dendritic branches all the way through the lamina, while L3 makes contacts with the afferent retinula fibers only in the upper third or at most half of the lamina. The bodies of the three fibers are markedly different in shape. L1 is thicker above, smoothly tapering downward, while L2 on the contrary becomes thicker toward the lower delimitation of the lamina. L3 has a swelling in the upper half of the lamina, rather abruptly going over into its thin segment. All three fibers are connected to cell bodies through thin necks.

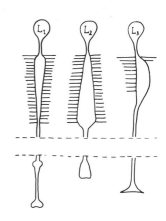

Fig. 7. Diagrammatic view of laminar
neurons L1, L2 and L3

In fig. 7, below are shown the terminations of these fibers in the medulla. They end at different
levels, L2 most superficially, L1 most deeply, however, with a swelling above the level of the
L2 termination. L3 has a termination at an intermediate level. The shape of the endfeet is also
different in the three fibers. These terminations have been described by CAJAL and SANCHEZ,
1915; the identification of the various types with fibers located in various positions in and around
the cartridge is due to TRUJILLO-CENOZ (1969) and in a more definitive form to CAMPOS-OR-
TEGA and STRAUSFELD (1972). The differences between these fibers are describable as different
relationships to different levels of the lamina. This would imply that the vertical dimension of the
lamina carries more meaning than just that of a parallel arrangement of many contacts between
the afferent retinula-cell fibers and the second order neurons. Dynamic aspects of the transmission
are probably important in this connection.

11. Peculiar characteristics of synaptic transmission in the lamina

It will be useful at this point to recall some recent results of microelectrode recording in the
lamina (SCHOLES, 1969; AUTRUM et al, 1970; ZETTLER and JAERVILEHTO, 1971; JAERVILEHTO
and ZETTLER, 1971). These authors recorded from retinula cell fibers as well as from laminar
L-neurons. A published photograph (JAERVILEHTO and ZETTLER, 1971) of an intracellularly
stained neuron from which they recorded shows it to be L1, but in other instances it may have
been L2 or L3, although no discrete types of electrical responses to light stimuli are evident. The
comparison of the potential deflections in first and second order neurons reveals an uncommon
type of transmission. No spikes are generated in the L-neuron and still the phenomenon transmitted, though graded, suffers no decrement along the axon. Apparently it is generated in the
L-fiber, at the arrival of the excitation in the R-fiber, by means of some kind of explosive process, since the peak of the excitation in the second order fibers is reached before the excitation
in the first order fiber arrives at the maximum amplitude. Also, the phenomenon in the second
order fibers has a higher amplitude and reaches saturation at levels at which the first order neuron
can still produce neatly graded potential deflections roughly proportionate to the logarithm of
the stimulus intensity. This information, however, is not lost in the second order fiber if the
rising slope of the signal there is the information carrier, since the slope, but not the amplitude,
increases all the way through the range tested. Finally, the excitation of the L-fiber takes the
form of a hyperpolarizing potential.

No doubt these peculiar properties of the R-L-synapse reflect the peculiar anatomy of this
junction. The explanation of one in terms of the other can perhaps be attempted once the differences of the postsynaptic phenomena in L1, L2 and L3 will be established and put in relation
with the different form of the three fibers.

12. Orderly arrangement of elements in the medulla

Besides the terminations of L1, L2 and L3, already mentioned, the endings of two more laminar neurons, as well as those of retinula cells 7 and 8 and in addition some recurrent fibers define an orderly one to one mapping of the compartments of the lamina onto as many compartments of the medulla. The front-to back order of the arrangement is, however, inverted due to the crossing of the fibers in each of the layers of the external chiasma. There is no exchange of fibers between various layers of the chiasma, so that the vertical order of the rows is preserved. Many of the neurons intrinsic to the medulla, third order neurons with respect to retinula cell axons 1 to 6 and (some of them) second order with respect to retinula axons 7 and 8, are arranged with the periodicity of these compartments.

13. Mapping of the medulla onto the lobuli (the inner chiasma)

In the lobular complex the visual field is again represented in an orderly way, judging from the structure of the inner chiasma which projects the inner surface of the medulla onto the two subdivisions of this region, the lobulus and the lobula. The caudal margin of the medulla is projected onto the lateral margin of the two ganglia, which thus comes to represent the anterior margin of the visual field (fig. 1). Again, as in the external chiasma, there is no fiber crossing in the vertical direction, but each of the layers of the inner chiasma carries fibers from two horizontal rows of elements in the medulla and therefore corresponds to two of the layers of the external chiasma (fig. 8).

The periodicity of the 3'000 compartments, so obvious in the retina and in the lamina and still very evident in the medulla, is not prominent in the lobular complex. There are periodically arranged elements in the anterior one of the two ganglia, the lobula, but spaced at distances much larger than would be expected considering the width of the original meshes of the network at the periphery. They seem to be arranged as the circles on the right hand diagram of fig. 8 (Lba), with the dots standing for the original unitary lattice. If this interpretation is correct, their number would be about 750 in the whole ganglion.

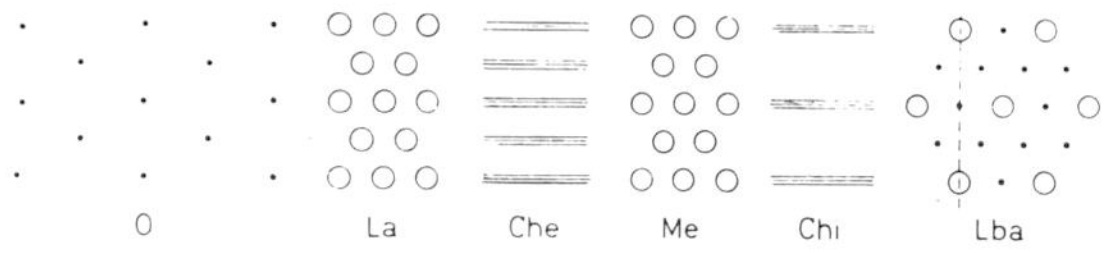

Fig. 8. Geometrical relation of the array of visual axes, of elements in the lamina, ext. chiasma, medulla, int. chiasma and lobula (from BRAITENBERG, 1970)

14. Elements in the ganglia coordinating large regions of the visual field

In each of the visual ganglia there are large fibers spanning several of the elementary columns described. (a) There is a network of rather irregularly oriented tangential fibers on the surface of the lamina. I have seen a single one of these elements bridge 10 rows of the elementary array. They seem to make contact with second order neurons in the lamina (see STRAUSFELD and CAMPOS-ORTEGA this volume). (b) In the medulla, a set of richly branching fibers about 15 μ below the surface are arranged vertically, spaced at fairly regular distances of about two elements of the array. Their length, in the vertical direction, exceed 20 rows. (c) Another set of fibers with a mainly vertical orientation, much thinner than the previous ones and without branches, termed H-fibers in our jargon, is evident in the medulla at a level corresponding to that of the termination of the L3 fiber. They span up to 20 rows of the elementary array. (d) Some more types of large fibers deeper in the medulla, some with a fairly regular, some with an apparently random course roaming the whole extent of the ganglion, are as yet poorly understood.

The very large fibers in the lobuli are interesting because of their small number. (e) A bundle of eight heavy (diameter over 5 μ) fibers enter the lobular complex from the medial side. Each divides into an ascending and a descending branch, the two branches of each fiber occupying successive vertical strips of the posterior surface of the lobulus. (f) Three more fibers of similar caliber are to be found near the anterior surface of the lobulus, which they traverse in a horizontal direction, dividing between themselves the ganglion in horizontal strips into which each fiber sends its branches. Thus very schematically (fig. 9) there is provision for integration in vertical strips extending the whole height of the visual field, as well as in horizontal bands, extending through its entire anteroposterior extent.

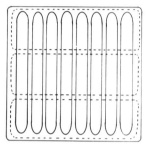

Fig. 9. Approximate distribution of giant fibers in the lobulus

15. Small range interactions

Experiments on optomotor reactions suggest the idea of sets of oriented movement detectors distributed throughout the whole eye. At least two sets with different orientation are required. Each detector is placed between neighbouring, or almost neighbouring channels. If the signals interacting in the movement detectors are conveyed by fibers, (at least) two sets of similar short fibers with different orientation should be found somewhere in the visual ganglia. Recent experiments by KIRSCHFELD (this volume) indicate that movement is seen between pairs of channels situated as shown in fig. 10a. Especially the vertical pair is surprising, since it implies interaction over

Fig. 10.
a: Arrows connect points of the ganglionic array which serve as inputs to a vertical and a horizontal movement detector (KIRSCH-FELD, this volume).
b: Points connected by L4 collaterals in the lamina.
c: Points connected by some fiber collaterals in the medulla

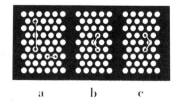

a   b   c

twice the range which would correspond to the most regular sets of short range inter-channel connections found in the lamina and in the medulla. The collaterals of L4 at the base of the lamina coordinate channels arranged as in fig. 10b, while a regular network in the medulla, at the level of the terminations of the L-fibers, is formed of branching fibers which connect sets of channels as in fig. 10c. Two of the patterns of fig. 10a or b, or a combination of the patterns of fig. 10a and b could mediate the vertical interaction across 5 rows. On the other hand, there are certainly neurons in the very complex network of the medulla which could be responsible for this as well as for the short range horizontal computation, but not enough is known about the precise geometrical relationships there (see the article by CAMPOS-ORTEGA and STRAUSFELD, this volume).

16. Gradients of size in the visual system

The sizes of some elements are not uniform throughout the layers at which they occur. Their variation has been described in a paper (BRAITENBERG and HAUSER, in press) from which the following summary is largely drawn. (a) The lenses are larger in front. Their diameter varies between 36 μ for the male and 30 μ for the female in front, and 21 μ in the back for both sexes. The largest lenses in the male are at the anterior border of the eye above the equator, in the female at the level of the equator or slightly below. (b) The thickness of the axons of the retinula cells follows the same pattern. Their diameter varies, in the male, by about a factor 2. (c) The thickness of the body of the L3 fiber shows the most striking variation of all elements, since it stays roughly proportionate, throughout the eye, to the square of the diameter of the corresponding lenses. (d) The thickness of L1 and L2 also varies monotonically with the variation of lens size, but this variation is less than that of L3. (e) The branches of L4 which in the superficial third of the lamina embrace the cartridge to which the fiber belongs, are longer and heftier in front (extending about 10 μ from the fiber, as compared to 5 μ in the back) perhaps simply because the whole cartridge is that much thicker in front. (f) L1 and L2 have about the same average thickness in the lamina near the anterior and posterior border of the ganglion. The ratio of their diameters increases to about 1:1,5 (L2 always being the thicker one) somewhere near the middle of the ganglion. The largest ratios were measured, in a female fly, below the equator at about 1/3 of the distance from front to back. (g) In the medulla, the terminations of L2 are thicker in the posterior part of the ganglion. The same is probably true for the terminations of L1 and L3, but precise measurements have not been made yet. (h) Also, in the medulla, the layer of vertically running fibers at the level of the termination of L3, the so-called H-fibers already mentioned, is much denser in the back and almost invisible near the anterior border of the ganglion.

Note that all of these gradients, except (f), follow the same general pattern, since the two gradients in the medulla, like all the others, considering the inversion due to the chiasma, also show a maximum with the elements referring to the front part of the visual field.

Fig. 11a illustrates the variation of L3 in the male fly.

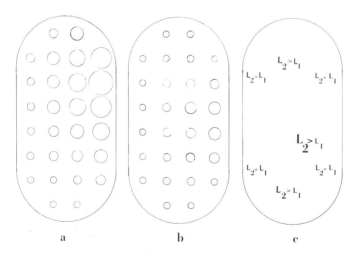

$$L_2 > L_1$$
$$L_2 = L_1 \qquad L_2 = L_1$$

$$L_2 > L_1$$

$$L_2 = L_1 \qquad L_2 = L_1$$
$$L_2 > L_1$$

a   b   c

Fig. 11. a: Circles of size proportionate to the cross-sections of L3 in different regions of the lamina in the male and b: in the female. c: Difference between L1 and L2. The forward direction is to the right on these diagrams

The diameters of the circles are proportionate to the diameters of the fibers. Fig. 11b shows the same for the female on the same scale. Fig. 11c shows in an approximate way the variation of the difference between L1 and L2 over the ganglion (the gradient f above).

17. Discussion of the gradients

The gradient of the ratio L1/L2 deserves a special discussion. For all the others the following argument may be valid, suggested by the proportionality of lens surface and fiber cross-section in the case of L3. The larger lenses produce a more intense light flux onto the retinula and there-fore a higher density of primary sensory events in the corresponding ommatidia. This would produce a stronger excitation of the associated nerve elements than in the ommatidia with smaller lenses, unless the excitation were distributed on membranes with surface area proportionate to the average light flux, i.e. to the surface of the lenses. For cylindrical elements of constant length, such as the L-fibers approximately are, this would require proportionality between surface of the lens and diameter of the fiber. A homogeneous optical environment would then again be represented by a constant level of excitation, and a superimposed modulation would affect the same part of the nonlinear characteristics of the transmission in all elements. The fact that the relatively small variation of the lenses, affecting the light flux at most by a factor 4 should require such elaborate compensation in a mechanism which can operate over an intensity range of several orders of magni-tude, is surprising but no more mysterious than the elaborate pupillary mechanism in the vertebrate eye, which also has a very low efficiency.

18. Role of L1 and L2

If we believe the electronmicroscopical evidence, L1 and L2 receive the same information from the afferent axons. It is possible, but not yet ascertained, that L3 is also connected in the same way to all retinula axons R1 to R6. Thus the information from one point of the visual environment in the cartridge is distributed on two, possibly 3 lines. From the different shape of L1, L2 and L3 one would predict different transmission properties and from the different localization of their endings different uses of this information in the functional context. We shall now explore the possibility that these fibers may be identified with some of the elements of the mechanism of movement perception, postulated on the basis of the analysis of optomotor reactions. If movement perception is done by means of elementary movement detectors, each sensitive to movement in one direction only, arranged with two different orientations (which for convenience we may call vertical and horizontal) between neighbouring channels, each channel will be connected to 8 movement detectors, namely (I) detectors for movement toward that channel and for movement away from that channel (II) on either side of the channel (III) in each of the two directions (fig. 12a). Is this distribution of the signal from one channel achieved by successive branching of the line, and if so, which is the sequence of the dichotomies? We should like to identify the first branching point with the cartridge, where the same information is relayed to L1, L2 and possibly L3.

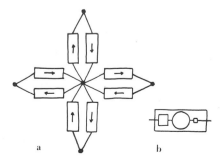

Fig. 12.
a: Connection of an ommatidial channel to 8 movement detectors.
b: A movement detector, see text

Fig. 12b shows a movement detector as containing a symmetrical element, the multiplier of the Reichardt–Hassenstein model (circle), flanked by two different elements (small and large rectangle) which give it the asymmetry necessary for the detection of the sense, as well as the magnitude of the movement.

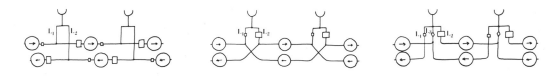

Fig. 13. a, b, c: Three interpretations of L-fibers in the context of the mechanism of movement detection

Fig. 13 shows three ways in which these elements may be inserted between neighbouring channels (say, along the horizontal coordinate). In fig. 13a, the information from the periphery is first divided between the detectors for movement in the two opposite directions. This would imply that the difference between L1 and L2 has something to do with the difference between the perception of movement, and therefore between the optomotor reaction elicited by movement in opposite directions, as it has recently been postulated by REICHARDT (this volume). Fig. 13b, on the contrary, divides the signal at first between detectors of movement directed toward that channel and those for movement away from it. Here L1 and L2 could be directly identified with the two different inputs of the multiplier (large and small rectangle of fig. 12b), and their different shape and size could be directly related to the different transmission characteristics which the model implies. Finally, fig. 13c makes use of L3 as well as L1 and L2 as inputs to the movement detectors. If the detectors for one direction are between L3 and L1 of neighbouring cartridges, and those for the opposite direction between L3 and L2, the variation of the difference between the optomotor reactions in opposite directions could be correlated with the variation of the difference of L1 and L2, while the average intensity of the reaction elicited in various parts of the visual field might depend on the size of L3. But the similarity with known facts is at present rather superficial. The scheme of fig. 13c is reminiscent of the fact that the L3 fibers terminate in the medulla at a level between the levels of the terminations of L1 and L3. The correlation of the differences between L1 and L2 with the postulated difference of the optomotor reactions in opposite directions (fig. 13a and c) is appealing because both differences disappear at the anterior and posterior borders of the visual system.

References

AUTRUM, H., ZETTLER, F., JAERVILEHTO, M.: Postsynaptic potentials from a single monopolar neuron of the ganglion opticum I of the blowfly Calliphora. Z. vergl. Physiol. 70, 414 – 424 (1970).

BOSCHEK, C.B.: On the fine structure of the peripheral retina and lamina ganglionaris of the fly Musca domestica. Z. Zellforsch. 118, 369 – 409 (1971).

BRAITENBERG, V.: Ordnung und Orientierung der Elemente im Sehsystem der Fliege. Kybernetik 7 (6), 235 – 242 (1970).

CAJAL, S.R. and SANCHEZ, D.: Contribucion al conocimiento de los centros nerviosos de los insectos. Parte 1, Retina y centros opticos. Trab. Lab. Invest. Biol. Univ. Madr. 13, 1 – 168 (1915).

CAMPOS-ORTEGA, J.A., STRAUSFELD, N.J.: The columnar organization of the second synaptic region of the visual system of Musca domestica L. I. Receptor terminals in the medulla. Z. Zellforsch. 124, 561 – 585 (1972).

HORRIDGE, G.A., MEINERTZHAGEN, I.A.: The accuracy of the patterns of connexions of the first – and second – order neurons of the visual system of Calliphora. Proc. Roy. Soc. Lond. B 175, 69 – 82 (1970).

JAERVILEHTO, M., ZETTLER, F.: Localized intracellular potentials from pre- and postsynaptic components in the external plexiform layer of an insect retina. Z. vergl. Physiol. 75, 422 – 449 (1971).

KIRSCHFELD, K.: Die Projektion der optischen Umwelt auf das Raster der Rhabdomere im Komplexauge von Musca. Exp. Brain Res. 3, 248 – 270 (1967).

KIRSCHFELD, K.: Absorption properties of photopigments in single rods, cones and rhabdomeres. Rendiconti S.I.F. XLIII, Ed. W. Reichardt, Academic Press, London and New York, 116 – 136 (1969).

SCHOLES, J.: Electrical responses of the retinal receptors and the lamina in the visual system of the fly Musca. Kybernetik 6 (4), 149 – 162 (1969).

STRAUSFELD, N.J.: The organization of the insect visual system (light microscopy). I. Projections and arrangements of neurons in the lamina ganglionaris of Diptera. Z. Zellforsch. 121, 377 – 441 (1971).

TRUJILLO-CENOZ, O.: Some aspects of the structural organization of the intermediate retina of Dipterans. J. Ultrastr. Res. 13, 1 – 33 (1965).

TRUJILLO-CENOZ, O., MELAMED, J.: Electron microscope observations on the peripheral and intermediate retinas of Dipterans. In: The functional organization of the compound eye, Ed. C.G. Bernhard, Pergamon Press, London (1966).

TRUJILLO-CENOZ, O.: Some aspects of the structural organization of the medulla in muscoid flies. J. Ultrastr. Res. 27, 533 – 553 (1969).

ZETTLER, F., JAERVILEHTO, M.: Decrement–free conduction of graded potentials along the axon of a monopolar neuron. Z. vergl. Physiol. 75, 402 – 421 (1971).

2. Synaptology of the Lamina ganglionaris in the Fly

C. B. Boschek

Max-Planck-Institute of Biological Cybernetics, Tübingen, Germany

Abstract.   In the fly's optical cartridge the types of neuronal elements as well as the types of synapses are described.

1. Introduction

A synapse is a point of contact which allows the transmission as biologically meaningful information between two nerve fibers, between a receptor and a nerve fiber or between a nerve fiber and an effector. Identifying such a point of contact from its fine structure is difficult for two reasons: Firstly, no single unique structural specialization is found in all synapses. Secondly, many characteristic structures such as membrane thickenings or accumulations of vesicles, found at synapses also occur in non-synaptic regions. In such instances where we are able to anatomically identify a synapse we would wish to be able to predict some of its functional properties from its appearance. For instance, we would like to know which of the contacting fibers is presynaptic and which is postsynaptic, if the synapse is excitatory or inhibitory, and if it functions electrically or through the discharge of a chemical transmitter substance. In regard to the last question, we might note that electrical synapses are generally recognized by so-called tight-junctions in which case the plasma membranes of the contacting cells are joined and at such points, no extracellular gap being recognizable. In the following discussion we will limit ourselves to those types of synapses which are generally considered to transmit by chemical means, as this is the only type which has been positively identified in the Dipteran retina.

No doubt the best understood synapse occurs at the vertebrate neuromuscular junction. Due to the ideal properties of this preparation, neurophysiologists have been able to obtain from it a great deal of information concerning the transmission characteristics of synapses in general. The large size of the branched endings of the motoneuron axons allowed the use of extracellular electrodes to positively identify these endings as the actual sites of synaptic activity. Electron microscopists have carefully studied the fine structural anatomy of the motor endplate and can easily recognize it from its structure. The motor-ending is elongated and branched, being separated from the axon by a layer of mitochondria. The terminal itself is filled with vesicles and is separated from the muscle fiber by a gap of 500 to 1000 Å. This gap is in turn filled with a fibrous basement membrane which sends processes into the invaginated plasma membrane of the muscle fiber. For a complete discussion of the physiology of this junction see KATZ (1966) and for details concerning the fine structure see v. DUERRING (1967).

Another synapse whose structure and function have been studied in great detail occurs in the vertebrate retina. As an example of such a study we might consider the mudpuppy retina, the synaptic structures of which have been described in DOWLING and WERBLIN (1969) and the electrophysiology in WERBLIN and DOWLING (1969). The rod and cone endings which are filled with synaptic vesicles may be seen to contain elongated ribbons at direct points of contact with bipolar and horizontal cells. At such points of contact the plasma membranes appear thickened and unusually electron dense. Due to their uniqueness and to the electrical events recorded in the contacting fibers, these structures are generally accepted as synaptic specializations.

For technical reasons, electrophysiologists have had difficulties in studies of insect visual systems. Primarily the problem lies in the miniscule dimensions of insect nerve fibers, many of which are smaller than .1 μ in diameter. Histologists also have problems when working with insects, due to the inherent difficulties of fixation of insect tissues as well as the task of embedding and successfully sectioning the tough, chitinous cuticle. In spite of these difficulties, insects have proven themselves to be especially suited to certain types of behavioral and physiological studies and as will become apparent throughout this symposium, a great number of such studies have already been undertaken.

Some cynical critics, most notably neurophysiologists, have been heard to criticize anatomists for chosing these well investigated insect nervous systems on grounds that the anatomists are only doing so in order to save themselves the effort involved in any creative thinking concerning possible function. Although this possibility cannot always be ruled out we should note that there are certain other valid reasons why such systems are also unusually well suited for neuroanatomical studies. As an example we will consider the Dipteran lamina ganglionaris, the first synaptic neuropile region of the compound eye. Because of its unique topography this region is a model system for the study of neuropile organization and synaptology. Its hemispherical shape, the periodic nature of its over 3'000 optical cartridges and the generally parallel course of fibers within this neuropile allow a thorough analysis of its cytoarchitecture by careful study of only a few survey electron micrographs. For a complete study of the synaptology, serial sectioning remains a necessary evil. However, due to the repetitive arrangement of the fiber bundles a high degree of certainty may be obtained from a relatively small number of low-power serial sections. In order to begin such a study, it is obvious that we must first be able to recognize points of synaptic contact. Although the functional data concerning this region is somewhat less satisfying than it is for similar vertebrate systems, we are still going to try to make some statements concerning such points of contact. TRUJILLO-CENOZ (1965) first described specializations within the axon terminals of the photoreceptor cells. These appeared as T-shaped structures which occured at points where the receptor terminals could be seen to be touching upon other fiber cross-sections.

Without going into detail concerning the spatial relationships of the various fibers within the lamina we will first take a look at the general organization of the optical cartridges as well as the fine structure of the T-shaped specializations and then we will attempt to visualize the form and possible interrelationships of the various fibers which have been observed in one optic cartridge. Finally we will discuss at some length why we feel justified in considering the T-shaped specializations as synaptic structures. For more details see BOSCHEK (1971).

2. Materials and methods

The compound eyes and optic ganglia of adult Musca domestica females were excised and fixes in phosphate buffered formaldehyde and post-fixed in veronal buffered osmium tetraoxide. After a rapid dehydration in an ethanol series they were embedded in Durcupan ACM. Gold or silver colored sections were cut with glass knives, mounted on grids and contrasted in uranium acetate and lead citrate. Serial sections were mounted on formvar films on copper one-hole apertures with help of a simple manipulator mounted on the stereo microscope of the ultramicrotome. A Zeiss EM-9A fitted with a magnification reducer was used at magnifications ranging from 800 to 37400X.

3. Results

In fig. 1 we see a section through an optical cartridge. A crown of six dense fiber cross-sections surrounds two lighter appearing fibers. Serial sectioning has shown that the fibers of the crown are axon terminals of the photoreceptor cells R_{1-6} from six different ommatidia in the peripheral retina. The central fibers are the primary axons of the two large monopolar neurons L_1 and L_2. Outside the crown of receptor axon terminals we see the axons of two smaller monopolar neurons L_3 and L_4 as well as six pairs of so-called centrifugal fibers which have been labelled α and β.

The centrifugal fibers can be seen to contain numerous small spherical structures. The cartridge is isolated from its neighbours on all sides by the collumnar epithelial glial cells.

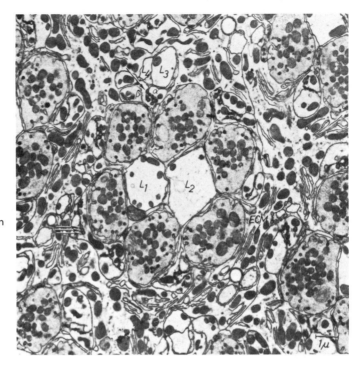

Fig. 1. Medial section through an optical cartridge. The crown of six fibers are terminals of the photoreceptor cells R_{1-6}. (L_{1-4}), primary axons of the monopolar neurons. (α, β), centrifugal fibers containing spherical structures (arrows). (ECM), membranous folds within the epithelial glial cells. X8000

Fig. 2 shows a high magnification of a portion of one of the receptor terminals as seen in fig. 1. Separated from the terminal by a gap of about 75 Å we see the cross-sections of two small fibers, which by means of serial sectioning were identified as side-branches of the central monopolar neurons L_1 and L_2. The receptor terminal contains numerous if somewhat poorly reserved vesicles and at the point of contact with the L_1 and L_2 side-branches a T-shaped ribbon. The L_1 and L_2 side branches are devoid of vesicles, but contain platelike membranous structures. Ribbons are also observed within receptor terminals at points of contact with one or both axis fibers of L_1 and L_2. Except at such points where ribbons are seen, the various fibers within the cartridges are isolated from one another by processes of the epithelial cell.

Ribbons have been observed at other points of contact between various fibers of the cartridges as indicated by the numbered arrows in fig. 3. In most instances, vesicles were seen surrounding the ribbons as well as plate-like structures within the opposing fiber or fibers. Contact was considered positively established after at least three independent observations in three different specimens.

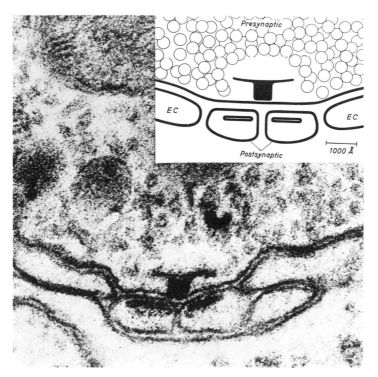

Fig. 2. A T-shaped ribbon within a receptor axon terminal at a point of contact with the side-branches of the monopolar neurons L_1 and L_2, each of which contains a membranous plate-like structure. (EC), processes of epithelial glial cells. X100000

4. Discussion

Let us first list the reasons why we feel justified in assuming that the T-shaped ribbons are of a synaptic nature: (1) in order to function, the receptors must have outputs to second order neurons. It is only at points where ribbons are seen that the receptor terminals make direct contact with the second order neurons. Since the receptor terminals must be presynaptic at such points of contact, we may assume that the fiber which contains the ribbon is always the presynaptic one. Thus we have an indication of polarity at such ribbon synapses. (2) It is only at points where ribbons are seen that the various fibers within the cartridges are not fully insulated from one another by epithelial cells. (3) Fibers containing ribbons are almost always found to be nearly filled with synaptic vesicles. (4) Similar ribbons have been observed at neuromuscular junctions of Dipterans, in segmental muscles by OSBORNE (1967) and at the Musculus orbitotentoralis by HENGSTENBERG (1971). (5) There is an apparent similarity of these structures to the well established synaptic junctions in the vertebrate retina.

Now looking at the diagram in fig. 3 let us consider some of the functional ramifications of the various synapses which have as yet been observed. The type 1 synapse is the structural basis for neural superposition as discussed by KIRSCHFELD, this volume. By means of this interconnection six receptors from six different ommatidia are coupled as a means of increasing the light gathering power in this type of compound eye.

The diagram in fig. 3 is still incomplete. There may be a number of synapses as well as fibers which have not yet been observed. In addition, a few fibers which have been seen to make synaptic contact have not yet been identified. We also know little about the fibers which leave the lamina and enter the medulla by way of the first optic chiasma (compare STRAUSFELD and

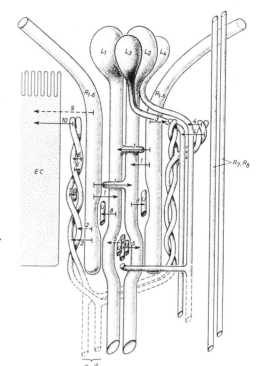

Fig. 3. "Wiring diagram" for an optical cartridge. For the sake of clarity only two of the six receptor terminals (R_{1-6}) and two of the six pairs of centrifugal fibers (α, β) have been drawn. (L_{1-4}) monopolar neurons. (R_7, R_8) axons of the central receptor cells which are in transit and have not been observed to make contact of any kind in the lamina. (EC) epithelial glial cell. (U) unidentified fiber fragments. Broken lines indicate areas of uncertainty

CAMPOS-ORTEGA, this volume). Due to this lack of evidence as well as to the sparseness of electrophysiological recordings it is difficult to even speculate on the functional consequences of most of the remaining synapses. Since L_3 provides a possible output from only one of the six photoreceptors it is possible that this output could contain information concerning the plane of polarized light. The type 9 and 10 synapses are very curious in light of the fact that the supposed postsynaptic elements are epithelial cells which are generally considered to be of a glial nature.

In spite of the incompleteness of fig. 3, however, all of the structures which have been drawn have been verified by at least three observations in at least three different specimens and it is therefore claimed that although some additions will need to be made to the diagram, no major alterations will be necessary.

References

BOSCHEK, C.B.: On the fine structure of the peripheral retina and lamina ganglionaris of the fly Musca domestica. Z. Zellforsch. 118, 369 – 409 (1971).
DOWLING, J.E., WERBLIN, F.S.: Organization of retina of the mudpuppy, Necturus maculosus. I. Synaptic structure. Z. Neurophysiol. 32, 315 – 338 (1969).
V. DUERRING, M.: Ueber die Feinstruktur der motorischen Endplatte von höheren Wirbeltieren. Z. Zellforsch. 81, 74 – 90 (1967).
HENGSTENBERG, R.: Das Augenmuskelsystem der Stubenfliege Musca domestica. I. Analyse der "Clock-Spikes und ihrer Quellen. Kybernetik 9, 56 – 77 (1971).
KATZ, B.: Nerve, Muscle and Synapse. McGraw-Hill Company. New York.
OSBORNE, M.P.: The fine structure of neuromuscular junctions in the segmental muscles of the blowfly larva. J. Insect Physiol. 13, 827 – 833 (1967).

TRUJILLO-CENOZ, O.: Some aspects of the structural organization of the intermediate retina of Dipterans. J. Ultrastruct. Res. 13, 1 - 33 (1965).

WERBLIN, F.S., DOWLING, J.E.: Organization of the retina of the mudpuppy, Necturus maculosus. II. Intracellular Recording. J. Neurophysiol. 32, 339 - 355 (1969).

3. Some Interrelationships between the First and Second Synaptic Regions of the Fly's (Musca domestica L.) Visual System

N. J. Strausfeld and J. A. Campos-Ortega
Max-Planck-Institute of Biological Cybernetics, Tübingen, Germany

Abstract. Electron microscopy of Golgi impregnated neurons has allowed the precise determination of the spatial relationships between identifiable profiles in optic cartridges and corresponding terminals in the medulla columns. Golgi-E.M. profiles can be correlated with
profiles in normal electron microscopy. Structural contacts of cells in the lamina as well as
their corresponding components in the medulla are described: functional contacts (synapses)
known up to the present time are listed in figure 5 (polarities arrowed). Possibly structural contact may be indicative of functional intimacy, but this is still to be determined. At least three
centripetal pathways (L1, L2, L3) convay information from R1 – R6 into the medulla. The other
monopolars L4 and L5 appear to be intimate with endings of centrifugal cells from the medulla.
The logic underlying the complexity of lamina organization seems to lie in the arrangement of
feed-back and feed forward loops in parallel with the centripetal cells.

1. Introduction

The first synaptic region of the fly's visual system (the lamina) is composed of columnar aggregates
of neural elements. The basic component of each column is a crown of six receptor terminals termed R1 – R6. In addition, there are at least ten types of non-receptor neural elements associated
with each crown: together these, plus the set of R1 – R6, constitute an "optic cartridge". Of the
non-receptor elements Golgi studies have allowed us to distinguish five forms of monopolar cells
(termed L1 – L5), that project from the lamina to the medulla, and at least three forms of cells
that project peripherally from the medulla into the lamina and have narrow field components in
the outermost synaptic region. In addition, there are two wide field elements, one of which links
the medulla to the lamina (the lamina tangential cell) and a second which resides exclusively in
the lamina (the lamina amacrine cell). There is also a set (a pair) of receptor prolongations
(termed R7 and R8), derived from the retina, that passes down beside each cartridge and finally
terminates at two levels in the medulla. Light microscopy studies have allowed reconstructions
of the topographical relationships of these elements (bar the amacrine cell) in cartridges. The
pathways of these neurons across the 1st optic chiasma and the forms and levels of their medullary
components are known from Golgi studies and the examination of degenerated fibres by electron
microscopy (STRAUSFELD, 1971a, b; CAMPOS-ORTEGA and STRAUSFELD, 1972).

The six retinula cell terminals in a cartridge and the corresponding "en passant" pair of R7 + R8
fibres sample the same point in the visual surround (KIRSCHFELD, 1967; BRAITENBERG, 1967).
The monopolar and medullato-lamina elements to or from the same cartridge terminate with the
corresponding R7 + R8 fibres at the same locus in the medulla. The map of cartridges in the lamina is convayed in an orderly fashion onto the medulla, being reversed through 180° along the
horizontal by the 1st optic chiasma. The set of L1–L5 and R7 + R8 constitute the implicator
elements of a medullary column.

A few of the synaptic relationships between R1 – R6 and L1 – L3 have been determined from
electron microscopical studies on normal material (TRUJILLO-CENOZ and MELAMED, 1966;
BOSCHEK, 1971). Other synaptic relationships have been seen between so-called α – β fibres,
R1 – R6, L4, unidentified profiles and the previously mentioned elements (BOSCHEK, 1971;

BRAITENBERG, personal communication): but until a profile can be safely identified as belonging to a specific cell type the function of each cartridge must remain enigmatic.

Our investigations have been concerned with the identification of cell types by the Golgi method and then sectioning impregnated cells for electron microscopy. Our method is simple: brains of flies are impregnated by the Golgi–Colonnier technique (COLONNIER, 1964) and post–treated with osmium and then phosphotungstic acid. Brains are subsequently embedded in araldite and serial sectioned at 80 μm. Once we have identified and mapped the position of a cell the section is washed in Xylol, re–embedded in araldite and sectioned for the electron microscope. In this way we have been able to study the dendrology of lamina and medulla elements and have been able to define their discrete relationships with one another.

2. The Matrix

A section cut through a cartridge, parallel with its long axis, shows that its matrix is not uniform in texture. If one plots the densities of profiles it is clear that different populations of dendrites reside at different levels (fig. 1).

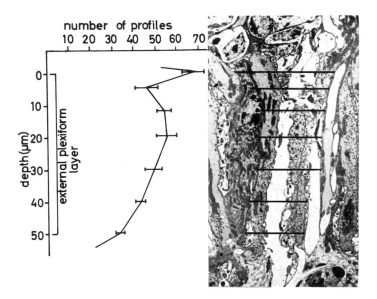

Fig. 1. The numbers of profiles counted at seven levels in the external plexiform layer of the lamina. Levels are indicated in the accompanying E.M. picture. The curve is drawn through the average number of profiles from six cartridges. Bars indicate standard deviation. Maximum number of profiles is at the outer face of the lamina. Majority of profiles reside in the outer 1/2 - 1/3 of the plexiform layer

The relative positions, in depth, of cell components (processes, spines, blebs, etc.) (fig. 2) show a discrete stratification according to density maxima–minima. This is our first indication that a cartridge may not be simply a relay station giving the same modes of outputs from R1 – R6 terminals to monopolar cells down its length. Rather, different kinds of integrative operations may occur in different segments of a cartridge.

Notwithstanding their different forms it was previously supposed that each of the five types of monopolar cells could be synaptically intimate with R1 – R6 terminals (STRAUSFELD, 1971 a).

However, this is now known not to be the case. Examination of Golgi impregnated profiles in the electron microscope has shown that only three of the five L—cells (fig. 2) contact R—fibres (fig. 3). The other two, namely L4 and L5, are primarily connected to elements derived from the medulla. We shall discuss, therefore, two main pathways between the lamina and the medulla. The first concerns L1, L2 and L3 with their corresponding medulla—to—lamina components, the second concerns L4 and L5 with medulla—to—lamina elements and the wide field tangential and amacrine cells.

3. L1, L2 and L3 pathways

L1 and L2 lie at the centre of the retinula cell crown. Both have tuberous swellings at the outer face of the external plexiform layer (fig. 2) and both give out short lateral processes down their length. These are radially arranged and extend to the inner and outer faces of retinula cells (fig. 3). The L1 and L2 pair is post—synaptic to all R1 – R6 terminals of a crown down its whole length. Also, L1 and L2 are post—synaptic to profiles that lie apposed to their axis fibres in the outer 1/3 of the lamina. Golgi—E.M. studies have shown that the branches of the tangential cell and one process from L4 lie between or against L1 + L2 at this level (fig. 2, 3). Lateral swellings from the medulla—to—lamina cell ending, C3, extend into the crown along its whole length between R1 and R6 (fig. 3). These terminal processes contact L1 + L2. The dimensions of L4, tangential and C3 processes in the crown are quite distinct. Those of C3 and the tangential match the profiles presynaptic to L1 + L2: those of L4 are similar to profiles post—synaptic to the axial monopolar cell fibres. Thus it seems that L1 and L2 are under the influence of a small and wide feed—back loop from the medulla (C3 and the tangential, respectively).

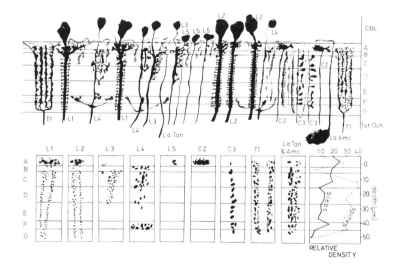

Fig. 2. The forms of monopolar cell components (L1 – L5), medulla—to—lamina cell components (T1, C2, C3, Lamina tangential —La Tan) and the lamina amacrine (La Amc) in the lamina. Note the relative levels of specializations: level a) is defined by the tuberous swellings of L1 and L2, the spines of L5 and the terminal specialization of C2; level b) almost lacks specializations and corresponds to the trough of the curve in fig. 1. Level c) is defined by tangential branches, levels c) and d) by processes of L3, level f) by the collateral terminals from L4. The relative densities of specializations of each cell type (spines and swollen specializations or blebs) are mapped below and plotted extreme right

A third centrifugal cell, C2, terminates at the extreme periphery of the lamina at the level of the tuberous specializations of L1 and L2 and the spines of L5. C2's axis fibre lies alongside that of L5 outside the retinula cell crown near R4 and R5 (fig. 3). The profile of the C2 terminal matches those of the tuberous swellings of the axial monopolars and the spines of L5 (fig. 2). In normal E.M. material L5 is clearly postsynaptic to C2. We have seen contact between C2 and L1, L2 and L5 in Golgi–E.M. material (fig. 3) but we cannot state with certainty the polarity of the medulla–to–lamina neuron, C2, with respect to L1 and L2: present evidence suggests that C2 is pre–synaptic to the two monopolars.

According to other authors L1, L2, L3 and L4 seem to be postsynaptic to one of the α - β basket processes. One of these, β , is derived from the T1 component in the lamina (see section 3). Our studies indicate that only α fibres are presynaptic.

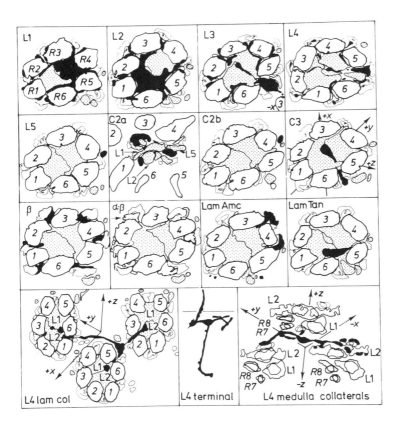

Fig. 3. Top three rows show locations and patterns of contact between elements examined with the Golgi–Electron microscopy technique and un–impregnated profiles. C2a indicates the mode of contact between the terminal specialization of C2 and L1, L2 and L5. C2b indicates the position of C2's axis relative to the R1 - R6 crown. β profiles are derived from the T1 cell. The origin of the α profiles lying beside each β fibre is still uncertain. They may be derived from T1a or perhaps the amacrine cell. T1a (fig. 4) certainly enters the 1st optic chiasma and may extend to the lamina. L4 processes contact between six and one α - β fibre pair depending on the location of the L4 cell across the lamina (see text). L4 has one constant feature, though; namely, its process between the L1 and L2 axis fibre. L4 contacts three pairs of L1/L2 via its collaterals (lower row, left hand diagram). Its terminal in the medulla (as far as the 40 μ m level in fig. 4) contacts three columns. The

longest branch contacts the L2 terminal in the -y column. The orientation of a retinula
cell crown relative to the three axes of the neuropil, x, y and z is shown in the extreme
right hand diagram, second row. The dimensions of these cartridge cross sections is approximately equal to 10 μm (diameter along y axis)

L1 and L2 are post-synaptic to L4 collaterals from two adjacent cartridges lying along the +x
and -y aces of the neuropil (STRAUSFELD and BRAITENBERG, 1970; BRAITENBERG, personal
communication). The inner processes of L4 to its parent cartridge at level F in fig. 2 contact
the axial monopolar cells and are probably pre-synaptic.

L3 is post-synaptic to all R1 - R6 terminals of its crown within the outer 1/2 of a cartridge.
According to BOSCHEK L3 is post-synaptic to one of the α - β fibres. L3 cells often have a
single posteriorly directed branch near the outer surface of the external plexiform layer (level
b, fig. 2). One of these, cut for electron microscopy, was followed to the R3 terminal of the
neighbouring cartridge lying along the -x axis (fig. 3, top row, third diagram from left). If this
pattern of connectivity is found to be ubiquitous then this neuron must interact with not one but
two optic cartridges in the lamina.

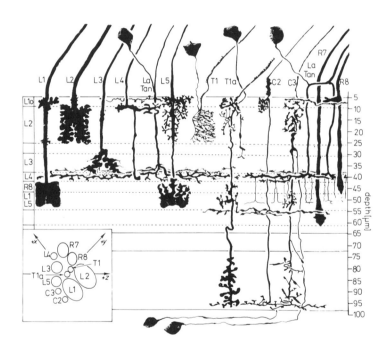

Fig. 4. The forms and levels of L- and R-endings with the medullary components of
medulla-to-lamina cells[1]. The tangential (La Tan) has two variants of medullary components, one of which extends as far as the level of L1. Note the ruffled specializations
from the T1 and T1a fibres in the 1st optic chiasma and the swollen component of C2
above the outer face of the medulla. Synaptic interaction cannot be excluded at these
levels. The inset shows the relative locations of axis-fibre profiles in an "average" column situated in the upper half of the left hand medulla. The position of L1 and L2 and
the other monopolar cells is the reverse of that in an optic cartridge

[1]Whether T1a is actually a medulla-to-lamina neuron is still to be determined

L1 and L2 appear to have identical connectivities with other elements in the lamina. However, we do not know their relationships with the amacrine cell. From physiological evidence ZETTLER and JAERVILEHTO (1972), JAERVILEHTO and ZETTLER (1972) have shown that under a limited set of conditions L1 and L2 appear to have identical response properties.

These neurons do, though, differ remarkably in form in the medulla (fig. 4). L1 is bistratified, L2 unistratified. The terminal of L3 resides between the ending of L2 and the innermost terminal component of L1. Golgi-E.M. studies show that a second form of T-cell (T1a), which enters the 1st optic chiasma, and the medulla-to-lamina cell C3 contact L1 and L2. C2 contacts the outer component of L1, and T1 is in contact with the ending of L2. L3 appears devoid of contact with a medulla-to-lamina neuron in the second synaptic region. At present we cannot state the pre- or post-synaptic nature of the other contacts mentioned in this paragraph.

4. L4 and L5 pathways and other elements

In the lamina L5 is post-synaptic to the terminal of C2. L4 is post-synaptic to one of the α - β baskets (BOSCHEK, 1971). It is probable that L4 is also post-synaptic to the axis-fibres of L1 and L2 within the outer 1/3 of the plexiform layer. The collaterals of L4 extend to L1 and L2 of adjacent cartridges (fig. 3) where they are pre-synaptic. In the medulla the terminal of L5 contacts, and is post-synaptic to the terminals of L1 and L2. The ending of L4 links three medullary columns in a pattern which is the reverse of that formed by its collaterals in the lamina. However, the polarity of branching is the same when considered with the horizontal cross-over of fibres via the 1st optic chiasma. Golgi-E.M. studies show the -y branch of L4 (in the upper half of the medulla, and the +x branch in the lower) to contact L2's ending. The other portions of the L4 terminal contact unidentified profiles.

At least one of the T-cells link the medulla to the lamina. It (T1) ends as a basket of processes around a cartridge. There are, however, two baskets pro cartridge, α and β (fig. 3). The basket endings of T1 correspond to the β fibres. TRUJILLO-CENOZ (1969), TRUJILLO-CENOZ and MELAMED (1970) and BOSCHEK (1971) have variously described α - β profiles as being pre- and post-synaptic to each other, presynaptic to L1, L2, L3 and L4 as well as to retinula cell fibres and epithelial cells. However, few criteria were evolved for distinguishing the two sets of profiles. In fact normal and Golgi-E.M. material show clearly that the T1 (= β) profiles are less electron dense than their mimetic partner profiles (TRUJILLO-CENOZ and MELAMED, 1970). Our studies have found that α profiles are electron dense, contain vesicles and ribbons whereas β profiles are post-synaptic and contain curious sausage shaped glia invaginations. The origin of α profiles is, as has been stressed earlier, uncertain. They may be derived from the amacrine cell or, perhaps, from the medulla element T1a.

L4 processes are post-synaptic to α fibres in the lamina and there is a gradient of L4 shape from front to back in the lamina: frontal L4s have many more outer processes than those posteriorly. Similarly, equatorial L4s have more processes than those situated at the poles of the lamina. Golgi-E.M. studies show that L4s with many processes contact all α fibres: those posteriorly contact as few as one α fibre. However, all L4s contact L1 and L2 via one branch in the outer 1/3 of the lamina. Thus there is a true gradient of connectivity defined by the numbers and lengths of fibres and hence their degree of contact with the α - β arborisations.

The last neuron to be considered is the lamina amacrine cell. It is complex in form, consisting of a large perikaryon beneath the lamina (fig. 2) and a set of processes through the external plexiform layer. They spread through a domain of several optic cartridges. Golgi-E.M. studies show that for one cartridge there are two, four or six amacrine processes situated alongside T1 fibres (β). Amacrine processes contact R-fibres and amacrine spines project into the crown to L1 and L2. We do not know the polarity of these contacts: possibly it is these components that were identified as centrifugal pre-synaptic components onto R-fibres by previous authors.

5. Discussion

The connections between a cartridge in the lamina and a column, or columns, in the medulla are exceedingly complicated. No monopolar cells can, theoretically, be assumed as being free from lateral interaction in the lamina either by lamina units or by feed-back loops from the medulla (e.g. via, possibly, C2 or the lamina tangential). L3 appears to be the simplest unit in terms of its connectivity, receiving only R1 – R6 inputs and a probable input from $\propto$ fibres (see also BOSCHEK, 1971). The output from L1 and L2 to medulla elements is, in part, through the ending of L5. This cell, it will be recalled, receives its sole input from the medulla-to-lamina cell, C2. Thus the output from the axial monopolars is indirectly, and in part, under wide field influence from the C2 component deep in the medulla at the level of L1 (fig. 4). They may be under complete C2 influence if this cell is, in fact, pre-synaptic to the outer segments of L1 and L2 in the lamina. A further source of wide field interaction could be via the tangential cell or via the collaterals of L4. The only two elements that appear to be free of feedback loops are R7 and R8. We have detected pre-synaptic contact to L2 from R8 in the medulla but R7 appears to be free of contact with L fibres. The schemes of connectivity are summarized in fig. 5.

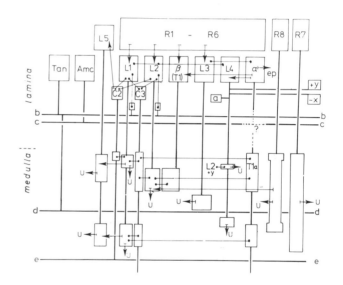

Fig. 5. The scheme of connectivity (synaptic interaction) and contact (apposition of fibres but mode of contact still undetermined) between elements of a cartridge and the corresponding medullary column. Dumbells indicate contact from Golgi–E.M. studies and arrows indicate synaptic intimacy (for instance, R1 – R6 to L1). a) denotes L4 presynaptic to L1 + L2 at the base of its parent cartridge. +y and −x indicate L4 collaterals presynaptic to L1 + L2 of the adjacent cartridges. b = extension of tangential component to other cartridges. c = extension of amacrine to other cartridges. d = extension of tangential to other medullary columns. e = extension of the medullary component of C2 to other columns. The contacts $\propto$ to e.p. indicates $\propto$ fibres presynaptic to the lamina epithelial cells (glia). The contacts $\propto$ presynaptic to β and to L4 are derived from BOSCHEK (1971). $\propto$ contacts L3 but synaptic intimacy is not proven. U in the medulla indicates unidentified profiles post-synaptic to retina-lamina inputs. The question mark means that the origin of $\propto$ fibres is still open to question. They may be derived from T1a or from the amacrine. The deep components of C2, C3 and T1a (see fig. 4) have not been included. Note also that the inputs to C2, C3 and the tangential from other elements are still unknown

The functional implications of these feed-back loops and the segregation of the various mono-polar cell pathways and the R7 / R8 pathways are various. We do not intend to discuss them here since at present the anatomy can produce many hypotheses but few functional explanations. Electron microscopy of Golgi labelled cells correlated with the identification of functional cell types by fluorescent dye injection may well allow us an optimal combination of techniques for the functional description of this intriguing and intricate visual system.

References

BOSCHEK, C.B.: On the fine structure of the peripheral retina and lamina ganglionaris of the fly Musca domestica. Z. Zellforsch. 118, 369 - 409 (1971).
BRAITENBERG, V.: Patterns of projection in the visual system of the fly. 1. Retina-lamina projection. Expl Brain Res. 3, 271 - 298 (1967).
BRAITENBERG, V.: Ordnung und Orientierung der Elemente im Sehsystem der Fliege. Kybernetik 7, 235 - 242 (1970).
CAMPOS-ORTEGA, J.A., STRAUSFELD, N.J.: The columnar organization of the second synaptic region of the visual system of Musca domestica L. I. Receptor terminals in the medulla. Z. Zellforsch. 124, 561 - 585 (1972).
COLONNIER, M.: The tangential organization of the visual cortex. J. Anat. (Lond) 98, 327 - 344 (1964).
JAERVILEHTO, M., ZETTLER, F.: Localized intracellular potentials from pre- and postsynaptic components in the external plexiform layer of an insect retina. Z. vergl. Physiol. 75, 422 - 449 (1971).
KIRSCHFELD, K.: Die Projektion der optischen Umwelt auf das Raster der Rhabdomere im Komplexauge von Musca. Exp. Brain Res. 3, 248 - 270 (1967).
STRAUSFELD, N.J.: The organization of the insect visual system (light microscopy). I. Projections and arrangements of neurons in the lamina ganglionaris of Diptera. Z. Zellforsch. 121, 442 - 454 (1971b).
STRAUSFELD, N.J.: The organization of the insect visual system (light microscopy). II. The projections of fibres across the first optic chiasma. Z. Zellforsch. 121, 442 - 454 (1971b).
STRAUSFELD, N.J., BRAITENBERG, V.: The compound eye of the fly (Musca domestica): connections between the cartridges of the lamina ganglionaris. Z. verg. Physiol. 70, 95 - 104 (1970).
TRUJILLO-CENOZ, O.: Some aspects of the structural organization of the medulla in Muscoid flies. J. Ultrastruct. Res. 27, 533 - 553 (1969).
TRUJILLO-CENOZ, O., MELAMED, J.: Electron microscope observations on the peripheral and intermediate retinas of Dipterans. In: The functional organization of the compound eye, e.d. C.G. Bernhard. London: Pergamon Press (1966).
TRUJILLO-CENOZ, O., MELAMED, J.: Light and electronmicroscope study of Muscoid flies. Z. Zellforsch. 110, 336 - 349 (1970).
ZETTLER, F., JAERVILEHTO, M.: Decrement-free conduction of graded potentials along the axon of a monopolar neuron. Z. vergl. Physiol. 75, 402 - 421 (1971).

4. Columns and Layers in the Second Synaptic Region of the Fly's Visual System: The Case for Two Superimposed Neuronal Architectures

J. A. Campos-Ortega and N. J. Strausfeld
Max-Planck-Institute of Biological Cybernetics, Tübingen, Germany

Abstract.   Combination of Golgi staining, reduced silver staining, normal and Golgi-E.M. has allowed estimates about the numbers of cells present in medullary columns. Up to the present time 120 cell types, classified according to orientation, depth, dendritic spread and projections, have been identified. Components of 34 cell types occur in each medullary column. In addition, each column contains the corresponding medullar components of elements destined for or derived from an optic cartridge, and the R7 / R8 endings. The basic three dimensional structure of the medulla, as defined by the 46 cells that contribute to each column, is columnar and stratified. Information flow must be both lateral and from periphery to centre or vice versa. This is in contrast with the lamina structure where most information flow must be along a cartridge. This structure is discussed with respect to the remaining cell types whose distribution per column is still unknown.

1. Introduction

In the previous paper we described the twelve kinds of neurons that connect the lamina with the medulla. The lamina components of eleven of these are contained within each optic cartridge. They project together in a bundle to the medulla where they form the basic elements of each medullary column. There is a twelfth type of neuron, the lamina tangential cell, whose axis-fibres are irregularly arranged with respect to the lamina mosaic; i.e. they neither have a syn-periodic relationship with cartridges (1 cell: 1 cartridge) nor a supraperiodic arrangement (1 cell to every n cartridges). Rather, they appear to be randomly distributed (aperiodic). Nevertheless, the subsequent specializations from the tangential fields in the lamina are arranged 1 per optic cartridge.

The cells that lead from optic cartridges in the lamina to the medulla, via the first optic chiasma, relay neural signals from the receptor level and the lamina to, presumably, a proportion of the nerve cells in the second synaptic region. Some of the cells link the medulla to the third and fourth synaptic regions, the lobula and the lobula plate. The problem that we are presently attempting to solve is the structural analysis of the medullary neuropil. We have examined over 1'000 optic lobes of the fly Musca domestica (L), impregnated with Colonnier's modification of the Golgi method. In total we have been able to distinguish 180 different nerve cell forms in the four synaptic regions of the optic lobes: of these 180, 120 forms are evident in the medulla alone. Each cell form is distinguishable from any other on the basis of its layer relationships, dendritic spreads and patterns of arborisations. Apart from two cells, each form has been seen on at least two separate occasions in more than one brain. The neurons have been classified according to the schema outlined in STRAUSFELD and BLEST (1970).

2. The medulla mosaic

The input bundles into the medulla from the lamina contain monopolar cell fibres and long visual cell terminals as well as components of the medulla-to-lamina neurons. These elements are asso-ciated with perpendicular axis-fibres of some transmedullary (Tm), T- and Y-cells that subse-quently prolongate to the lobula, lobula plate and to both lobula + lobula plate, respectively.

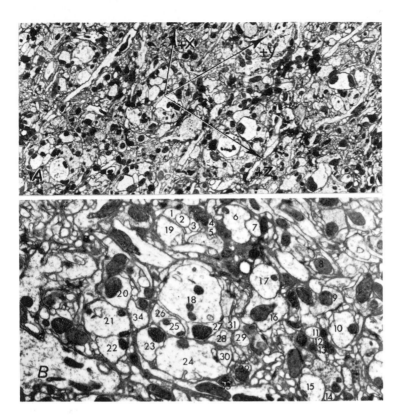

Fig. 1. A: a tangential section of the medulla, cut parallel to its outer face showing clearly delineated sets of elements (column cross-sections) aligned along the three axes of the neuropil (x, y and z: z points posteriorly).
B: a cross section of one column showing 34 profiles identified as axis-fibres of perpendicular elements

Together these fibres give the medulla its columnar appearance. The columns are hexagonally arranged along the three axes of the neuropil (BRAITENBERG, 1970). There are as many columns as optic cartridges and as many cartridges as ommatidia.

The borders of each column are impossible to define with an absolute degree of certainty. They are not separated by glial barriers like those formed around optic cartridges by epithelial cells. Nevertheless, the cores of the columns can be relatively easily defined within the outer two thirds of the medulla, peripherally to the serpentine layer, due to the presence of closely packed bundles of axis- (or mother-) fibres (fig. 1). But at the presumed edge of a column, where most of the synaptic interaction takes place, the clearest feature that allows us to approximately distinguish one column from another is the presence of laterally oriented processes from tangential and amacrine cells. These do not, however, occur with the same abundance at all levels throug the region and the perimeter of one column may, at certain depths, be confused with those of its surrounding neighbours.

3. The cross layers of the medulla

Even in unstained medullae, viewed with phase contrast illumination, the neuropil appears
distinctly laminated. These layers are even better resolved in reduced silver preparations: they
can be primarily related to levels of a few amacrine and tangential cells as well as the in-depth
relationships of a few perpendicular neurons. The medulla is, therefore, composed of columnar and
layered aggregates of neural components. This situation is in contrast with the lamina where, with
the exception of three cell types, through the majority of its depth the direction of information
transfer is perpendicular and is restricted to single columnar units, the optic cartridges.

4. The blue print

The architectural blue print of these two synaptic regions, the lamina and the medulla, cannot,
however, be considered as totally different since both have neural elements in common: it is
these which form the implicator elements of both cartridges and columns. Visual information
reaches the medulla through at least five channels: either directly from the retina via R7 and R8
or, after relay, from R1 - R6 via L1, L2 and L3. The endings of R7 and R8 and the three monopolar
cells define six coarse levels in the outer two thirds of the second synaptic region. These coincide
with sets of strata, each or which is qualitatively different from the next, that are formed by
tangentially oriented amacrine and tangential cells.

5. Strata

Using the term "stratum" we mean discrete concentrations of synaptic specializations, arranged
as a tangentially oriented plexus across the medulla's lateral extent, that are derived from two
or more neural elements (an element can be a whole cell or a fraction of it). Each element is
characteristically found at a particular depth in the neuropil. Strata are not to be confused with
the gross-layers described in section 3. Rather, a stratum is defined by the in-depth coincidence
of amacrine and/or tangential cell processes and specializations as well as processes from some
of the perpendicularly arranged cell types whose lateral spreads exceed the limits of a column
in which their axis-fibre resides. Per definition, a stratum forms a continuous system of specializations throughout the medulla's lateral extent: the planar plexus of processes lies parallel to
the region's outer or inner face. It follows that a stratum must be composed of cells whose specializations, at least, are synperiodic.

We can assume, therefore, that there are two directional modalities of information transfer: the
first being from the region's periphery to its centre (or vice versa), the second being in a lateral
direction via strata and laterally arranged processes of all cell classes.

The inputs from the lamina/retina coincide with many such strata. Each stratum is qualitatively
different from the next and it might be supposed that each represents a structural basis for a
different kind of integration. Naturally with the methods presently at hand we cannot state
whether this arrangement represents a hierarchical disposition of the elaboration of input signals
or whether each stratum serves to extract one meaningful parameter alone from the input and to
transfer this to specific cells in the lobula complex and the pre-motor command centres. However,
there are intrinsic cells in the medulla that project from one stratum to another and it is reasonable
to suppose that the activity of one level is interdependent on that of another.

6. Strata versus columns

As far as we can tell each stratum appears to extend as a planar sheet through the whole of the
medulla's lateral extent. However, it would be rash to jump to any conclusions that in the fly
this region consists solely of planar networks.

Due to the extraordinarily precise repetitive organization of the lamina elements, where each
optic cartridge is qualitatively like its neighbour, and where each can most probably function

as a unit on its own, we might expect that the precise mapping of the lamina onto the medulla and the transfer of this map through its whole depth implied a similarity of functional organization. Each column, as defined by the input bundle from the lamina/retina, would have the capacity to perform the same set of operations as any other column. But we are faced with an enoumous number of neural elements as well as a multitude of strata. This staggering numerical complexity poses in its own right the following problems and questions:

1. Do all the perpendicular elements occur in each column?
Most probably they do not: sections taken across columns in the medulla show that the numbers of profiles that could safely be identified as axis fibres from electron microscopic criteria varies between 11, close to the 1st optic chiasma, and 44 just proximal to the level of the R7 terminals. Counts from several columns at this level indicated between 17 and 48 axis fibres (figs. 1 and 2).

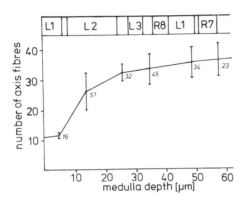

Fig. 2. Numbers of axis fibre profiles (diameter > 0,3 μm) per column at six levels in the medulla (l1 – R7). Curve is drawn through mean number of fibre counted in between 16 and 51 columns (numbers given in italics). Horizontal bars indicate standard deviation for each sample level. The maximum number of recognisable axis fibres is at the R7 level

Notwithstanding the difficulties encountered in making such counts (and the obvious subjective bias) it seems that fibre densities can vary from column to column. One further word of caution: there are 38 perpendicular cell types whose fibre diameters as far as the R7 level are axceedingly small (< 0,3 μm): these could easily be confused with some profiles of lateral processes having the same diameter and electron microscopical characteristics.

2. How many cells, apart from the 11 elements linking each column in the medulla with each cartridge in the lamina, occur in each column?
To answer this question we have used the following criteria: a) some cell types, as identified with the Golgi method, can be recognized in Bodian preparations by virtue of their characteristic reduced silver Gestalts; b) pairs or triplets of cells (of the same type) can be fortuitously impregnated with the Golgi method in adjacent columns on many occasions; c) cells may be impregnated in an "en masse" fashion by the Golgi method. These last two criteria demand a fourth: cells impregnated in pairs or triplets or "en masse" must be stainable at any location in the medulla's horizontal or vertical extent. Each of the criteria cited above is considered sufficient indication of synperiodicity. From these criteria we have arrived at a minimal figure of thirteen types of perpendicular neurons in each column in addition to L, R and medulla-lamina neurons. There are also thirteen forms of amacrine cells that have lateral spreads through many columns and which send specializations to each. These, therefore, are considered as having a synperiodic arrangement irrespective of the spacing of their mother fibres. There are also eight forms of tangential arbori-

zations, derived from or leading to the ipsi- or contralateral brain, whose components invade each column. In summary, every column, irrespective of its position in the medulla, contains the same minimal set of 46 neural elements.

As has been outlined in section 5 the in-depth coincidence of tangential and amacrine cell specializations form the discrete strata. These are shown in fig. 3. This illustration also figures the forms and dispositions of perpendicular cells as well as the distribution of their fields and specializations, from a single column, through the medulla's depth.

3. How are the remaining 74 forms of neurons in the medulla related to the periodic structure of this region?
As yet there is no conclusive answer to this question. Instead we prefer to manipulate two possibilities and treat the following as our working hypothesis.

1st hypothesis. The minimal set of all elements is not a single column as in the lamina, but a group of columns, the shape and size of a group being determined by the number and geometrical arrangement of the minimal number of columns containing components of all 120 medullary elements. This hypothesis implies that all elements should occur more than once per medulla and in a characteristic pattern; i.e. each column of a group would be a sub-unit whose basic components were of 46 cell types whereas between 1 and n of the remaining 74 cell types would be distributed among all or some of the columns in a group. With this hypothesis we could treat the medulla of the fly as having several groups of cell arrangements where in each there are two superimposed systems. The one is restricted to each column of the group; the other is more complex and is localized in the whole group as well as defining its shape with respect to the medulla mosaic.

2nd hypothesis. Given that members of the remaining 74 cell types are randomly distributed throughout the medulla then there can no longer be particulate groups, as in the first hypothesis but instead columns whose basic 46 components are synperiodic but, taken together, whose compositions (derived from 46 synperiodic elements plus combinations of n of 74 cell types) are heterogenous. It is possible, too, that some cell types occur only once in the whole medulla. This means that the medulla may have a heterogenous structure some of whose parts function differently from others.

How can we resolve the dilemma of randomness versus order? If we take a section through the depth of the medulla one has the overriding impression that at any level the neuropil is homogenous across its lateral extent. There are, it is true, differences of fibre densities from front to back (BRAITENBERG, 1970) but these are probably merely a consequence of the cartesian distortions of cells due to the medulla's curvature and the subsequent differences in columnar packing.

The skeleton of the medulla is composed of columns which, in some respects, are equivalent to each other: the axes of the 24 perpendicular synperiodic elements form the homogenous columnar structure whereas the tangential and amacrine cells define the stratification and form its lattice work. This is the principle feature of medullary organization. Superimposed upon the elemental lattice are the remaining 74 cell types whose distribution relative to the medulla's hexagonal mosaic is at present questionable. If they are periodically arranged (either syn- or supraperiodic) then we can treat the whole region as a homogenous three-dimensional system of nerve nets. If, however, some cells are randomly arranged then it may be that no portion of the medulla can be isolated as a minimal structural unit, or medullon, containing representatives of all cell types.

References

BRAITENBERG, V.: Ordnung und Orientierung der Elemente im Sehsystem der Fliege. Kybernetik 7, 235 - 242 (1970).
STRAUSFELD, N.J., BLEST, A.D.: Golgi studies on insects. Part I. The optic lobes of Lepidoptera. Phil. Trans. Roy. Soc. Lond. B 258, 81 - 134 (1970).

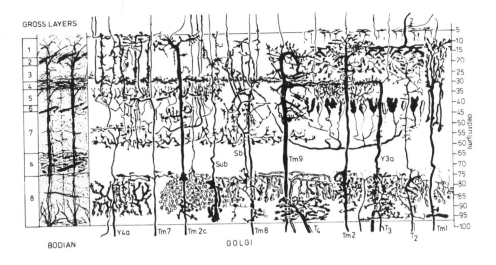

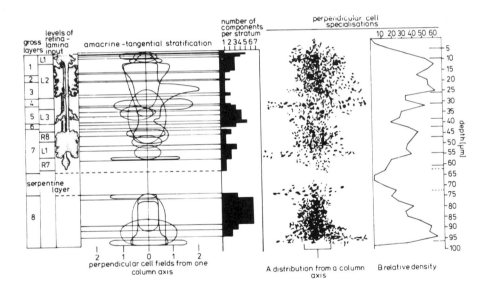

Fig. 3. Upper diagram shows gross layering of medulla from Bodian preparations. To the right are drawn the 13 synperiodic perpendicular elements, Tm, T- and Y-cells against 11 of the 13 amacrine cell types whose specializations occur synperiodically and which, with 8 tangential cell arborizations, form the strata. The amacrines arrowed 2 and 4 define gross layers 2 and 4. The lower diagram shows the gross layering against the levels of the lamina-retina inputs and the strata. Fields of all synperiodic perpendicular elements, around one column axis, are drawn against the strata. The histogram shows relative number of strata components. To the right are shown the distribution of perpendicular cell specializations from one column. The relative densities through the medulla are plotted extreme right. Maximum amacrine and tangential components occur beneath the serpentine layer; the majority of perpendicular components occur at the L2 level, coinciding with the minimum of stratum elements. In the upper two thirds of the medulla the majority of the stratum elements occur at the L1 and L3 level

5. The Fine Structure of the Compound Eye of Formica polyctena – Functional Morphology of a Hymenopterean Eye

R. Menzel

Department of Zoology, Institute of Technology, Darmstadt, Germany

Abstract.   Three aspects of the functional organization of the wood ant's compound eye are described. (1) The fine structure of the ommatidium: In addition to the dioptric apparatus of the cuticular lense and crystalline cone, the 140 μ long ommatidium consists of 8 retinula cells (6 large and 2 small ones), which extend over the whole length of the ommatidium. In the pro-ximal third part a nineth retinula cell appears, which also forms a rhabdomere. The fused rhabdom immediately joins the crystalline cone and consists of 6 rhabdomeres whose microvilli have a direction of their own. The retinula cells are numbered clockwise beginning with the small one pointing to the dorsal side of the eye. (2) Arrangement of the ommatidia in the eye: After descrip-tion of the pattern of cuticular lenses, the changing location of the ommatidia relative to a system of spatial coordinates is described, and the importance of the irregular pattern for an eye which is able to analyse the polarization plane of light are discusses. (3) Fine structural changes of light and chromatic adaptation: In exposure to light the intracellular vacuoles around the rhabdom split and move toward the outer part of the cell, where as the pigment granules of the retinula cells wander closely to the rhabdom. Eyes exposed to uv light (337 nm) show an adaptation in the cells nos. 1 and 5 (the two small cells); the six large cells show adaptation in yellow light (591 nm). Dependence on intensity of chromatic exposure is separately stated for all 8 retinula cells. The importance of this selective process of adaptation is discussed with regard to the con-trol of pigment movement.

The Red Wood Ant is among the few insects whose optical orientation has been thoroughly investi-gated. Early studies (survey: BRUN 1914) and later ones (VOWLES, 1954; JANDER, 1957; VOSS, 1967; KIEPENHEUER, 1968) showed that wood ants use the sun and the polarization pattern of the sky as a compass, learn landmarks, which they remember for a long time, and they have colour vision. They perceive form patterns by analysing their black areas, vertical edges and disruption. VOSS (1967) proved that the limit of physiological visual accuity is about 0.5°, and also found that in the frontal visual area resolution in a horizontal direction is different from that in a ver-tical one.

The wood ant is able to perform all these optical tasks with a compound eye which is in comparison to other Hymenoptera relatively small: it contains 750 ommatidia, in contrast the eyes of bees and wasps have more than 4'000 ommatidia. The small number of ommatidia, their specific structure, which makes identification of each retinula cell possible, and the great variety of behaviour patterns (spontaneous and trained runs) make the wood ant a suitable ocject for experiments aiming to clarify the interdependence of structure and function of the eye. I would like to go into details about three aspects of our problem: The fine structure of the ommatidium, the arrangement of the ommatidia in the eye, and changes in the fine structure of retinula cells caused by light and chro-matic adaptation.

1. The fine structure of the ommatidium

The compound eye of the wood ant is of the apposition type whose ommatidia form a central fused rhabdom. For the investigation the ommatidia of the central area of the dark adapted right eye were examined. Here the cuticular lenses have a medium diameter of 18.4 μ. The crystalline cone (25 μ long) is formed intracellularly by 4 Semper cells, whose curved nuclei are situated in the distal corner. The two kidney-shaped principal pigment cells surround the crystalline cone; they contain large pigment granules (0.66 μ Ø) accumulated in their proximal ends. The nuclei of retinula cells are arranged in two levels (fig. 1): 20 μ and 40 μ proximal to the end of the cone; the retinula cells, too, contain pigment granules which, with 0.37 μ Ø, are significantly smaller than those in the principal and long pigment cells.

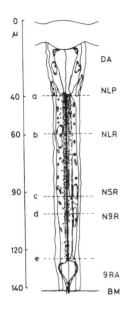

Fig. 1. Diagram of ommatidium in the central area of the ant's eye. DA, dioptric apparatus; NLP, nucleus area of long pigment cells; NLR, nucleus area of the 6 large retinula cells; NSR, nucleus area of the small retinula cells; N9R, nucleus of the nineth retinula cell; 9RA, formation of 9 retinula axons; BM, basement membrane; a, shows level of cross-section in fig. 2a; b, the same in fig. 2c

At the proximal end of the ommatidium the pigment granules are accumulated in basal pigment cells. Here the retinula cells form axons which after penetrating the basal membrane extend to the lamina lying 120 - 140 μ below. In its entire length from the cuticular lense to the basal membrane, the ommatidium is surrounded by 8 long pigment cells, which contain pigment granules with 0.45 μ Ø. The nuclei of the long pigment cells lie at the level of the border between the crystalline cone and the rhabdom. At the same level a cross section (fig. 2a) shows that the omma-tidium consists of 8 retinula cells, 3 large and one small pair. The small dorsal cell was numbered one, and the following cells were numbered clockwise; the line arbitrarily drawn through the center of the rhabdom and small retinula cells is called the ommatidium axis. Four fibrillar processes extend from the crystalline cone to the basal membrane closely along the rhabdom. Desmosomes can always be found in the near vicinity of the rhabdom connecting all 8 retinula cells.

During dark adaptation the rhabdom is surrounded by a system of large cavities formed by the endoplasmatic reticulum; the numerous pigment granules lie in the outer part of the retinula cells in dense plasma filled with many large mitochondria.

In section 3 I shall elucidate the changes in the fine structure by light adaptation, which occur in the distal parts (5 - 10 μ) of the retinula cells. Fig. 3 shows the exact spatial construction of that area: the 6 large retinula cells protrude in a distal direction between principal and long pigment cells as well as along the crystalline cone. The principal pigment cells form an identation in the underlying retinula cells. The two small retinula cells nos. 1 and 5 begin a few μ below as thin processes and attenuate downward. Thus only the rhabdomeres of the 6 large cells contact the crystalline cone directly.

The pigment granules accumulate in these places: in the principal pigment cells their number decreases in the distal direction, in the retinula cells in the proximal direction.

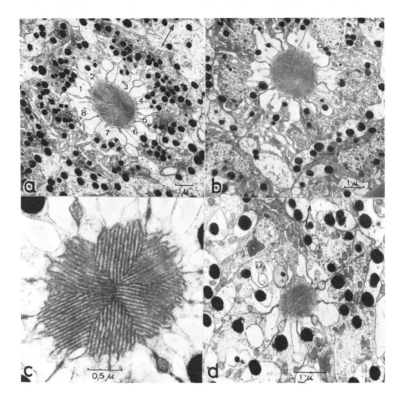

Fig. 2. Cross-sections through an ommatidium at various depths. a: a few μ below end of crystalline cone at the level of nuclei of long pigment cells (see fig. 1a); b: cross-section through rhabdom at the same level as in a; c: cross-section at level of six retinula nuclei (see fig. 1b); d: cross-section through end of rhabdom. Arrows point to the dorsal direction of the eye

As shown in the comparison of figures 2a and 2c this structure of the retinula cells is the same in the distal half of the ommatidium. Below the level of the nuclei of 6 large retinula cells, the small retinula cells widen, until their width become proportional to the other six and contain a nucleus each.

The fused rhabdom consists of microvilli 550 $\overset{\circ}{\text{A}}$ wide, which are arranged in six packages (fig. 2b): the few short microvilli of the small cells have an arrangement which is parallel to one of the two neighboring large cells. As a rule, the microvilli of each package have a direction of their own (see below). From cross-sections of the microvilli we can see that each microvillus contains an opaque central fibre (for details see MENZEL, 1972).

Below the nuclei of the two "small" retinula cells, a nineth retinula cell makes its appearance (MENZEL, 1972, fig. 7). Its beginning protrudes along cell no. 1 or no. 5; then it becomes wider without forming microvilli. Some μ below it contributes a few microvilli to the rhabdom. The microvilli pattern here is quite different from that in the distal half of rhabdom. The rhabdom consists only 4 or 5 microvilli packages. The rhabdomere of the nineth cell forms a separate microvilli package.

Thus the ommatidium of the wood ant is similar to that of the honey bee drone. PERRELET and BAUMANN (1969) also found 9 retinula cells, 6 large ones and 2 small ones along the entire length, and an additional nineth in the proximal third part. The worker bee, however, has only 8 retinula cells; one of them extends in two axons (VARELA and PORTER, 1969).

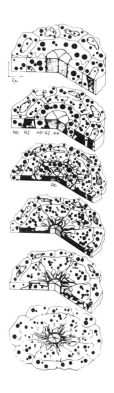

Fig. 3. Spatial diagram of area of transition from crystalline cone to rhabdom (see text). NP, long pigment cell; RZ, retinula cell; HP, principal pigment cells; KK, crystalline cone; Rb, rhabdom

The end of the rhabdom where the microvilli run only in one direction is about 1 μ long. Four basal pigment cells densely filled with pigment granules join the rhabdom. The axons, at their origin surround the basal pigment cells, then protrude between them and combine in a bundle at the level of the basement membrane. The diameters of the nine axons increase below the basement membrane up to 0.5 - 1.2 μ ; two or three axons are always thinner than the others. All axons are surrounded by basal pigment cells or glial cells over their entire length.

2. Arrangement of the ommatidia in the ant's eye

The long axis of the compound eye builds an inclination angle of 30° to the sagittal frontal axis of the head (fig. 4). The eye is not symmetrical; it bulges on the median side. Its dorsal and median parts are flatter. In the ventral area the ommatidia are arranged in regular rows and have a typical honeycomb form. In the dorsal half of the eye these rows ramify causing irregular forms of the cuticular lenses to be more frequent. The diameter of the cuticular lenses decreases from the ventral to the dorsal side: Central part of the 2. - 4. row (from ventral): 20 μ, 10. - 12. row: 19.4 μ, 17. - 19. row: 18.5μ, 23. - 25. row: 16,8 μ .In view of the fact that the dorso-medium part is flatter, we can expect a better spatial resolution in this part of the eye. This improved spatial resolution of the dorsal and frontal ommatidia can be viewed in its relationship to the accurate orientation with sky and land marks. The resulting diminished high speed of the dioptric apparatus is compensated by the higher light intensity from above. But more detailed analysis is still outstanding.

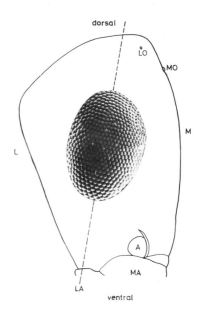

Fig. 4. Stereoscan picture of the right eye true to position and oriented in a diagram of the head seen laterally from the front. L, lateral; M, median; LO, lateral ocellus; MO, median ocellus; A, antenna; MA, mandible. Two rows around a medio-dorsal center are marked with black dots

We are aquainted with extremely regular patterned eyes in diptera (BRAITENBERG, this volume) and bees (BAUMGAERTNER, 1928). Such regularity is not to be found in the ant's eye, neither by the arrangement of the cuticular lenses (see above) or the retinula cells in the ommatidia. In cross-sections at the level of the nuclei of the long pigment cells the relative location of the ommatidia changes even where there are no ramifications in their rows (fig. 5): The deviation of the angles between the long axis of the eye and the axis of the various ommatidia is large: the median angle is 36°, the standard deviation $\pm$ 49° (n=200).

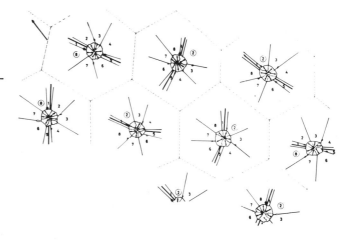

Fig. 5. Diagram of arrangement of 9 neighboring ommatidia in a cross-section at the level of nuclei of long pigment cells. Arrow in upper left-hand corner points in dorsal direction of the eye's long axis. The ommatidium axis in retinula cells nos. 1 and 5 is shown by a bolder line. The circles around nos. 2 or 8 indicate that the rhabdomere of cell 1 with that of cell 2 or 8 forms a common package

Some measurements are shown in the histogram (fig. 6). Besides a maximum at 10 - 20° there is another one at 110°. From this distribution we can conclude that the axis of the ommatidia differ widely, but two directions are preferred, which are perpandicular to each other.

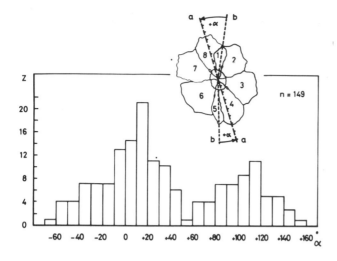

Fig. 6. Histogram of angle α between the eye's long axis and the axis of the ommatidium. Ordinate: numbers of ommatidia; abscissa: angle α in groups of 10° (see text)

Because the varying directions of the ommatidia, the microvilli in the six rhabdomeres, too, are arranged at different angles relative to a spatial system of coordinates. However, if the direction of the microvilli is measured in relation to the axis of their own ommatidium, the deviation is smaller: Retinula cell no. 2: $26^{\circ} \pm 10,5^{\circ}$, no. 3: $90^{\circ} \pm 12,5^{\circ}$, nos. 4 and 5: $159^{\circ} \pm 13,9^{\circ}$, no. 6: $210^{\circ} \pm 13,2^{\circ}$, no. 7: $270^{\circ} \pm 14,8^{\circ}$, nos. 8 and 1: $329^{\circ} \pm 14,1^{\circ}$.

To all those measurements the objection could be made that in cross-sections of the convex eye the neighboring ommatidia are cut at different depths. If they twist around their long axis the directions of the ommatidia and of the microvilli would differ in cross-sections. To disprove this the ommatidia were examined in series of cross-sections. An example is shown in fig. 7, three cross-sections of the same ommatidia at a distance of $2\,\mu$ are illustrated. Within those $4-5\,\mu$ no twisting can be found: ommatidia axis and microvilli direction each are the same at the various depths.

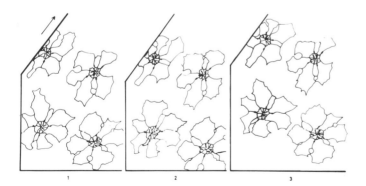

Fig. 7. Three cross-sections (1, 2, 3) through 4 ommatidia at three different levels (distance between each cross-section is $2\,\mu$). Arrow in upper left-hand corner points in dorsal direction. The direction of microvilli in each rhabdomere is indicated by a double arrow

Since ants perceive the direction of polarized light (JANDER, 1957) and as far we know the direction of the microvilli is in a constant relation to the polarization plane (KIRSCHFELD, 1969), the changing position of the retinula cells must be of consequence for the nervous connections of the retinula axons. How this pattern of connections is organized is unknown. There are principally two possibilities: (1) The position of the ommatidium is taken in such a way into account that in various ommatidia not the same pattern of connection prevails, but only the "right" retinula cells relative to a spatial system of coordinates reach the respective second neurons. (2) All ommatidia have the same pattern of connection, but the differences in the facilitation of synapses causes the "right" retinula cells to depolarize the respective second neurons. As the microvilli of the nineth retinula cells also have different directions (MENZEL, 1972, fig. 8), the same considerations will apply for them too. The second alternative could be decided upon, if we could find out whether ants must learn orientation by the polarization pattern; as yet we only know that ants (JANDER, 1957) and bees (LINDAUER, 1959) must learn compass orientation to the sun, whereas menotaxis is innate.

3. Changes in the fine structure of retinula cells caused by light and chromatic adaptation

With appropriate electron-microscopic methods it is possible for us to analyse the dynamic processes in the retinula cells caused by light. Here I would like to handle with the changes in the position of pigment granules and the control of this adaptation mechanism. When compairing fig. 8 with fig. 2a, are at once see the changes caused by exposure to light. The intracellular

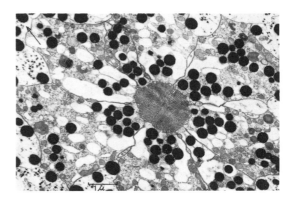

Fig. 8. Cross-section through light adapted ommatidium at level of nuclei of long pigment cells (40'000 Lx, Xenon arc)

cisternae surrounding the dark adapted rhabdom split into smaller vacuoles, which move into the inner part of the cell; some of the pigment granules wander closely to the rhabdom. Further changes in the fine structure have been described in another paper (MENZEL and LANGE, 1971). The effects of such pigment movement were investigated by KIRSCHFELD and FRANCESCHINI (1969): In Musca as well as in Formica the light is conducted through the rhabdom by total reflection; if diffraction changes in the vicinity of the rhabdom, the change of the angle of total reflection causes the light to emerge from the rhabdom, which then is absorbed or reflected by the pigment granules. KIRSCHFELD and FRANCESCHINI deduced that changes in the surrounding medium up to the width of about a wave length (0.5μ) control the light flux in the rhabdom. It is logical that the pigment movement takes place in the distal area of the rhabdom, for here the whole rhabdom is protected from too strong exposure.

How is the movement of the pigment granules controlled? Firstly, the light absorbed by the visual pigments might regulate this mechanism, possibly by the excitation of the sense cells, a hypothesis which KIRSCHFELD and FRANCESCHINI derived from their observations of the Musca eye. Secondly, a pigment dissolved in the cytoplasma (e.g. a riboflavine) or the ommochrome of the pigment granules (BUTENANDT, BIEKERT and LINZEN, 1958; DUSTMANN, 1966; HOEGLUND, LANGE, STRUWE and THORELL, 1970) might activate the movement. Since ants have ultraviolet and green receptors (see ROTH, this volume), we should expect that an analysis of pigment distribution after selective spectral adaptation will be helpful in solving the problem.

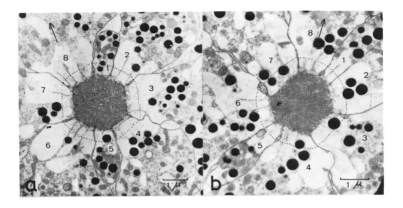

Fig. 9. Cross-section through (a) uv - and (b) yellow-adapted ommatidium at the level of nuclei of long pigment cells. The dotted line indicates a distance of 0.5 μ around the rhabdom

Fig. 9 shows cross-sections through ommatidia adapted to uv or yellow light. After adaptation to 337 nm the pigment granules accumulate closely to the rhabdom only in the small cells nos. 1 and 5; after adaptation to 591 nm the small cells show dark adaptation, whereas in the six large ones the pigment granules lie close to the rhabdom. A great number of oriented cross-sections at the level of 2 - 5 μ below the end of the crystalline cone (level of nuclei of long pigment cells) were examined to decide whether one or more pigment granules were lying 0.5 μ or nearer to the rhabdom; thus each cell was separately determined to be either adapted or not adapted. Fig. 10 shows dependence on intensity of exposure to 337 nm. For comparison, on the left side (D) the condition after a 12-hour dark adaptation is shown, and on the right side (L) after an adaptation to 40'000 Lx (light of a Xenon arc). By definition of adaptation (see above) the maximum of adaptation in the small cells is about 40%, in the large cells about 90%. The selective effect on cells 1 and 5 after uv exposure is clearly seen: With 0,018 W/m^2 the small cells reach the maximum, whilest the other cells are adapted only after a 20 times stronger exposure. If we compare this dependance of intensity with adaptation to 591 nm (fig. 10b), the selective effect on the 6 large cells is evident. Besides, it follows that the cells sensitive to uv (nos. 1 and 5) are at least twenty times more sensitive to 337 nm than the other cells to 591 nm. The high sensitivity of the small receptors is obviously the cause why their pigment granules move in exposure to yellow, while the large retinula cells do not yet reach their maximum.

The difference in the single retinula cells becomes more distinct, if in the area of medium intensity (337 nm: 0,018 - 0,056 W/m^2, 591 nm: 0,11 - 0,3 W/m^2) the number of adapted cells is entered in a histogram (fig. 11). The small cells are to a high degree selectively sensitive to uv light, the large ones selectively to yellow light. Between the 6 large cells there are no significant differences (Student-test). At the moment we are investigating in series of adaptation tests with blue light (447 nm) whether there are differences between the 6 large cells.

From these findings of selective pigment movement the following conclusions can be drawn:
1. The movement of the pigment granules is controlled by the visual pigments, and not by some pigments dissolved in the cytoplasma, or by the ommochromes of the pigment granules; in the latter cases there would be no differences between the 8 retinula cells.
2. Every ommatidium in the central area of the eye contains two cells sensitive to ultraviolet (uv-receptors) and 6 cells sensitive to yellow (green-receptors). In this way the basis of a visual system at least a dichromatic one is ascertained (see ROTH, this volume).
3. Since the microvilli contain the visual pigments (LANGER and THORELL, 1966), the uv visual pigment is to be expected in the microvilli of the two small cells. In the distal half of the rhabdom every small cell carries fewer and shorter microvilli than the six large cells; in the proximal half the rhabdomeres of the small cells are about the same size as those of the other cells. The content

of the uv visual pigment is supposedly less than that of the pigments sensitive to yellow. Never-theless sensitivity to uv is twenty times greater than to yellow, which can be derived from fig.10 and from experiments in spectral phototaxis (own experiments, unpublished).

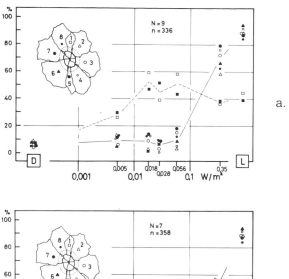

Fig. 10. Dependence of pigment movement on the intensity of spect-ral light in each of the 8 retinula cells (symboles see inserted diagram). (a) 337 nm, (b) 591 nm. D, percent-age of adapted cells after 12 hours' dark adaptation; L, percentage of adapted cells after exposure to Xenon light (40'000 Lx); N, number of fixation series; n, number of evaluated cross-sections. The me-dian data for cells nos. 1 and 5 are connected by a interrupted line, that for cells nos. 2, 3, 4, 6, 7, 8 by an uninterrupted line

4. The uv retinula cells presumably have a less active metabolism because of their smaller volume (in the distal half it measures one fifth of a large cell) and the smaller number of mitochondria (on an average 1,7 mitochondria per cross-section, 14,4 mitochondria in one of the large cells per cross-section; n=80). This might indicate that the sensitivity of the uv receptor depends on photoreisomerization, where as the resynthesis by metabolic processes is of lesser importance (cp. HAMDORF, GOGALA and SCHWEMER, 1971).

5. It remains an open question whether the pigment movement is caused by the depolarization of the receptor cell. If this can be proven, it would mean that the large and the small receptors are electrically isolated from each other; some current experiments seem to point in this direction. BUTLER (1971) also found that coupling between individual receptors must be relatively small. On the other hand SHAW (1969) measured interreceptor coupling in the ommatidia of honeybee and locust compound eye.

6. Changes in the structure of the microvilli as such GRIBAKIN (1969) found in bees could not be ascertained in exposure to white nor to chromatic light. In some series of experiments the microvilli did not contain opaque central fibres after strong light adaptation (more than 50'000 Lx); this effect, however, could not be reliably reproduced.

As yet there has been no investigation whether the ratio of two uv and six green receptors holds true for the whole eye. Also the question of the time-course of adaptation remains open, as well as the question what influence adaptation may exert on the dynamic properties and spectral

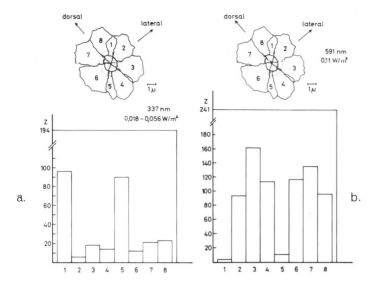

Fig. 11. Histograms for the number (Z) of adapted cells after exposure to (a) 337 nm (0,018 - 0,056 W/m^2) and (b) 591 nm (0,11 W/m^2), noted separately for each of the 8 retinula cells (abscissa). The sum of all evaluated ommatidia is given as uppermost number on the ordinate

sensitivity of the receptor, not to mention the whole eye. In the future those problems will be studied in behavioral experiments and by electrophysiological methods. We expect to benefit the small number of ommatidia in the ant's eye and shall try to determine eventual differences in the functional abilities of various areas of the eye.

References

BAUMGARTNER, H.: Der Formensinn und die Sehschärfe der Bienen. Z. vergl. Physiol. 7, 56 - 143 (1928).

BRUN, R.: Die Raumorientierung der Ameisen und das Orientierungsproblem im Allgemeinen. Jena: G. Fischer (1914).

BUTENANDT, A., BIEKERT, E. und LINZEN, B.: Ueber Ommochrome XIV. Zur Verbreitung der Ommine im Tierreich. Hoppe-Seyler's Z. physiol. Chemie 313, 251 - 258 (1958).

BUTLER, R.: The identification and mapping of spectral cell types in the retina of Periplaneta americana. Z. vergl. Physiol. 72, 67 - 80 (1971).

DUSTMANN, J.H.: Ueber Pigmentuntersuchungen an den Augen der Honigbiene Apis mellifica. Naturwissenschaften 53, 208 (1966).

GRIBAKIN, F.G.: Cellular basis of colour vision in the honey bee. Nature 223, 634 - 635 (1969).

HAMDORF, K., GOGALA, M. and SCHWEMER, J.: Beschleunigung der "Dunkeladaptation" eines UV-Rezeptors durch sichtbare Strahlung. Z. vergl. Physiol. 75, 189 - 199 (1971).

HOEGLUND, G., LANGER, H., STRUWE, G. and THORELL, B.: Spectral absorption by screening pigment granules in the compound eye of a moth and a wasp. Z. vergl. Physiol. 67, 238 - 242 (1970).

JANDER, R.: Die optische Richtungsorientierung der roten Waldameise, Formica polyctena. Z. vergl. Physiol. 40, 162 - 238 (1957).

KIEPENHEUER, J.: Farbunterscheidungsvermögen bei der roten Waldameise Formica polyctena. Z. vergl. Physiol. 57, 409 - 411 (1968).

KIRSCHFELD, K.: Absorption properties of photopigments in single rods, cones and rhabdomeres. In: Processing of optical data by organisma and maschines. Ed. W. Reichardt. New York, London: Academic Press, 116 - 136 (1969).

KIRSCHFELD, K. and FRANCESCHINI, N.: Ein Mechanismus zur Steuerung des Lichtflusses in den Rhabdomeren des Komplexauges von Musca. Kybernetik 6, 13 – 22 (1969).

LINDAUER, M.: Angeborene und erlernte Komponenten in der Sonnenorientierung der Bienen. Z. vergl. Physiol. 42, 43 – 62 (1959).

MENZEL, R.: Feinstruktur des Komplexauges der Roten Waldameise, Formica polyctena (Hymenoptera, Formicidae). Z. Zellforsch. 127, 356 – 373 (1972).

MENZEL, R. and LANGE, G.: Aenderungen der Feinstruktur im Komplexauge von Formica polyctena bei der Helladaptation. Z. Naturforsch. 26b, 357 – 359 (1971).

PERRELET, A. and BAUMANN, F.: Presence of three small retinula cells in the ommatidium of the honeybee drone eye. J. Microscopie 8, 497 – 502 (1969).

SHAW, S.R.: Interreceptor coupling in ommatidia of drone honeybee and locust compound eye. Vision Res. 9, 999 – 1029 (1969).

VARELA, F.G. and PORTER, K.R.: Fine structure of the visual system of the honeybee (Apis mellifera). I. The retina. J. Ultrastruct. Res. 29, 236 – 259 (1969).

VOSS, C.: Ueber das Formensehen der Roten Waldameise (Formica rufa-Gruppe). Z. vergl. Physiol. 55, 225 – 254 (1967).

6. Measurements on the Arrangement of Ommatidial Structures in the Retina of Cataglyphis bicolor (Formicidae, Hymenoptera)

P. L. Herrling
Department of Zoology, University of Zurich, Switzerland

Abstract. The orientation of the ommatidial axes of the central region of the eye of Cataglyphis bicolor shows a small range of variation. The microvilli of the 8 retinular cells present in the distal part of the ommatidia in the central eye region seem to have each a prefered orientation of their long axis which is important with respect to the perception of polarized light. The receptor cells for ultraviolet must be looked for amongst the four slim cells of the distal retinular part in the central eye region.

1. Introduction

The desert ant Cataglyphis bicolor is known to use visual stimuli while hunting over distances exceeding 150 m from the nest entrance (WEHNER, 1972; WEHNER and DUELLI, 1971; see also the contributions of DUELLI; WEILER; WEHNER and FLATT; BURKHALTER; TOGGWEILER in this volume). Parameters of this optical orientation are for instance the pattern of polarized light and horizon landmarks as deduced from experiments in the natural environment, and spectral colours, tested in the laboratory. This paper reports the results of preliminary measurements performed to show which are the possible receptor elements, able to perceive the visual stimuli mentioned above. It has been shown that the e-vector of polarized light is maximally efficient in those retinular cells that bear microvilli, the long axes of which extend in the same direction (LANGER, 1967; KIRSCHFELD, 1969). For the perception of polarized light one must presume a certain order of the microvilli throughout the retina or at least in some of its regions. An alternative would be an order to the same effect in the neuronal connections of retinular cells with equally oriented microvilli. Therefore we would like to present evidence from ultrastructural work showing whether such an order exists at the level of the retina.

2. Methods

The measurements have been performed on electronmicrographs of dark adapted, Xenon-arc (XBO, 450W) adapted or ultraviolet (341nm) adapted eyes (central region), fixed in an acrolein-glutardialdehyd, cacodylat-buffered (pH 7.2) mixture, postfixed in osmiumtetroxide, cut on a LKB Ultramikrotom and stained with uranylacetate and leadcitrate. As a spacial reference system we used the x, y, z axes described by BRAITENBERG (1967) for Musca and WEHNER, EHEIM and HERRLING (1971) for Cataglyphis and determined its orientation on ultrathin sections by means of a doubly asymmetrical pyramid.

3. Results

a) Structure of the ommatidium. As previous studies on the fine structure of the ommatidia have shown (WEHNER, HERRLING, BRUNNERT and KLEIN, 1972; BRUNNERT and WEHNER, in press), one ommatidium consists distally of 8, proximally of 9 retinular cells. If we first refer to the distal part of the retinula (the region distal of the nuclei) it can be subdivided in two parts: a distal portion connected with the tip of the crystalline cone, consisting of four large cells and four slim ones inserted between the large ones (4 + 4 type, see fig. 1). Just distal of the nuclei region (they are in the middle region of the retinula) two of the slim cells lying opposite to each other grow larger so that a 6 + 2 arrangement emerges (fig. 2).

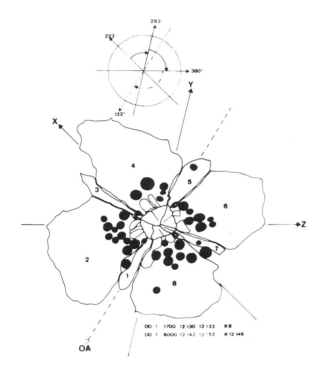

Fig. 1. Halfschematic drawing of an ommatidium cut perpendicular to its long axis just under the crystalline cone

On that level the two slim cells have been numbered 1 and 5. Number 1 beeing the small cell making an angle at the nearest ninety degrees to the z-axis in the manner shown in fig. 1 and 2. The sense of increasing numbers was determined after the +(-) x, + (-) y, + (-) z direction (see arrows in fig. 1 and 2). A line going through cells 1 and 5 and through the middle of the rhabdom defined as the ommatidial axis (OA in figs. 1 and 2). All angle measurements have been done in the x, y, z direction mentioned above, the + z axis beeing defined as the zero direction. This give comparable measurements in different eyes as this system is eye-constant. The ninth cell in the proximal retinula is situated next to cell No. 1 or 5, but it can not be distinguished from them, so that depending on where it appears, 1 and 9 or 5 and 9 are called 1A and 1B respectively 5A and 5B.

b) The orientation of the ommatidial axis. Within a region of 20-40 ommatidia the ommatidial axes are very constant referring to the coordinates of the eye. In some examples the direction and length of the mean vectors are: $\propto = 107.35^\circ$, r = 0.98 (XBO-3, 11 896-11 905); $\propto = 120.21^\circ$, r = 0.98 (XBO-3,11 931-11 935); $\propto = 76.99^\circ$, r = 0.93 (DD-1,12 130-12 133). The angles of the eye named XBO-3 are all within a narrow range. Measurements of the eye DD-1 are not exactly from the central region but from a region above it (dorsally). Here we notice a somewhat greater inhomogenity of the values (r = 0.93) and another mean direction. In contrast to Formica rufa (MENZEL, this volume), the retina of Cataglyphis is characterized in respect to the ommatidial axes by a high degree of order at least within the central region of the eye. Fig. 3 shows a plan from which measurements of the ommatidial axes were taken.

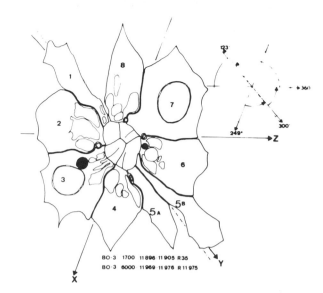

Fig. 2. Halfschematic drawing of an ommatidium cut as
in fig. 1, but more proximally
in the upper nuclei-region.
Note the ninth cell (5A or
5B)

c) The orientation of the microvilli long axes.   The next step in our investigation was to determine the degree of order in the microvilli directions (see introduction). As for the ommatidial axes,
the orientation of the microvilli were measured in angles to the +z axis. It is obvious that only
angles from 0° to 180° are needed for these calculations. The results are given in fig. 4. In the
upper graph (uppermost trace) we see that the values of all retinular cells are widely distributed
over the whole range with slight peaks at 60° and around 120°. From the traces 1a-8 in the upper
graph it becomes obvious that these directions seem to be preferred directions for single retinular
cells. Calculating the arithmetic mean angle for each cell results in the curve shown in the lower
graph of fig. 4. The question now arises, whether there are phase shifts in this curve, if it is calculated from a number of different eyes and if the tendencies shown here can be statistically
confirmed.

d) Spectral sensitivity types of retinular cells.   To determine which retinular cells are sensitive
to ultraviolet, we referred to the method of selective adaptation as it was used by BUTLER (1972)
for Periplaneta. From BRUNNERT and WEHNER (in press) and WEHNER, HERRLING, BRUNNERT
and KLEIN (1972) we know that in the light adapted retinula there is a migration of the pigment
granules in the cells 2, 4, 6, 8 (the four large cells) in the distal third of the retinula to the rhabdom and away from it in the dark adapted condition. If any of these cells should be sensitive to
ultraviolet then they should be in the light adapted state after several hours (6 - 12 h) of dark
adaptation and subsequent illumination by monochromatic light of 341 nm. This wavelength has
been chosen, as it lies near the maximum of the sensitivity distribution found by WEHNER and
TOGGWEILER (1972). The electronmicrographs show that these four large cells are in a dark
adapted condition after the above mentioned treatment. This indicates that the ultraviolet receptors must be looked for amongst the slim retinular cells (1, 3, 5, 7). This finding corresponds with
fig. 9b in WEHNER, HERRLING, BRUNNERT and KLEIN (1972) as well as with the results of
MENZEL reported in this volume for Formica rufa. There is now work in course to precise the
position of ultraviolet sensitive cells and of cells sensitive to other wavelengths found by WEH-
NER and TOGGWEILER (1972) to be lying near the maxima of the sensitivity distribution.

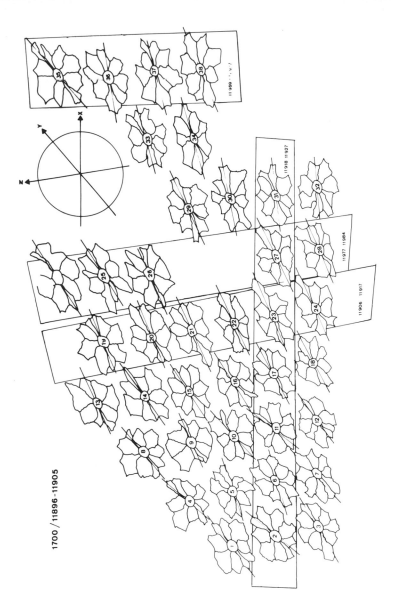

1700 / 11896–11905

Fig. 3. Plans on which the measurements of the angle of the ommatidial axes were taken

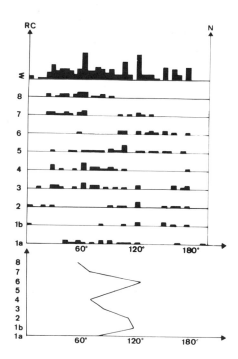

Fig. 4. The orientation of microvilli. For explanation see text

Acknowledgement. The work was financially supported by a grant offered to RUDIGER WEHNER by the Fonds National Suisse de la Recherche Scientifique, No. 3.315.70.

References

BRAITENBERG, V.: Patterns of projection in the visual system of the fly. I. The retina-lamina
   projection. Exp. Brain Res. 3, 271 - 298 (1967).
BRUNNERT, A., WEHNER, R.: Fine structure of light and dark adapted eyes of desert ants, Cata-
   glyphis bicolor (Formicidae, Hymenoptera). J. Morph., in press (1972).
BUTLER, R.: The identification and mapping of spectral cell types in the retina of Periplaneta
   americana. Z. vergl. Physiol. 72, 67 - 80 (1971).
KIRSCHFELD, K.: Absorption properties of single rods, cones and rhabdomeres. Proc. Internat.
   School of Physics ENRICO FERMI: Processing of optical data by organisms and machines.
   New York, Academic Press (1969).
LANGER, H.: Wahrnehmung von Wellenlänge und Schwingungsebene des Lichtes. Verh. Dtsch.
   Zool. Ges., Zool. Anz. Suppl. 30, 195 - 233 (1967).
WEHNER, R., DUELLI, P.: The spatial orientation of desert ants, Cataglyphis bicolor, before
   sunrise and after sunset. Experientia 27, 1364 - 1366 (1971).
WEHNER, R., EHEIM, P., HERRLING, P.L.: Die Rastereigenschaften des Komplexauges von Cata-
   glyphis bicolor (Formicidae, Hymenoptera). Rev. Suisse Zool. 78, 722 - 737 (1971).
WEHNER, R., HERRLING, P.L., BRUNNERT, A., KLEIN, R.: Periphere Adaptation und zentral-
   nervöse Umstimmung im optischen System von Cataglyphis bicolor (Formicidae, Hymenoptera).
   Z. vergl. Physiol. 79, 197 - 228 (1972).
WEHNER, R., TOGGWEILER, F.: Verhaltensphysiologischer Nachweis des Farbensehens bei Cata-
   glyphis bicolor (Formicidae, Hymenoptera). Z. vergl. Physiol. 77, 239 - 255 (1972).

7. The Ultrastructure of the Photosensible Elements in the Eyes of Collembola and Their Orientation (Insecta)

H. F. Paulus
Department of Zoology, University of Vienna, Austria

Abstract.   Collembola maximally have eight ommatidia on each side of the head, forming a so-called "atypical complex eye". Their constructions are remarkably different within the order. In respect to their symmetries the rhabdoms of the known species can be distinguished in (1) completely irregular rhabdoms (Orchesella), (2) radial symmetrical rhabdoms (Tomocerus, Sminthurides), (3) bilateral symmetrical rhabdoms (Podura, Neanura, Entomobrya, Allacma (Sminthurus). All rhabdoms consist of eight retinula cells (except in Sminthurides) which form a photosensible part in a very different manner. In Orchesella, Allacma and Entomobrya you may find eight, in Tomocerus seven, in Podura four or six (double eye!), and in Neanura six rhabdomeres. In the compound eye of Podura all rhabdoms considered as cross-sections (fig. 3) lie parallel to one another (disregarding the curvature of the complex). In that of Entomobrya they lie in distinct angles of 30°, 60°, 90°, and 120° to one another (fig. 4).

Investigations of ommatidia or of the Arthropoda are very popular. The main interest lies on the photosensible part which forms a rhabdom in very different manners. Naturally the authors mostly investigate such species whose behaviour is best known. Here it is easiest to draw conclusions from structure and function. The present investigation on Collembola primary dealt with phylogenetical aspects. Nevertheless the question of function of these ommatidia seems of great interest The more as these animals are very primitive, their optical capacity is of a very low degree. But this seems to me a chance to approach to the very complex mechanism of the composed eyes of the higher insects. Characteristically Collembola have eight relatively big and round ommatidia which form a so-called "atypical compound eye". Mostly they are of different size within one complex, they often are reduced, sometimes they are missing completely. If present, they can be seen as two big black pigmented areas lying either on the lateral parts of the head just behind the antennae or dorsally between antennae (so in Sminthuridae).

For a long time the lateral eyes of Collembola were supposed to be simple ocelli as they are typical for the Myriapoda. WILLEM (1887) was the first to show that these eyes are ommatidia of the eucone type. HESSE (1901) gave us the first more detailed analysis of this ommatidia whose result can be found as the only known scheme of the eye of Collembola in the manuals (last in SCHALLER 1970). According to HESSE, the ommatidia of Orchesella consists of an eucone crystalline cone, two corneagean cells and a two-layered retina composed of seven rhabdomeres. I showed the retina of this genus to be of one layer composed of eight cells (PAULUS, 1970, 1972).

Analysis of the construction of the rhabdoms shows great differences between the investigated species so that a single scheme of the Collembolan eye cannot be given. According to their symmetry the following types can be distinguished: (1) a completely irregular rhabdom in Orchesella villosa (fig. 1 A), (2) radial symmetrical rhabdoms in Tomocerus (fig. 1 B) and Sminthurides (fig. 1 E), (3) bilateral symmetrical rhabdoms in Podura (fig. 1 D), Neanura, Entomobrya (fig. 1 C) and probably in Allacma fusca (fig. 1 F). Except in Sminthurides all ommatidia are composed of eight sensory cells, but not necessarily all those cells contribute to the forming of a rhabdom.

1. Results

a) Orchesella villosa. The photosensible part consists of eight rhabdomeres. The composition
to a rhabdom is very irregular. The picture in fig. 1 A shows only one possible composition of
the rhabdomeres. The single rhabdomeres are often interlocked, sometimes the microvilli-borders
are interrupted alongside of the cell. These microvillis are unusually big. Their diameter is 540 –
570 Å, their length varies greatly, they can be as long as 5 μ. The direction of corresponding
rhabdomeres also varies so that diverging angles can be found in the different ommatidias. The
rhabdom is lying just beneath the crystalline cone and reaches up to the sides of the cone. In
the distal two third the rhabdom is completely compact, but proximally the microvilli-borders
separate from each other. As a whole the diamter of the rhabdom is in the central part about
8 – 10 μ and the length is 20 – 25 μ. In respect to rhabdoms of other insects these ommatidia
have a considerable volume which varies greatly from ommatidium to ommatidium because the
microvilli-borders often are shortened, interrupted or altogether missing.

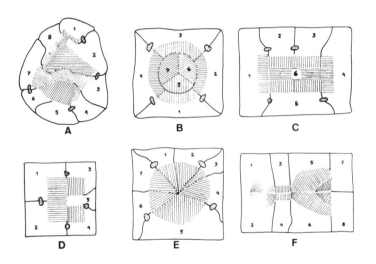

Fig. 1. Central cross-sections
through the rhabdoms of different species of Collembola.
A. Orchesella villosa, B. Tomocerus vulgaris, C. Entomobrya sp., D. Podura aquatica,
E. Sminthurides aquaticus,
F. Allacma (Sminthurus) fusca

b) Tomocerus vulgaris and T. longicornis. In these two species the rhabdom consist of seven
rhabdomeres. The eighth retinula cell is only rudimentary with no microvilli-borders. In contrast
to the rhabdom of Orchesella the rhabdomeres of this genus form a more regular radial symmetrical rhabdom. But even here the composition to a rhabdom varies. First there is a ring of four
cells whose microvillis build right angles to each other. This ring reaches far up to the side of
the crystalline cone. They extend to the whole length of the rhabdom, but in the last third they
separate from each other. In the central part are three further retinula cells. Their microvilli-
borders form a compact cone and angles of about 120° to each other (fig. 1 B). This central part
can touch the crystalline cone at its base or may be located more proximal. In this case the four
rhabdomeres of the outer ring touch each other in the middle. Sometimes the three central rhabdomeres are separated from the four of the ring either of their whole length or in the basal one-
or two thirds. In the first case we may call it an "open rhabdom".

c) Sminthurides aquaticus. This springtail is living on the water surface of small pools. The
ommatidium of this species contain a typical radial symmetrical compact rhabdom similar to that
of Orthoptera (HORRIDGE, 1966; WACHMANN, 1969) or of Camponotus (Formicidae) (WOLKEN
1971). The circular rhabdom consists of six to seven rhabdomeres as can be seen in fig. 1 E. The
whole rhabdom is uniformly built in this way. A more detailed description will be given in a
later paper.

d) Podura aquatica and Neanura sp.   These two species belong to the primitive representatives
of the Collembola (Poduromorpha). Whereas Podura also lives on the surface of small pools and
has eight ommatidia, Neanura lives hidden under barks and usually has only three ommatidia.
The rhabdom of both species are built not all of them being rhabdomeres. In Podura you will find
four microvilli-borders in the distal part, two of them forming right anles to each other (fig. 1 D).
The cells 3 and 4 are parted by a process of cell 5 lying beneath, so that there is an open space
between these two rhabdomeres. These two microvilli-borders regularly are lacking in the upper
two ommatidia of the complex. That means we find here a sort of a double eye (PAULUS, 1970).
In the proximal part of the rhabdom only these two cells lying parallel under the cells 1 and 2
have microvilli-borders. The rhabdom of the dorsal eye contains four, that of the ventral eyes
six rhabdomeres. But all of the eight retinula cells have axones. The Neanura rhabdom is almost
identical to that of Podura, only cells 3 and 4 touch one another to form a rectangular or square
rhabdom.

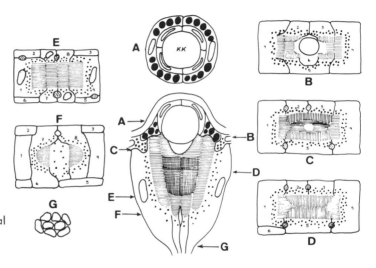

Fig. 2. Reconstruction of the
ommatidia of Entomobrya sp.
As exception of all known
Arthropleona this genus contains in its ommatidia primary
pigment cells. The rhabdom is
extremely bilateral consisting
of eight retinula cells. In the
middle is a central longitudinal
section, the fig. A - G are
cross sections (arrows)

e) Entomobrya sp.   This species has a long extremely bilaterally built rhabdom consisting of eight
rhabdomeres (figs. 1 C, 2). Because of insufficient material and an error in the sequence of sections
my interpretation of the rhabdom of this species was not quite right (PAULUS, 1972). After reexami-
nation of the retinula cells the rhabdom can be presented in the following short description:

In its length the rhabdom consists of three layers. In the distal third the microvilli-borders of six
cells reach far up to two opposite sides of the crystalline cone (fig. 2 B). Only in the distal part
of the rhabdom one cell is lying in the center with rhabdomeres on two opposite sides. More
proximal this cell is disappeared. Here the rhabdomeres of the cells 2, 3 and 5 touch each other
in the middle (fig. 2 D) building right angles with their microvillis. In the basal part also these
cells are ending. In place of these cells we now find two further cells (no. 7 and 8) which are
lying parallel with their microvillis between the cells no. 1 and 4 (fig. 2 E). In fig. 2 F there
can only be seen the rhabdomeres of the cells 7 and 8.

f) Allacma fusca.   The construction of this rhabdom is not completely known. In cross-sections
it is a longish rhabdom consisting of seven or eight rhabdomeres. Three different types of rhabdomeres have been found, but probably they show three different levels of only one ommatidium as
seen in Entomobrya. Fig. 1 F shows one of these possibilities.

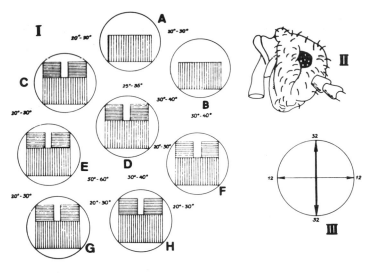

Fig. 3. Distribution of the eight ommatidia in the "double eye" of Podura aquatica. I. The rhabdoms within the circles are shown as cross sections. The divergence angles are declared between the ommatidia. By neglecting these angles the rhabdoms are lying parallel to each other. II. The head of Podura in natural position (lateral). III. The microvilli directions of all rhabdomeres are noted in this circle. There are 32 cells in vertical and 12 cells in horizontal direction

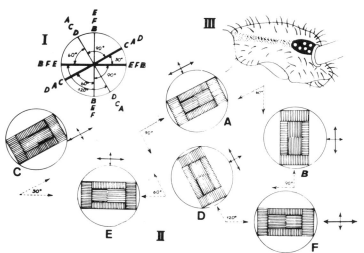

Fig. 4. Distribution of the six ommatidia of an undetermined species of Entomobrya. The head in natural position (III). The rhabdoms within the circles are shown as cross section Here the rhabdoms build distinct angles to each other as declared between the circles (II). In I the microvilli directions of all rhabdomeres are noted in this circle. There can be seen two main directions (big axes). One of them lies parallel to the long axe of the animal, the other is twisted by 30°

2. Discussion

It seems to me of great importance for a functional analysis to clear up not only the dioptric apparatus and the innervation of the retinula cells but also the orientation of the rhabdomeres within the whole composed eye. Hardly any author has tried to do that, not least because of the multitude of facets and the curvature of the complex. Nevertheless, those few publications dealing with these questions show that at least in the central part of the complex the rhabdoms are parallel to one another (disregarding the curvature of the complex, i.e. the diverging angles) (SCHNEIDER and LANGER, 1969; DANNEEL and ZEUTSCHEL, 1957). Only between dorsal and ventral part of the composed eye you find symmetrical distortions by 180°, or as in the double eye of Gerris the rhabdom can here be seen as the so-called "ventral-type". But here also the rhabdoms lie in exactly the same position to one another within a district. In Collembola an analysis of this kind is difficult as here the diverging angles of the ommatidia lie between 40° to 60°. In Podura (fig. 3) these angles are smaller (20° to 30°). Here the rhabdomes similar to the higher insects lie parallel to one another. Summing up all microvilli-directions you find only two directions, the vertical one (vertical to the watersurface by natural position of the insect's head) consisting of 32 and the horizontal one of 12 rhabdomeres. The distribution of angles of rhabdoms in Entomobrya (fig. 4) seems very interesting. Here you can find six ommatidia in a typical arrangement. In ommatidia A and C or E and F the extremely bilateral rhabdoms lie parallel to each other, whereas those of A and C are distorted by 30° to those of E and F. Ommatidium D with its rhabdom is distorted to that of A by 90°, to that of F by 120° and to that of E by 60°. B is distorted to A by 60°, to F by 90°, which means that D is vertical to A and C and B to E and F. Summing up all microvilli-directions (fig. 4, I) you find two preferred directions again, one parallel to the longitudinal axis of the animal, the other distorted by 30°. I cannot say what that means. But it seems important to me that these rhabdomes reach far up to two opposite sides of the crystalline cone so that obliquely incident light also may be perceived, but only light falling in the directions of the diameters shown in fig. 4, I. Whether this is a directive component in the optical reactions or this is an analysing system of polarized light, by division of labour between or coordination of all ommatidia has to be studied. On the whole it would be interesting to find out if there is any correlation between the construction of a rhabdom and its optical capacity. But here Entomobrya and also the other known Collembola seem to show that it is not so, as the optical capacities known to us until now are no more than a first step in the distinguishing of shapes (SCHALLER, 1969). But possibly you can find very different kinds of rhabdoms in different species of Collembola. Such a variability is not known in any insect order.

References

DANNEEL, R., ZEUTZSCHEL, B.: Ueber den Feinbau der Retinula bei Drosophila melanogaster. Z. Naturforsch. 12 B, 580 – 584 (1957).

HESSE, R.: Untersuchungen über die Organe der Lichtempfindungen bei niederen Tieren. VII. Von den Arthropodenaugen. Z. wiss. Zool. 70, 347 – 473 (1901).

HORRIDGE, G.A.: The retina of the locust. in "The functional organization of the compound eye" (BERNHARD, C.G. edit.) Pergamon Press Oxford, 513 – 541 (1966).

PAULUS, H.F.: Zur Feinstruktur des zusammengesetzten Auges von Orchesella. Z. Naturforsch. 25 B, 380 – 381 (1970a).

PAULUS, H.F.: Das Komplexauge von Podura aquatica, ein primitives Doppelauge. Naturwiss. 57, 502 (1970b).

PAULUS, H.F.: Der Feinbau der Komplexaugen einiger Collembolen, eine vergleichend anatomische Untersuchung. Zool. Jb. Anat. 89 (1), 1 – 116 (1972).

SCHALLER, F.: Zur Frage des Formsehens bei Collembolen. Verh. Dt. Zool. Ges. 1968 (Kiel), Zool. Anz. Suppl. 32, 368 – 375 (1969).

SCHALLER, F.: Collembola. in Handb. Zool. 4 (2), 2/1, 72 pp., Walter de Gryter, Berlin (1970).

SCHNEIDER, L., LANGER, H.: Die Struktur des Rhabdoms im Doppelauge des Wasserläufers Gerris lacustris. Z. Zellf. 99, 538 – 559 (1969).

WACHMANN, E.: Zum Feinbau der Ommatidien von Pteronemobius heydeni (Orthoptera, Gryllidae). Z. Zellf. 99, 263 – 276 (1969).

WILLEM, V.: Recherches sur les Collemboles et les Thysanoures. Mem. C. et Mem. Sav. Etrang. Acad. Roy. Belg. 58, 1 – 144 (1887).

WOLKEN, J.J.: Invertebrate photoreceptors, a comparative analysis. Acad. Press, New York, 179 pp. (1971).

II. Optics of the Compound Eye

1. The Visual System of Musca: Studies on Optics, Structure and Function

K. Kirschfeld
Max-Planck-Institute of Biological Cybernetics, Tübingen, Germany

Abstract.   Experiments were performed in order to obtain information on the function of receptive and neural structures, and to relate this to the histology. The study of the optical properties of the ommatidia, and the "wiring diagram" of the first optic ganglion as evaluated by means of histological methods (BRAITENBERG, 1967; TRUJILLO–CENOZ and MELAMED, 1966), lead to the hypothesis that the unfused rhabdoms of the Dipteran eye, combined with the special neural connexions between retina and lamina are a means of increasing the light gathering power in this type of compound eye by "neural superposition" of the output of receptors nos. 1 to 6. The concept of neural superposition predicts functional differences between the system composed of receptors 1 to 6 and the other composed of receptors 7 and 8. These differences, concerning absolute and spectral sensitivity, sensitivity to polarized light, and contrast transfer, have been demonstrated by means of optomotor experiments. One other consequence of the concept of neural superposition is that one ommatidium alone should be able to analyze movement. This has been verified by stimulation of pairs of receptors within one single ommatidium. By means of this technique it has been shown, furthermore, that movement detectors in the upper front region of the eye are arranged in two orthogonal directions. This information has been used in order to draw a network of connexions which are the least necessary and sufficient to explain the experimental results. These connexions are also the minimal ones that must be found in the histology of the information carrying channels.

1. Introduction

An analysis of the manner in which nervous systems function may be carried out by means of various methods. This paper gives an account of experiments which were made in order to obtain information on the function of the Musca–eye and the optic ganglia, and if possible to relate these functions to structures that can be described histologically. In the case that the present state of anatomical knowledge is insufficient to allow comparison with obtained physiological data, the experiments should give as precise information as possible as to a required histological basis.

Besides for this aim – of attributing to certain structures the capacity for taking up information and processing it – a further aim was to make use of the precise knowledge of the functional capacity of individual structures in order to predict consequences in behaviour and orientation, which could then be tested by specifically designed experiments. Furthermore, it is hoped that by the application of these methods, the system being investigated can be more thoroughly understood than would be possible by the use of methods which do not allow a correlation of function and structure.

The experimental methods used were optical investigations on the compound eye of Musca and experiments, in which optically stimulated turning reactions of experimental animals were measured. Furthermore considerations on a necessary and adequate structure had to be taken into account, which started from the considerations of HASSENSTEIN and REICHARDT (1953), and which had the aim of making statements on the required histological structure.

Of course, it is true that electrophysiology would be one of the most direct methods of obtaining information on the function of individual cells; however, results of electrophysiological experiments will not be reported here. Electrophysiology - however ideal it would seem at first - has certain difficulties of interpretation inherent in it; it has several times led us astray in our conceptions of the characteristics of the Dipteran eye. As an example, electrophysiological measurements of sensitivity-distributions of receptors in the Musca eye (KIRSCHFELD, 1965) gave half-width values which are, as we now know, too high by about a factor of two (see SCHOLES, 1969). This was because the methods available at that time did not allow us to determine the position of the electrode tip with sufficient accuracy. Moreover, changes in the dioptrics as a result of opening the head capsule could not be controlled. - But even with improved techniques, electrophysiology has not been able to show that the Musca eye contains two receptor types, each with a different width of sensitivity distribution, as has clearly been shown by optical methods (KIRSCHFELD and FRANCESCHINI, 1968) and in behavioural experiments (KIRSCHFELD and REICHARDT, 1970; ECKERT, 1971). Other authors have drawn the conclusion from electrophysiological data that the Calliphora eye contains 3 types of photo-receptors, each with a different spectral sensitivity; this is as we know now almost certainly not true; it was also concluded that about half of the photo-receptors in the Calliphora eye are insensitive to a change in the position of the E-vector of linearly polarized light, which is not the case, they are all polarization sensitive. Further examples could be cited concerning this matter.

The chief advantage in electrophysiological technique is that with the aid of micro-electrodes it is possible to record the activity of individual cells and so to gain an insight into the "black box" that is the interior of the nervous system being examined. Certain peculiarities of the Musca compound eye, particularly the fact that the rhabdomeres of the sensory cells of each ommatidium are unfused, allow the application of another technique, by means of which it is also possible to "open" the visual system: by means of stimulating separately individual photo-receptors of the ommatidia, individual elements can be isolated and thus information on the functional and structural interactions between single cells can be obtained.

2. The "neural superposition" in the lamina

The visual system of Musca is composed of the ocelli, the compound eyes and the 3 optical ganglia (lamina, medulla and lobuli). Each compound eye consists of about 3000 similarly constructed subunits, the ommatidia. The sensory cells of the ommatidia form the retina. Between the retina and the lamina there is a network of nerve-fibres with a remarkable interweaving of the fibres. The fact that these structures are obviously arranged according to a strict set of rules has constantly inspired anatomists and physiologists to put forward hypotheses on their possible function. Some of these hypotheses may at least be mentioned here (for more detailed information see KIRSCHFELD, 1971). For instance, these structures have been related to the ability to distinguish colours, forms and movements, as well as to the ability to analyse polarized light; it has also been suggested that lateral inhibitory processes may proceed along these fibres.

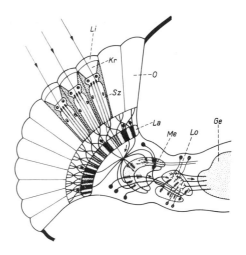

Fig.1.   Horizontal section through the compound eye of Musca, with the corresponding brain parts. Li - corneal lens, Kr - crystalline cone, Sz - photo-receptors, O - ommatidium, La - lamina, Me - medulla, Lo - lobuli, Ge - brain (from KIRSCHFELD, 1971)

As we have already seen, a peculiarity of the Musca eye lies in the fact that the rhabdomeres of the photo-receptor cells are unfused, and this, as AUTRUM and WIEDEMANN (1962) and de VRIES and KUIPER (1958) were able to show, has the consequence that the optical axes of the 7 thabdomeres of an ommatidium are not parallel, but diverge. A more detailed analysis of the situation has shown that - with exception of certain eye areas - 7 rhabdomeres from 7 different ommatidia are always directed to one and the same point in the optical surroundings (KIRSCHFELD, 1967). Starting with the assumption that the axons of these receptors are brought together respectively in the subunits of the lamina called the cartridges, a suitable oriented section will show the connecting scheme between the retina and the lamina with its typical fibre interweaving, as in fig. 1. An imaginary section cut approximately at right angles to the axes of the ommatidia shows a relatively complicated network of nerve fibres between the retina and the lamina (fig. 2). Histological examinations have shown that this connecting scheme between retina and lamina is realized in the Dipteran eye (BRAITENBERG, 1967; TRUJILLO-CENOZ and MELAMED, 1966).

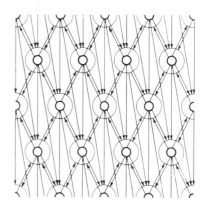

Fig.2.   If the axons of those retinula cells 1 - 6, which "look" at one and the same point in the optical surroundings are drawn converging on one and the same cartridge of the lamina the result will be the scheme of connexions shown here between ommatidia of the retina (small circles) and cartridges of the lamina (large circles) (from KIRSCHFELD, 1967)

A special part is played by the axons of the receptor cells 7 and 8, which form the rhabdomere no. 7: they bypass the lamina without synaptic contact and are finally connected synaptically to second order neurons in the medulla.

Thus 6 axons from 6 receptor cells out of 6 different ommatidia converge onto one single cartridge. Since the optical axes of these receptor cells are all directed to one and the same point in the surroundings, the signals entering one cartridge represent one point of the optical surround.

After the facts had been clarified so far, the question arose as to the functional meaning of the relatively complicated projection, first of the surroundings onto the retina, and second of the retine onto the lamina. To answer this question it was first of all necessary to take into account the functional characteristics of the photo-receptors. During the last years investigations carried out on Dipteran eyes by means of different methods have shown: (1) The rhabdomeres of the receptor-cells 1 to 6 all have the same spectral extinction, its maximum lying at approximately 515 nm; in contrast, the extinction maximum of rhabdomere no. 7 is at approximately 470 nm (LANGER, 1966). (2) The rhabdomeres of all the receptor cells nos. 1 to 7 absorb dichroically, as do probably also those of the receptor-cells no. 8; the directions of maximum extinction are set parallel to the direction of the microvilli (receptors 1 - 6) or vertically to them (receptor 7) (LANGER, 1966; KIRSCHFELD, 1969). (3) The receptors 1 - 6 have a sensitivity distribution which extends further than that of receptors nos. 7 and 8 (KIRSCHFELD and FRANCESCHINI, 1968; ECKERT, 1972).

The knowledge of the projections first of the optical surroundings onto the retina and second of the retina onto the lamina, and the knowledge of the receptor characteristics just described, already allow one to exclude the possibilities, earlier formulated as hypotheses, on the question of the function of the retina-lamina projection. If all the known facts are taken into consideration, it can be shown that these complicated arrangements are a means in order to provide an eye with an increased light gathering power. This can easily be explained by means of fig.1: parallel light entering the eye illuminated not only one rhabdomere of one ommatidium optimally, as in an apposition eye, but it also illuminates 7 rhabdomeres of 7 ommatidia three of which are shown in the section fig.1. This means that the efficient aperture of this eye, and thus its light gathering power, is increased by a factor of 7 in comparison with the classical apposition eye. This increase in light gathering power becomes especially efficient for the visual process as far as signal to noise ratio is concerned when the signals triggered in the axons of the receptor cells are superimposed in the cartridges of the lamina. The structural requirement necessary for this superposition is at least given for the two second class neurons (L_1 and L_2) which are each synaptically connected with the 6 receptor-cell axons entering the cartridge (TRUJILLO-CENOZ, 1965; BOSCHEK, 1971). In analogy to the optical superposition eye described by EXNER (1891), the Musca eye may be described as a "neural superposition" eye.

Neural superposition however can be valid only for the signals of the receptor cells nos. 1 to 6, since only their axons are synaptically connected in the cartridges with second order neurons. This means that the Musca eye is composed of two sub-systems, formed by the receptors 1 - 6 on the one hand and those of the receptor cells 7 and 8 on the other (KIRSCHFELD and FRANCESCHINI, 1968).

The concept of neural superposition of the signals of receptors 1 - 6 in the lamina has been developed from optical and histological findings and must remain hypothetical until the superposition itself as well as the functional consequences of such a superposition are proven. The following paragraphs are intended to draw from this concept such consequences as can be tested experimentally.

3. Optomotor reactions to stimulation under "normal" conditions

As we have seen, the Musca eye is composed of two sub-systems, the system of the receptor cells 1 - 6 and the system of the receptors 7 and 8. In many respects these two systems are expected to have different characteristics: (1) Because of the different spectral extinction of the rhabdomeres involved, the spectral sensitivities of the two systems are expected to be different. (2) Because of the smaller sensitivity distribution the system of the receptor cells 7 and 8 should transmit smaller spatial wavelengths with sufficient contrast than is possible for the system of the receptor cells 1 to 6. (3) On the other hand, the system of receptor cells 1 - 6 should have a higher absolute sensitivity than that of the receptor cells 7 and 8 for several reasons. First, because the neural superposition can be effective only in this system (gain: factor 6); then, because in it the sensitivity distribution is extended about twice as far (gain: factor 4) compared with system 7, 8; and moreover, the rhabdomeres 1 - 6 are longer than the rhabdomeres 7 and 8 (gain: factor 1 - 2). Altogether therefore, an absolute sensitivity gain of a factor of 24 - 48 should result (KIRSCHFELD, 1971; ECKERT, 1971). (4) The system of the receptor cells 1 - 6 should not be sensitive to linearly polarized light if the signals entering the cartridges are really superposed. This is because the angular orientation of the analyzers of the six receptor cells, whose axons enter a cartridge, is evenly spread over $360°$ in angular distances of $120°$.

These 4 conclusions may be tested by optomotor experiments under "normal" conditions, in which relatively large areas of the eye are stimulated by moving striped patterns.

First, concerning the question of polarization sensitivity. It can be shown in optomotor experiments that a polarization-sensitive and a polarization-insensitive system contribute to the reaction as long as the mean light intensities are not too low and as long as stripe patterns with large stripes are used for stimulation. If stimulation is caused by stripe cylinders made up of very narrow stripes (spatial wavelenght about $3°$) however, the signals being then transmitted with sufficient contrast only by the receptors 7 and 8, then only the polarization-sensitive system contributes to the reaction. If stimulation is caused by stripe cylinders with stripes of large spatial wavelength, and if light of longer wavelengths is used ($\lambda > 550$ nm) which particularly stimulates the receptors 1 - 6, it is found that the reaction is independent of the E-vector direction of linearly polarized light (KIRSCHFELD and REICHARDT, 1970).

If spectral sensitivity is determined at very low intensities (at the intensity threshold) by means of stripe cylinders of relatively large stripes, the signals of which therefore can be well transmitted even by the receptors 1 to 6, a spectral sensitivity is found with a maximum at about 490 nm, and a second one in the ultraviolet at about 360 nm. But if cylinders are used with very narrow stripes ($\lambda \approx 3°$), which can stimulate only the receptor cells 7 and 8, then higher mean light intensities are necessary to get any measurable reaction at all; moreover now a reaction maximum is found at about 460 nm, and no further maximum in the ultraviolet spectral range (ECKERT, 1971). - It is true that the maxima found in the behaviour experiment do not correspond exactly with the maxima determined microspectro-photometrically, and moreover the sensitivity of the system composed of receptor cells 1 - 6 is considerably greater in the ultraviolet spectral range than might be expected from the extinction measurements. But at least an agreement exists in so far as both methods show a displacement of the sensitivity maximum towards shorter wavelengths in the visible spectral range at the transition from the 1 - 6 system to the 7 and 8 one.

In optomotor experiments McCANN and McGINITIE (1965) had found that the contrast transmission in the Musca eye is better at high light intensities that in low ones, which means that the sensitivity distributions must be smaller at high mean light intensities. They interpreted this findings as the result of a change of the sensitivity distribution of the receptors dependent on the average brightness. But from what has been said above, we should rather expect that it is not the sensitivity distribution of the individual receptor that changes, but that a transfer takes place from one set of receptor cells to another.

In comprehensive optomotor behaviour experiments ECKERT (1972) has determined the dependence of the sensitivity distribution (characterized by their halfwidth $\Delta\varphi$) on the mean luminosity. The value found for $\Delta\varphi$ was about $4°$ at a luminance of about 10^{-4} apostilb (asb), which is the lowest at which optomotor reactions can be induced. The angle $\Delta\varphi$ remains at about this value up to a luminance of about $5 . 10^{-3}$ asb. With increasing luminances $\Delta\varphi$ decreases very rapidly and at $3 . 10^{-2}$ asb it has reached a value of about $1.7°$, at which it remains up to the highest luminances applied. The conception that this decrease in $\Delta\varphi$ is really caused by other receptors coming into action is supported by the fact that the spectral sensitivity does change in the anticipated manner. However, the possibility cannot be ignored that as the eye becomes bright-adapted both the value of $\Delta \varphi$ and the spectral sensitivity change. But that the Musca eye possesses two systems which might be effective simultaneously, independent of each other, distinguished in the required way by sensitivity distribution and polarization sensitivity has been shown in the experiments with polarized light as described above. The observed changes in the characteristics therefore cannot be changes affecting one type of receptor; there must be two types of receptors with different characteristics present in the Musca eye.

The experiments carried out so far have thus confirmed the concept of neural superposition and have shown that in the Musca eye there are two systems with the required characteristics as regards contrast transmission, spectral and polarization sensitivity.

4. Stimulation of individual photo-receptors

The experiments under "normal" conditions have confirmed some predictions arising from the concept of neural superposition. In order to answer a good many questions which are important particularly with regard to the question of which histological structure produces the functional characteristics, the system has to be "opened". Such questions are e.g. whether one single ommatidium and the neural apparatus following it are really sufficient for analysing movement, or whether it is enough to activate a single receptor-cell axon entering the cartridge in order to produce a signal at the output of the cartridge exit – both essential consequences of the concept of neural superposition. Furthermore it is not clear whether each receptor cell or each cartridge form an input for several correlators as would be necessary for movement-perception, e.g. in a row of ommatidia, or whether only one correlator is activated from the exit of a single cartridge. Finally it is not known how the movement detectors are arranged in the array of ommatidia, nor how the effective scanning basis for seeing movement comes into play, what qualitative and quantitative characteristics are to be attributed to the individual movement detectors, if there is an interaction between different movement detectors or whether there are interactions between signals derived from the receptor cells of the system 1 – 6 and those from system 7 and 8.

The "opening" of the system has been achieved by stimulating individual sensory cells separately. The analysis of the dioptrics of the individual ommatidia (KIRSCHFELD, 1967; KIRSCHFELD and FRANCESCHINI, 1968; FRANCESCHINI and KIRSCHFELD, 1971 a, b) had shown that individual photoreceptor cells in the intact Dipteran eye can be stimulated according to two principally different methods: either the corneal lenses can be used as imaging elements, or the refractive power of the corneal lenses can be optically compensated for, i.e. by the application of immersion oil or laquer. In the experiments, the results of which will be described here, the corneal lenses were used as short focal distance "objectives" ($f \approx 50 \mu$, corresponding to a magnification of approx. 5000 x) in order to illuminate individual rhabdomeres.

The ray path of the stimulating light (fig.3) produces an image of diaphragm B_2 on the corneal surface, as well as an image of a second diaphragm B_1, which is formed by a double diaphragm, on the center of the eye in which the "deep pseudo-pupil" is found. The deep pseudo-pupil consists of virtual images of the distal rhabdomere endings (cf. FRANCESCHINI, this volume). If a sufficiently small diaphragm aperture B_1 is imaged onto the virtual image, e.g. of rhabdomere no. 1 in the deep pseudo-pupil, then, as long as diaphragm B_2 is large, this rhabdomere no. 1 will be illuminated in a great number of ommatidia: the corneal lenses in each ommatidium cause the radiation path to converge onto the rhabdomere in question. Moreover, if diaphragm B_2 is reduced in diameter so that light can only fall into the eye through one single facet, only one receptor will be illuminated in one single ommatidium. The photographs in fig.4 show what can be observed through the oculars of the experimental set-up at various diaphragm adjustments of B_1 and B_2. In the experiments B_1 is formed by a double diaphragm, so that two receptors could be illuminated. An electromechanical shutter was fixed in front of each of the two apertures in the double diaphragm. Both apertures could be opened and closed independently. By means of pre-set phase angles between the periodic opening times of both shutters, imaginary movements to the right and to the left could be programmed.

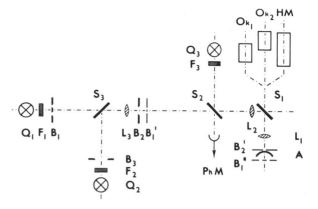

Fig.3. Ray path of stimulating light for the illumination of individual receptors in the Musca eye. A – eye of animal, B – diaphragms, or their images, L – lenses, S – partially reflecting mirrors, F – filters, Q – sources of light, Ok – oculars, HM – telescope, PhM – photo-multiplier

The experiments which are to be described here were confined to stimulation of pairs of receptors
in single ommatidia, and to measuring the optomotor turning tendencies induced by this stimulation as behaviour reactions. For this, the experimental animals whose heads were fixed to the
thorax with glue-wax could be adjusted in the ray path and walk on a y-maze globe (HASSEN-
STEIN, 1958 a). The reaction $\Delta W/2$ is defined as one half of the difference between the turning
tendencies induced by movement stimuli to the right and to the left.

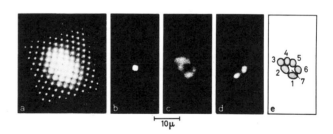

Fig.4. Corneal surface of the experimental animal with a square
diaphragm a) open and b) closed to a dimension that only one facet
is illuminated. c) plane of the deep pseudo-pupil (B_1'' in fig.3) without diaphragm B_1 in the beam path. The receptors 1 - 7 may be recognized. e) receptors 1 - 7 diagrammatically. d) with diaphragm
B_1 in place, only receptors 1 and 6 are illuminated. - All photographs taken with incident-light

If the experiment is confined to stimulation of receptors in one single ommatidium, we are able
to activate, as we know from histology (cf. part I), a set of 6 cartridges in the lamina, arranged
as shown in fig.5 c.

First of all the question had to be answered whether any measurable reaction at all could be elicited when only one single pair of the approx. 50,000 receptors of both compound eyes was stimulated (= 0.04‰!). The histogramme in fig.6 shows that this really is the case. The reactions were released by stimulating the receptors 1 + 6, whereby the light intensity of the stimulating light apparatus was reduced to 10 % of the maximum available. On the average the reactions were 0.13, and reactions of this magnitude are still clearly measurable. Thus a basic condition for carrying out experiments with stimulation of single receptors is fulfilled.

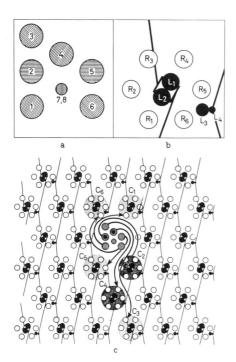

Fig.5.a) Cross-section through the rhabdomeres 1 - 7 of an ommatidium. The shading shows the direction of the microvilli, which in rhabdomeres 1 - 6 is also the direction for maximum extinction of linearly polarized light. b) cross-section through a cartridge. R_1 - 6 receptor axons; L_1 - L_4 second class neurons (after BRAITENBERG, 1970). c) connexions between the axons of an ommatidium and different cartridges C_1 - 6 of the lamina. The cartridges that get an input from the receptors of one ommatidium are shown dotted

If the reaction is measured with different light intensities we find a reaction maximum at about 10 % of the maximum intensity. But even 1 % and 0.1 % elicit measurable reactions.

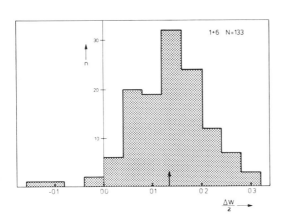

Fig.6. Frequencies n, with which in N = 133 experiments reactions of the strength $\Delta W/2$ occured when receptors 1 + 6 of one single ommatidium were stimulated. For the experiments more than 65 different flies were used

-In order to determine which pairs of cartridges that can be activated by stimulation of receptors within one single ommatidium (fig.5) produce an input to a movement detector, all 15 combinations of pairs of receptors 1 – 6 in single ommatidia were stimulated. The result was, that with a relative light intensity of 10 %, all combinations except 1 + 2, 2 + 3 and 5 + 6 yielded reactions greater than 0.05 which are therefore statistically significant. Nevertheless it cannot be concluded from these results that an interaction on a neuronal basis between the axons or cartridges leading from the receptors in question actually is taking place. For it must be taken into account that bad imaging conditions or stray light may allow light to affect not only those receptors on which the diaphragm apertures of B_1 are imaged but also others, particularly neighbouring ones, and that their signals may induce optomotor turning reactions.

Optical crosstalk between different receptors of individual ommatidia can be directly measured on the intact animal (KIRSCHFELD and FRANCESCHINI, 1972). It has been found that, given comparable optical conditions as they are realized in the apparatus for individual cell stimulation (fig.3), there is transmission from 1 % to 10 % between one rhabdomere and a neighbourought one. As we have seen, intensities between 1 – 10 % of the optimal effective intensity can still elicit measurable reactions. Therefore a method had to be found to eliminate the influence of this "stray light".

This can be done as follows. If e.g. we want to test whether the receptors 1 and 6 form the input to a movement detector, the double diaphragm B_1 will first be imaged onto the receptors 1 and 6 of one single ommatidium (experimental condition a, fig.7) and a reaction to stimulated movement is measured. Then a diaphragm B_3 (fig.3) will be imaged onto the deep pseudo-pupil. This must be adjusted in such a way that from it, all receptors are illuminated with a DC adaptation light intensity la except for receptor 1 (experimental condition b, fig.7). Then the adaptation intensity la is increased gradually and at each step the reaction to stimulation of receptors 1 and 6 by light passing through diaphragm B_1 is measured. The result is that the reaction to stimulated movement is reduced as la increases. This is a consequence of at least two facts: first, for every possible partner of receptor 1 the degree of modulation of the stimulation light is reduced by the adaptation light from diaphragm B_3. And second, the degree of modulation of the stimulation light on receptor 1 is reduced by the influence of stray light: both must lead to a reduction of the measured reaction.

Fig.7. Adjustment of diaphragms in different experimental conditions to prove a neural interaction between the signals released by receptors 1 and 6

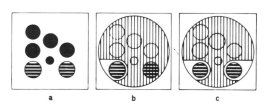

In the next experiment, which was performed in order to determine if the receptors 1 and 6 are the input to a movement detector, a diaphragm B_3 was applied which had a shape that allowed adaptation with a DC light of all the receptors except nos. 1 and 6 (experimental condition c, fig.7). An intensity la was used, at which the reaction in experimental condition b was noticeably reduced. If the receptor pair 1 + 6 contributed to the reaction measured in experimental condition a, then in experimental condition c an increase in the reaction is to be expected, as compared with experimental condition b; but if the reaction measured in experimental condition a was caused by receptor 1 and any other than receptor 6, then no increase in the reaction in experimental condition c is expected. At the end of each series of experiments the reaction in experimental condition a was measured once more as a control.

Fig.8 gives the individual values of the measurements of several different pairs of receptors. The circles and dots show the reaction in experimental condition a, at the beginning and at the end of the series; the triangles are points measured in experimental condition b and the crosses those in experimental condition c. The bars show the change in reaction at the transition from experimental condition b to c, and their length indicates whether the tested combination makes a real contribution or not.

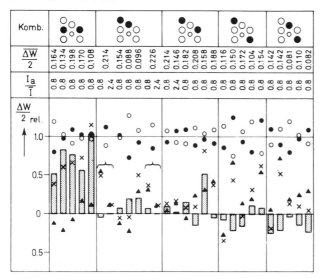

Fig.8. Results of experiments for proving neuronal interaction between different pairs of receptors. row 1: stimulated receptor pairs. row 2: absolute mean reactions ΔW/2 taken in exp. cond. a (cf. fig.7). row 3: relative adaptation-light intensity Ia/I in exp. conditions b + c. row 4: reactions measured in exp. condition a at the beginning (o) or end (●) of a series of experiments. ▲ – reactions for exp. condition b. x – for exp. condition c. – Vertical bars show the increase in reactions at the transition from exp. cond. b to c. Each vertical bar gives the values of one animal at a time

Fig.8 shows, that there are apparently positive reaction changes in the mean when the combination 2 + 5 is stimulated, whereas in the other combinations the mean reaction change is negligible. The results of the experiments shown in fig.8, and those of further receptor combinations tested, are statistically evaluated in fig.9. The black bars give the confidence limits for P = 0.05. Only the combinations 1 +3, 1+ 6, 2 + 5, and 4 + 6 give any values significantly different from 0; all the others are nil. The reaction change of the combination 4 + 6 is significant, but very small, and will not be further discussed here. Large contributions to the reaction around the vertical axis are thus given in the direction 1 – 6 = 2 – 4 by pairs of neighbouring receptors, and in the

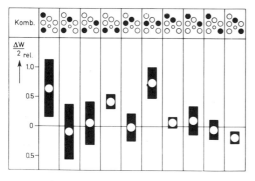

Fig.9. Mean reaction increase measured in experimental condition c compared with condition b (cf. fig.7) for several pairs of receptors. The vertical bars give the confidence limits for P = 0.05

direction 1 – 3 by pairs of receptors neighboured but one. This result agrees with the assumption that there are two sets of movement detectors, orthogonally arranged in the direction 1 – 6 and 1 – 3, both influencing turning around the vertical axis. More evidence for this interpretation is given elsewhere (KIRSCHFELD, 1972).

5. Discussion of the experiments with stimulation of individual receptors

a) Conclusions drawn from the concept of neural superposition. The concept of neural superposition postulates that one ommatidium should be enough to enable movement perception. The experiments show that this is actually the case. Furthermore if a superposition takes place in the cartridges it should be enough that one axon entering a cartridge is stimulated in order to activate the output of the cartridge. This is also the case as we have seen. If it is found that the receptors 1 + 6 form the input to a movement detector, then, finally, as can easily be seen in fig.5, the receptors approach have the same orientation and the same distance as those approached by receptors 1 + 6; they are therefore homologous. Here, too, the results of the experiments fulfill the theoretical expectations.

The finding that movement–detectors which induce turning around the vertical axis are arranged in the two orthogonal directions 1 – 6 and 1 – 3 has been surprising for several reasons. The experiments done by GOETZ (1968) certainly have shown clearly that movement detectors which are arranged in at least two different directions are present in the Musca eye; but only one of the two directions required was presumed necessary for turning around the vertical axis, the other was assumed to guide the thrust of the animal. The findings of histology up to now were interpreted in such a way that movement detectors were rather to be expected in the directions 2 – 4 or 3 – 4 (BRAITENBERG, 1970). The same conclusion, that movement detectors are arranged along these diagonal directions, was drawn from optomotor experiments in which the scanning basis $\Delta \varphi$ of the eye, which with simple assumptions may be interpreted as the divergence angle between the ommatidia, has been determined (McCANN and MacGINITIE, 1965; ECKERT, 1971, 1972). A more detailed consideration of the data however shows that if there are movement detectors aligned in different directions, a scanning basis $\Delta \varphi$ will be effective which cannot simply be interpreted anatomically (THORSON, 1966). On the basis of such lines of thought, the findings of the authors quoted above may be brought into harmony with the finding, that two movement-detector systems are realized, arranged in the directions 1 – 6 and 1 – 3.

The question as to a possible advantage of two orthogonally oriented movement–detector systems, both of them affecting turning on the vertical axis, cannot yet be answered. Therefore only a few consequences should be indicated here: it can be shown that the effective scanning basis of the eye becomes largely independent both of the direction of the movement of a pattern in relation to the array of ommatidia and also of the alignment of the rows of ommatidia in the eye. Moreover if there are two orthogonal movement detector systems, negative reactions which may arise due to geometric interference may under certain conditions be suppressed. And finally it must be considered that due to the definite alignment of the ommatidia in the Musca eye, a turning reaction on the vertical axis will even occur when a pattern is moved upside down or reversed in front of the eye.

b)   The minimum histological structure necessary. Finally, the question must be posed as to the histological structure which might be responsible for the investigated reactions. When a "block diagram" is being worked out it is essential that it be adequate and necessary (HASSENSTEIN and REICHARDT, 1953); it is also essential if one postulates a histological structure, that this postulated structure should be adequate and necessary too. When setting up a block diagram, elements with certain basic mathematical functions can be specified in the beginning; this is not possible, however, when working out a necessary and sufficient histological structure such as is required by the results of input–output analyses, since our knowledge of the performances of the individual nerve cells is still too limited. Therefore at present it is reasonable to limit pronouncements concerning histology to the minimal number of channels conveying information.

The minimum block diagram of a movement detector, necessary and sufficient for the basic performances, indication of speed and direction of movement, is shown in fig.10 (KIRSCHFELD, 1972). It is based on the principle of autocorrelation (HASSENSTEIN, 1958 b; REICHARDT, 1957; REICHARDT and VARJU, 1959). The general proof that such a structure is the simplest possible cannot be presented at present. But if, as in the minimum model shown in fig.10, the channel cross-section does not exceed the minimum number of independent entrance channels necessary, then of course, as far as the channel cross-section is concerned, it is clear that no simpler structure can be found.

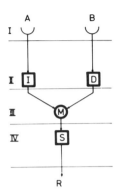

Fig.10. Necessary and sufficient structure for a movement detector allowing the basic performances: indication of speed and direction. I – inputs, II – filters, III – non-linear interaction, IV – filter, R – reaction

Starting from the minimum block diagram we can now draw the necessary histological structure, as required from the individual cell stimulation experiments (fig.11). Starting from an entrance drawn as a semi-circle there is a connexion on the horizontal plane (direction 1 – 6) each time

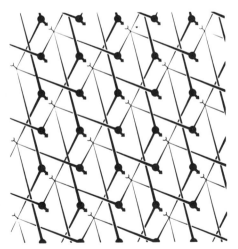

Fig.11. Necessary and sufficient histological structure required from the results of the experiments with individual receptor stimulation. The thickening of lines indicates that they come from the lower level of the inputs (half-circles) into the higher level of the correlators (filled circles). While the arrangement of the entrances is fixed by the array of ommatidia (from l. to r.: direction 1 – 6) nothing can be said as yet on the position of the correlators; the drawing merely explains the required connexions

to the neighbours on the right and left, and upwards and downwards each time to the next-but-one neighbours. The arrows mark the exits of the correlators which contain the information on movement. It is not by chance that each of the inputs is drawn leading into 4 correlators. For in principle it could well be that each input was connected with only one correlator, or with 2 or 3. But the individual cell stimulation experiments have shown that in fact there is a split at least into 4 information channels leading to 4 different correlators (KIRSCHFELD, 1972). This means that there are twice as many correlators as inputs and therefore cartridges in the lamina.

While the existence and localization of the connecting scheme shown in fig.2 has been localized histologically, the localization of the network shown in fig.11 is as yet unknown. However it is to be expected that the inputs correspond to the cartridge outputs, and that the correlators will be localized in the medulla.

Acknowledgements. I would like to thank Dr. C.B. BOSCHEK for reading the manuscript.

References

AUTRUM, H.J., WIEDEMANN, I.: Versuche über den Strahlengang im Insektenauge (Appositionsauge). Z. Naturforsch. 17b, 480-482 (1962).

BOSCHEK, B.: On the Fine Structure of the Peripheral Retina and Lamina ganglionaris of the Fly, Musca domestica. Z. Zellforsch. 118, 369-409 (1971).

BRAITENBERG, V.: Patterns of Projection in the Visual System of the Fly. I. Retina – Lamina Projections. Exp. Brain Res. 3, 271-298 (1967).

BRAITENBERG, V.: Ordnung und Orientierung der Elemente im Sehsystem der Fliege. Kybernetik 7, 235-242 (1970).

DE VRIES, H., KUIPER, J.W.: Optics of the Insect Eye. Ann. NY. Acad. Sci., 74, 196-203 (1958).

ECKERT, H.: Die spektrale Empfindlichkeit des Komplexauges von Musca (Bestimmung aus Messungen der optomotorischen Reaktion). Kybernetik 9, 145-156 (1971).

ECKERT, H.: Optomotorische Untersuchungen am visuellen System der Stubenfliege Musca domestica. Kybernetik, in preparation.

EXNER, S.: Die Physiologie der fazettierten Augen von Krebsen und Insekten. Leipzig – Wien: Deuticke 1891.

FRANCESCHINI, N., KIRSCHFELD, K.: Etude optique in vivo des éléments photorécepteurs dans l'oeil composé de Drosophila. Kybernetik 8, 1-13 (1971a).

FRANCESCHINI, N., KIRSCHFELD, K.: Les phénomènes de pseudopupille dans l'oeil composé de Drosophila. Kybernetik 9, 159-182 (1971b).

GOETZ, K.G.: Flight Control in Drosophila by Visual Perception of Motion. Kybernetik 4, 199-208 (1968).

HASSENSTEIN, B.: Die Stärke von optokinetischen Reaktionen auf verschiedene Mustergeschwindigkeiten. Z. Naturforsch. 13b, 1-6 (1958a).

HASSENSTEIN, B.: Ueber die Wahrnehmung der Bewegung von Figuren und unregelmässigen Helligkeitsmusters (Nach verhaltensphysiologischen Untersuchungen an dem Rüsselkäfer Chlorophanus viridis). Z. vergl. Physiol. 40, 556-592 (1958b).

HASSENSTEIN, B., REICHARDT, W.: Der Schluss von Reiz-Reaktions-Funktionen auf System-Strukturen. Z. Naturforsch. 8b, 518 (1953).

KIRSCHFELD, K.: Das anatomische und das physiologische Sehfeld der Ommatidien im Komplexauge von Musca. Kybernetik 2, 249-257 (1965).

KIRSCHFELD, K.: Die Projektion der optischen Umwelt von Musca. Exp. Brain Res. 3, 248-270 (1967).

KIRSCHFELD, K.: Absorption Properties of Photopigments in Single Rods, Cones and Rhabdomeres. Rendiconti SIF XLIII, Ed. W. Reichart, Proc. Int. School Physics Enrico Fermi: Processing of Optical Data by Organisms and by Machines. New York: Acedemic Press 1969.

KIRSCHFELD, K.: Aufnahme und Verarbeitung optischer Daten im Komplexauge von Insekten. Naturwiss. 58, 201-209 (1971).

KIRSCHFELD, K.: in preparation (1972).

KIRSCHFELD, K., FRANCESCHINI, N.: Optische Eigenschaften der Ommatidien im Komplexauge von Musca. Kybernetik 5, 47-52 (1968).

KIRSCHFELD, K., FRANCESCHINI, N.: in preparation (1972).

KIRSCHFELD, K., REICHARDT, W.: Optomotorische Versuche an Musca mit linear polarisiertem Licht. Z. Naturforsch. 25b, 228 (1970).

LANGER, H.: Grundlagen der Wahrnehmung von Wellenlänge und Schwingungsebene des Lichtes. Verh. Dt. Zool. Ges., Göttingen, 195-233 (1966).

McCANN, G.D., MacGINITIE, G.F.: Optomotor Response Studies of Insect Vision. Proc. Roy. Soc. B, 163, 369-401 (1965).

REICHARDT, W.: Autokorrelations-Auswertung als Funktionsprinzip des Zentralnervensystems (bei der optischen Bewegungswahrnehmung eines Insektes). Z. Naturforsch. 12b, 448-457 (1957).

REICHARDT, W., VARJU, D.: Uebertragungseigenschaften im Auswertesystem für das Bewegungs-sehen. Z. Naturforsch. 14b, 674-689 (1959).

SCHOLES, J.: The Electrical Responses of the Retinal Receptors and the Lamina in the Visual System of the Fly Musca. Kybernetik 6, 149-162 (1969).

THORSON, J.: Small-Signal Analysis of a Visual Reflex in the Locust. I. Input Parameters. Kybernetik 3, 41-53 (1966).

TRUJILLO-CENOZ, O.: Some Aspects of the Structural Organization of the Intermediate Retina of Dipterans. J. Ultrastructure Res. 13, 1-33 (1965).

TRUJILLO-CENOZ, O., MELAMED, J.: Compound Eye of Dipterans: Anatomical Basis for Integration - An Electron Microscope Study. J. Ultrastructure Res. 16, 395-398 (1966).

2. Pupil and Pseudopupil in the Compound Eye of Drosophila

N. Franceschini
Max-Planck-Institute for Biological Cybernetics, Tübingen, Germany

Abstract. An explanation is given of the phenomenon of "deep pseudopupil" in the compound eye of Drosophila. This phenomenon is here exploited as a technique which, combined with that of optical neutralization of the corneal lenslets, permits the real time analysis of the rapid pigment migration occuring in receptor cells 1 to 6 under bright adaptation.

1. Introduction

The ommatidia of Diptera share with those of some Hemiptera and Orthoptera the pecularity of having an open rhabdom. This is in fact a hopeful situation for studying the processing of information in a compound eye because it makes it possible – at least in principle – to stimulate a single receptor cell without simultaneously exciting its neighbours. Such an unitary stimulation technique is doubtless more difficult to achieve in other types of compound eyes in which the fused rhabdom, axially built by several receptor cells, acts as a common waveguide (VARELA and WIITANEN, 1970) so that the only possibility left to excite different receptors to different extents is to exploit their different spectral or polarization sensitivities.

The present contribution first summarizes some results obtained concerning the optical properties of the Drosophila eye. In fact it turns out that these results yield some methods for analysing and stimulating well identified receptor cells in the eyes of live insects. A first application of these methods will be presented in the second part of this paper, in connection with the pigment migration mechanism of the dipterous eye.

The technique used here is an in-vivo microscopical observation of the eye, the illumination being either orthodromic or antidromic (see fig.1).

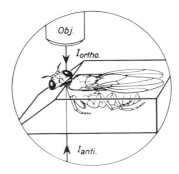

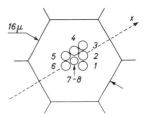

Fig.1. Antidromic and orthodromic illumination techniques of a Drosophila eye. The living insect lies comfortably in a specially built transparent object-holder

Fig.2. Arrangement and numbering of the rhabdomes within a dorsal ommatidium of a right Drosophila eye (as seen from outside)

2. The formation of a "deep pseudopupil"

A geometric optical model of the dipterous ommatidium is shown in fig.3a (P and P' = principal planes of the dioptric system). Three rhabdomere distal endings A, B, C are seen in this longitudinal section oriented along axis X of fig.2. As shown by previous experiments with Musca (KIRSCHFELD and FRANCESCHINI, 1968) these receptor terminals are located in the inner focal plane of the dioptric apparatus, and virtual images of them appear in the depth of the ommatidium. It could be shown that sharp images are seen only at a considerable distance from the cornea ($1 \text{ mm} \simeq \infty$) but that "pseudo-images" (i.e. not perfectly sharp) appear at any intermediary plane proximal to the focal plane (the deeper the plane, the sharper and larger the images). In model of fig.3a, A', B', C' ... A^n, B^n, C^n represent the centers of such pseudo-images which would be observed at the four arbitrarely choosen levels.

Now the question is what should occur if this optical property of the single ommatidium is applied to the eye as a whole. Fig.3b shows that the presence of two ommatidia diverging by an angle θ (interommatidial angle) necessarily leads to the formation of superpositions of virtual images in the depth of the eye. Three levels of superposition can thus be predicted (I, II, III in fig.3b). The formation of each is explained by the following example: C'_1, C''_1, C'''_1, images of rhabdomere tip C_1 thrown by ommatidium (1), should respectively superimpose on A'_2, B''_2, C'''_2 which are images of A_2, B_2, C_2 thrown by ommatidium (2).

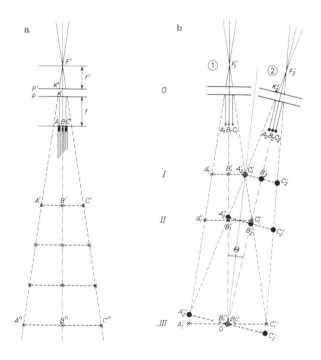

Fig.3.   Schematic longitudinal sections of ommatidia along the X axis of fig.2

Fig.3b makes it clear that the first superposition level should occur on the bisector of each interommatidial angle θ, that the second one should occur on the optical axis of each ommatidium, and that the third should appear at the center of curvature of the eye. Three image levels and no others, are actually observed at different distances from the cornea in the depth of an antidromically illuminated eye (FRANCESCHINI and KIRSCHFELD, 1971b), and it could be shown that the dimensions, the quality, the brightness, and the depth of the three kinds of "images" observed are in good agreement with the expectation made on the basis of the geometric optical model designed in fig.3b. In particular we have shown by using another in vivo technique, that the distance of 180 μm separating the third image level from the cornea does correspond to the radius of curvature R of the eye.

Let us now focus our attention on this third superposition of virtual images. It is clear that at this level (III on fig.3b) images of homologous receptor endings, belonging to neighbouring ommatidia, should superimpose. For example A'''_1, image of an ending A from ommatidium (1), should superimpose on A'''_2 (the image of another ending A) belonging to ommatidium (2). For this reason, there should appear, at the level of the center of curvature of the eye, a group of 7 images, arranged in a pattern similar to that of the rhabdomere distal endings within each ommatidium (fig.2). This is actually the case, as shown in fig.4. We have called this phenomenon a deep pseudopupil because it could be shown that the principle of its formation explains a lot of so-called "pseudopupils", which had been observed for more than a century in the compound eye of various Arthropods (LEYDIG, 1855; EXNER, 1891).

As can be seen on fig.4, the deep pseudopupil of Drosophila is very large, each of the 7 light spots being about 15 μm in diameter (its magnification with respect to the receptor endings themselves is thus about 10x). For this reason it can be observed with a microscope objective of relatively low power (in the present case: 13x) and long working distance (14 mm). The deep pseudopupil is brighter, the larger the aperture of the objective, i.e. the more ommatidia participate in its formation. In the photograph of fig.4, taken with an aperture of sin u = 0,22, about 25 virtual and magnified images of rhabdomere patterns identical to that of fig.2 are superimposing at the level of the center of curvature of the eye.

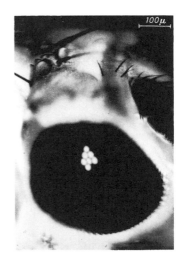

Fig.4. Deep pseudopupil, as observed in the eye of a living Drosophila (se mutant) under antidromic illumination. Microscope (sin u = 0,22) focused at about 180 μm proximal to the corneal surface

One must confess that such a precise superposition of so many images simply attests the formidable regularity in the architecture of the compound eye, since the appearance of a deep pseudopupil as sharp as that of fig.4 implies that the following constancies are maintained: (1) constancy of the center-to-center distance between the 7 rhabdomere distal endings, within the 25 ommatidia concerned, (2) constancy of the orientation of the rhabdomere patterns in the focal plane, (3) constancy of the power of the corneal lenslets.

No doubt the deep pseudopupil is a matter of indifference to the insect, which obviously cannot perceive it! But - and here appears its scientific beauty - this phenomenon is not indifferent to the experimenter because it enables him to observe, directly, the distal part of the retina while the animal is living. Possible applications of this phenomenon have already been described (see chapt. 6-3 of FRANCESCHINI and KIRSCHFELD, 1971b). Above all, the technique of the deep pseudopupil makes it possible to easily analyse, and/or stimulate, well identified sense cells in a living animal. Combined with relevant behavioural and electrophysiological experiments, this technique might help us investigate one of the main problems we are concerned with, namely how the visual information is processed by the neurons of the optic ganglions.

3. The light attenuation in rhabdomeres 1 to 6

A drastic change of reflectance is observed when looking at the deep pseudopupil of a dark adapted Drosophila in incident light (orthodromic illumination, see fig.1). Within a few seconds, it switches from state (a) to state (c) of fig.5. In the latter state – which we call "bright adapted" – the pseudopupil selectively reflects a glittering green light, even if the illumination is done with white light. Correspondingly the deep pseudopupil shows a drastic change of transmittance when observed with an antidromic blue-green light: within a few seconds it switches from a dark adapted state (fig.5d) to a bright adapted state (fig.5f), in which the transmittance of rhabdomeres 1 to 6 is considerably decreased. The time course of these reflectance and transmittance changes, as given in fig.5b and 5e respectively, have been directly recorded by a photomultiplier mounted on the ocular (after placing a field stop in the intermediate image plane, so as to intercept only the light coming from the deep pseudopupil).

Certainly one of the most striking results is that both observed changed concern only the six peripheral sense cells (1 to 6) and not the central ones (7-8). Thus whatever the mechanism responsible for these changes might be, this result indicates the existence of two functionally different receptor classes in the ommatidia of Drosophila.

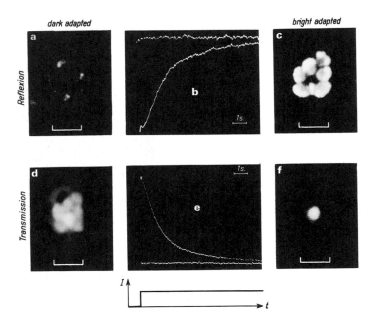

Fig.5.   Reflexion change (a → c) and transmission change (d → f) of the rhabdomeres as observed on the deep pseudopupil of mutant w^a (weakly pigmented eye) and corresponding time courses of these changes in response to a light step of high intensity (orthodromic in b, antidromic in e). In both records the straight line is the continuation of the curve. Calibration: 50 μm

The observed phenomena are apparently analogous to those previously described for the eye of Musca (KIRSCHFELD and FRANCESCHINI, 1969). The method presently used is different in that the deep pseudopupil gives a direct insight into what occurs at the level of the rhabdomere tips themselves, whereas the previously used method (which makes use of what we now call the reduced corneal pseudopupil, i.e. a pseudopupil observed at the level of the cornea with a reduced aperture of the microscope) only enables us to see how much light comes out from 7 facets, each of which is illuminated by a single rhabdomere. However it is important to note that the reduced corneal pseudopupil permits analysis of the transmission or reflection properties of single receptor cells. This is not true for the deep pseudopupil which obsiously brings out the average properties of a lot of homologous rhabdomeres. Fortunately this is not an inconvenience in the present situation because it can be verified, by using another in vivo technique, that the phenomena described in fig.5 simultaneously occur within each illuminated ommatidium and not sporadically among the ommatidia (see fig.6). Indeed, the spatial summation process naturally occuring in the deep pseudopupil leads not only to an amplification of the optical signals available for recording but also to an improvement of the signal to noise ratio.

Since the changes reported in figs.5 a,c,d,f are seen in the deep pseudopupil of Musca, as well (the selective reflection being yellow rather than green) it is likely that they owe their origin to a pigment migration within retinula cells 1 to 6, like that already demonstrated for Musca (KIRSCHFELD and FRANCESCHINI, 1969). Actually fine pigment granules, similar to those recently found by BOSCHEK (1971) in the retinula cells of Musca, have been observed in Drosophila as well (YASUZUMI and DEGUCHI, 1958). As in Musca, their diameter lies around 0,1 μm and, according to later authors, their chromophore may be an ommochrome (NOLTE, 1961; FUGE, 1967). Possible changes in the location of these granules with respect to light/dark adaptation have not been studied in Drosophila as they were in Musca (BOSCHEK, 1971), and the two former authors describe the granules only as clustered against the wall of the rhabdomeres (especially at their distal tip), that is to say, in what we would now call the bright adapted condition.

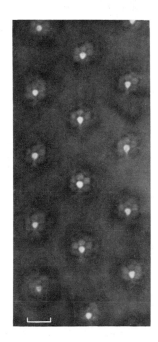

Fig.6.　In-vivo observation of the extinction of rhabdomeres 1 to 6 in the bright adapted state (antidromic illumination of the eye: λ = 500 nm; Drosophila w^a). Technique of optical neutralization of the cornea (see FRANCESCHINI and KIRSCHFELD, 1971a). Calibration: 5 μm

An interpretation of the interaction between light and pigment has been given in our 1969 paper on Musca. Although the granules pile up against the rhabdomere without penetrating it, they can interact with the evanescent waves produced outside the rhabdomere which acts as a waveguide. A frustrated total reflection thus occurs, due to the (probably) high refraction and extinction indices of the pigment.

An important question which might help the understanding of this mechanism is the following: at which luminance does it come into play? One can answer this question by adapting the insect eye to a series of luminances and by photometrically recording, for each of them, the degree of reflection or transmission of the deep pseudopupil in response to a test flash of constant intensity. The result is shown in fig.7 for the mutant w^a of Drosophila. In both cases, the adaptation of the eye was performed orthodromically in white light, whereas the test flash was either orthodromic (a) or antidromic (b). It appears that the pigment migration has a threshold and that it reaches a saturation level for a luminance about 100 times higher (two decades).

The location of the "adaptation range" on the luminance axis has been calibrated within the rotating drum used for testing the insect optomotor response in order to make comparisons easier. First it comes out that the threshold of pigment migration (about -3 log relative units on fig.7) correspond to a luminance of about 0,3 cd/m^2 for Drosophila (wild-type) and 3 cd/m^2 for Musca (wild-type). Secondly, taking into account the results of the optomotor experiments made by ECKERT (1972) in exactly the same rotating drum, it appears that the threshold of pigment migration in Musca is about 100'000 times higher than the absolute threshold of movement perception.

Using again the weakly pigmented eye mutant w^a of Drosophila (the head of which is transparent enough to allow an antidromic illumination down to about 400 nm) we have shown that the extinction brought out in rhabdomeres 1 to 6 by the pigment migration is higher the shorter the wavelength, within the range 600 nm to 420 nm (figs.5f and 6 just illustrate this finding for λ = 500 nm. Considering then the high sensitivity of receptor cells 1 to 6 for blue-green light in the ommatidia of Diptera (BURKHARDT, 1962; LANGER, 1966; ECKERT, 1971) it appears that the spectral range in which the granules are efficient covers - partly at least - the spectral range which is of interest for the visual pigment.

This mechanism thus appears as an ingenious throttle reducing the amount of photons to be absorbed for the sake of vision. This is in a sense a "longitudinal pupil" not unlike that postulated by KUIPER (1962) for the optical superposition eye. Whether such a mechanism has developed simply for protecting sense cells 1 to 6 aginast overstimulation is still an open question. In this respect it is worth noting that it comes into play within a quite physiological range of luminance. Optomotor experiments performed with the eye mutant w^a of Drosophila show that this insect can still perceive movements remarkably well at a luminance about 100 times higher than the saturation points of its pigment migration. The problem is whether sense cells 7-8 alone would be responsible for vision at such high intensities.

Three findings have lead us to the conclusion that the photosensor triggering the pigment migration is the visual pigment itself (FRANCESCHINI and KIRSCHFELD, in prep.). First a microstimulation of single receptor cells 1 to 6 shows that this sensor must be located within each of them. Secondly the sensor has exactly the same polarization sensitivity as the visual pigment in each rhabdomere 1 to 6; and thirdly the action spectrum for the threshold of pigment migration shows the two famous maxima (blue-green and UV) which characterize the absorption of the visual pigment in receptors 1 to 6.

From several histological studies it might be expected that such a light control mechanism is quite common amongst arthropods. Photon and electron microscopy have attested the existence of pigment movement within retinula cells in Limulus (MILLER, 1958), in a crustacean (TUURALA, LEHTINEN and NYHOLM, 1966) and in several orders of insects (Orthoptera: JOERSCHKE, 1914; TUURALA and LEHTINEN, 1967; Hymenoptera: MENZEL and LANGE, 1971; MENZEL, this volume; Hemiptera: LUEDTKE, 1953; Blattoides: BUTLER, 1971); but the static and dynamic properties of these pigment migrations are unknown. The technique of the deep pseudopupil, combined with that of optical neutralization of the corneal lenslets, might be a fruitful help for the real time analysis of these phenomena in the eye of live insects.

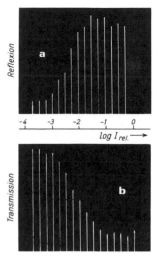

Fig.7. Statical behaviour of the pigment migration system. Each test flash immediately follows a 15 sec. adaptation period to the desired luminance (I [relative units]) and is followed by 90 sec. dark adaptation

Acknowledgements. I am indebted to Dr. K. GOETZ and Dr. K. KIRSCHFELD for help and discussions throughout the experimental work, and to Dr. STRAUSFELD and Dr. ROSSER for help with the English manuscript.

References

BOSCHEK, B.: On the fine structure of the peripheral retina and lamina of the fly Musca. Z.Zellforsch. 118, 369-409 (1971).

BURKHARDT, D.: Spectral sensitivity and other response characteristics of single visual cells in the arthropod eye. Symp. Soc. exp. Biol. 16, 86-109 (1962).

BUTLER, R.: The identification and mapping of spectral cell types in the retina of Periplaneta americana. Z. vergl. Physiol. 72, 67-80 (1971).

ECKERT, H.: Die spektrale Empfindlichkeit des Komplexauges von Musca. Kybernetik 9, 145-156 (1971).

ECKERT, H.: Optomotorische Untersuchungen am visuellen System der Stubenfliege Musca. Kybernetik (in prep., 1972).

EXNER, S.: Die Physiologie der fazettierten Augen von Krebsen und Insekten. Leipzig und Wien: Deutike, (1891).

FRANCESCHINI, N., KIRSCHFELD, K.: Etude optique in vivo des éléments photorécepteurs dans l'oeil composé de Drosophila. Kybernetik 8, 1-13 (1971a).

FRANCESCHINI, N., KIRSCHFELD, K.: Les phénomènes de pseudopupille dans l'oeil composé de Drosophila. Kybernetik 9, 159-182 (1971b).

FRANCESCHINI, N., KIRSCHFELD, K.: An automatic gain control in the photoreceptors of Drosophila (in prep.).

FUGE, H.: Die Pigmentbildung im Auge von Drosophila und ihre Beeinflussung durch den white[+] locus. Z. Zellforsch. 83, 468-507 (1967).

JOERSCHKE, H.: Die Facettenaugen der Orthopteren und Termiten. Z. wiss. Zool. 111, 153-280 (1914).

KIRSCHFELD, K., FRANCESCHINI, N.: Optische Eigenschaften der Ommatidien im Komplexauge von Musca. Kybernetik 5, 47-52 (1968).

KIRSCHFELD, K., FRANCESCHINI, N.: Ein Mechanismus zur Steuerung des Lichtflusses in den Rhabdomeren des Komplexauges von Musca. Kybernetik 6, 13–21 (1969).

KUIPER, J.W.: The optics of the compound eye. Symp. Soc. exp. Biol. 16, 58 (1962).

LANGER, H.: Spektrometrische Untersuchung der Absorptionseigenschaften einzelner Rhabdomere im Facettenauge. Zool. Anzeig. Suppl. 29, 329–338 (1966).

LEYDIG, F.: Zum feineren Bau der Arthropoden. Müller's Arch. Physiol.; p. 431 (1855).

LUEDTKE, H.: Retinomotorik und Adaptationsvorgänge im Auge des Rückenschwimmers Notonecta glauca. Z. vergl. Physiol. 35, 129–152 (1953).

MENZEL, R., LANGE, G.: Aenderungen der Feinstruktur im Komplexauge von Formica polyctena bei der Helladaptation. Z. Naturforsch. 26b, 357–359 (1971).

MILLER, W.H.: Fine structure of some invertebrate photoreceptors. Ann. New-York Acad. Sc., 174, 204–209 (1958).

NOLTE, D.J.: The pigment granules in the compound eye of Drosophila. Hered. 16, 25–38 (1961).

TUURALA, O., LEHTINEN, A.: Zu den photomechanischen Erscheinungen im Auge der Gewächshausheuschrecke Tachycines asynamorus A. Comment. biol. Soc. Sc. fenn. 30, 6 (1967).

TUURALA, O., LEHTINEN, A., NYHOLM, M.: Zu den photomechanischen Erscheinungen im Auge einer Asselart Oniscus asellus L. Ann. Acad. Sc. fenn. A IV, 99, 1–8 (1966).

VARELA, F.G., WIITANEN, W.: The optics of the compound eye of the honeybee (Apis mellifera). J. gen. Physiol. 55, 336–358 (1970).

YASUZUMI, G., DEGUCHI, N.: Submicroscopic structure of the compound eye as relealed by electron microscopy. J. Ultrastruct. Res. 1, 259–270 (1958).

3. Directional Intensity Distribution in Single Ommatidia of the Honeybee, Apis mellifera

W. P. Eheim
Department of Zoology, University of Zurich, Switzerland

Abstract. Intensity distribution curves in single ommatidia of worker bees have been recorded. Measurements were carried out in vertical (x/y-) and horizontal (z-) directions in light and dark adapted ommatidia from which similar Gaussian curves resulted. The intensity distribution in bees' ommatidia may be approximated by the function $I(\varphi) = \exp(-0.41 [\text{degree}^{-2}] \varphi^2)$ with a half-width of $2.60°$ (acceptance angle $\Delta\varrho$). These findings are discussed in relation to earlier results and considering adaptation mechanisms.

Hymenoptera have apposition eyes of the closed-rhabdom type. In these eyes the optical transmission properties are determined by the angle $\Delta\varphi$ between the optical axes of neighbouring ommatidia on the one hand and by the acceptance angle $\Delta\varrho$ of single ommatidia on the other hand. Studying the visual abilities of the honeybee, knowledge of $\Delta\varphi$ and $\Delta\varrho$ is therefore important.

In the bee's eye being composed of about 4000 ommatidia with a corneal diameter of about $30 \mu m$ (VARELA and PORTER, 1969; VARELA and WIITANEN, 1970), the interommatidial angle $\Delta\varphi$ was measured histologically by BAUMGARTNER (1928) whose results were revised in 1936 by DEL PORTILLO. AUTRUM and WIEDEMANN (1962) and WIEDEMANN (1965) reported from optical experiments the width of the visual field of single ommatidia in worker bees to be $6.8°$ in vertical (x/y-) and $7.2°$ in horizontal (z-) direction. In 1962 KUIPER published a bell-shaped intensity distribution curve of bees' ommatidia with a full half-width of about $6°$. KUIPER's result was confirmed by VARELA and WIITANEN (1970) who calculated a Gaussian "lens admittance function" that determined an acceptance angle of $5.6°$.

Incompatible with these values are the electrophysiological findings of SHAW (1969) and LAUGHLIN and HORRIDGE (1971). SHAW determined an acceptance angle of about $2°$ from measurements of directional sensitivity in the honeybee drone, a value close to the horizontal acceptance angle of $2.5°$ and the vertical acceptance angle of $2.7°$ reported by LAUGHLIN and HORRIDGE from dark adapted ommatidia of worker bees.

In 1970/71 we developed experimental equipment to make possible continuous recordings of intensity distribution curves as a function of angle of incidence of light in single ommatidia of arthropod compound eyes in order to ascertain these data in the worker bee.

Having been described elsewhere in detail (EHEIM and WEHNER, 1972), the experimental procedure will be briefly commented on here. All experiments were performed on animals which had been kept in total darkness for 12 hours. The eyes were prepared either immediately after this dark period using illumination of 652 nm or after having been exposed 30 - 90 min to an intensity of illumination of 15'000 lx from a 450 W Xenon lamp. In the beam of this lamp the preparation of the light adapted eyes was accomplished.

Following the method described by KIRSCHFELD (1967) a calotte was cut off the central part of the eye by means of the splinter of a razor blade oscillating on the membrane of a loud-speaker. Cut side to a cover-slip moistened with insect Ringer's solution the calotte was placed in the hole of a light-impervious plexiglass mount and instantly brought under the microscope so that the preparation could be watched from proximal.

The microscope without condenser and foot-plate is fixed on a table top with a large hole below the stage-plate to allow the beam from the test-light to hit the cut-off eye from distal. The test-light on the cross-bar of a pendulum with its axis of rotation through the axis of the objective on the level of the preparation can be moved along a circular arch through the visual field of certain ommatidia in the preparation (comp. BURKHARDT and STRECK, 1965).

In this stage of completion pretests were made with normally adapted eyes using as light source a 6V/5A filament lamp at a distance of 35 cm from the preparation. As the light source moved through the visual field of the observed ommatidium, the angular position of the light source on both sides of the visual field at which the rhabdom turned dark was determined without a light measuring instrument. This method is similar to that used by AUTRUM and WIEDEMANN (1962) and WIEDEMANN (1965).

We determined a vertical width of the visual field of $6.68^{+}_{-}0.14°$ (n=20) and a horizontal width of $7.42^{+}_{-}0.24°$ (n=20) (angular width of the lamp's filament of $0.25°$ included). These values correspond with those published by AUTRUM and WIEDEMANN. The difference between the two measurement directions is not significant, on the other hand, the agreement in the larger horizontal width in both investigations makes a little difference probable.

To record intensity distribution curves, the experimental arrangement was furnished with additional equipment. The driving axle of the pendulum was directly connected to the axis of a 10-turn potentiometer, whereas the photo-tube of the microscope was equiped with a fast photomultiplier (diameter of measurement field $5.6 \mu m$; spectral sensitivity from 380 – 700 nm, maximum at 420 nm). A 200 W mercury-vapour high-pressure lamp was used as light source installed in a box with a diaphragm aperture of 3 mm in a distance of 70 cm from the preparation (angular width $0.25°$). Comparison of the spectral sensitivity of the photo-cathode with the emission spectrum of the light source shows that wavelengths between 400 and 450 nm form the main part of the measurements. The output of the potentiometer, being a standard of the angular position of the light source, was connected to the X-input of a plotter, the Y-input of the plotter depended on the output of the photomultiplier.

Before presenting the results some remarks concerning the evaluation of the curves have to be made. For reasons not discussed here there is probably no standard distribution of the widths of the intensity distribution curves. Therefore means have been calculated, but the best estimation of the most probable values is given by the modal values of the frequency-distributions and the curves determined thereby. Consequently the nonparametric Wilcoxon test was used for statistics.

As expected, any successful recording showed a bell-shaped intensity distribution curve. Any half degree on both sides of maximum amplitude (optical axis) the hights of the curves were measured and all these values standardized to amplitude one. Fig. 1a shows as an example the means from vertical and horizontal measurements in dark adapted eyes. Beside that the widths of the curves on 95, 75, 50, 25, 15, 10 and 5% hight were measured and modal values were calculated from the respective frequency-distributions.

No significant difference, neither between the results from vertical or horizontal measurements nor between the results from light and dark adapted ommatidia, could be found. On the contrary the modal values of the widths of the curves from the four measurements nearly coincide (fig. 1b). Following examination of the reciprocal dependence of angular width (mode) and standardized relative intensity, the curve given by the modal values may well be approximated by the Gaussian function

$$I(\varphi) = \exp(-0.41 \ [\text{degree}^{-2}] \ \varphi^{2}).$$

The full width on half maximum of this function is $2.60°$ whereas acceptance angles calculated separately for differently treated ommatidia are $2.53°$ (vertical) and $2.56°$ (horizontal) for light adapted ommatidia and $2.75°$ (vertical) and $2.78°$ (horizontal) for dark adapted ommatidia. Again considering only the means and the modal values, the visual field of an ommatidium seems to be a little bit wider in horizontal direction. But a difference can not be ensured, not even taking together the values from one measurement direction from light and dark adapted eyes.

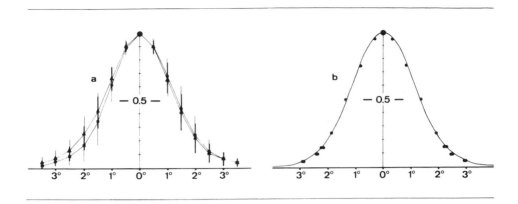

Fig. 1. Standardized intensity distribution curves from dark adapted ommatidia. Means of measurements in vertical (▲) and horizontal (■) direction and the appertaining standard deviations are shown in fig. 1a (from 10 and 12 recorded curves respectively). In fig. 1b a Gaussian function is fitted in the modal values (●) of the widths of the intensity distribution curves shown in fig. 1a. Abscissa: angle of incidence of light

With the given experimental conditions no significant difference between the results from light and dark adapted ommatidia could be ascertained. In both cases the identical input was motified by the dioptric apparatus and its surrounding in the same way. After neither pigment migration in the pigment cells nor changes in the fine structure of the ommatidial cells had been found comparing light and dark adapted ommatidia, this had to be expected (GOLDSMITH, 1963, note added in proof; VARELA and PORTER, 1969).

Our measurements lasted no longer than 1 - 3 min each. According to GOLDSMITH (1963) light adaptation is completed in a few seconds. Therefore adaptive processes in the retinula cells themselves as described by MENZEL in Formica and by HERRLING in Cataglyphis (this volume; see also WEHNER, HERRLING, BRUNNERT and KLEIN, 1972; BRUNNERT and WEHNER, in press) might occur remaining undetected with the present method. But first, the distribution of the retinula pigment in the worker bee differs from that described for example in Formica and Cataglyphis in that considerable numbers of granules are found only in the distal sixth of the hole length of the retinula cells. Secondly VARELA and PORTER (1969) could neither find any dislocation of the retinula pigment nor any change in the fine structure of the retinula cells. To sum up these results, in the worker bee there seems to be no stimulus regulation neither in the dioptric apparatus nor in the rhabdom.

The last statement is confirmed by the following indication. We don't know at which depth the one rhabdom chosen for measuring was cut off. The only criterion for the choice of an ommatidium was the direction of its optical axis, and thus it is probable that the different measurements were taken along at least a big part of the hole length of the rhabdom . Consequently we tried to correlate the widths of the single intensity distribution curves with the respective maximum intensities. No positive results were obtained. With the present experimental conditions this allows no definitive conclusions, particularly because the amplitude is very sensitive to deviations of the plane of the way of the light source from the optical axis of the measured ommatidium. But still it confirms once more the rhabdom being a light guide, a very efficient light guide in that at maximum amplitude some 10^8 quanta·sec^{-1} passing the bleached rhabdom   no considerable extinction could be figured out (comp. HAMDORF and LANGER, 1965) and no dependence of the shape of the intensity distribution curves from the length of the rhabdom   could be shown.

Nevertheless, different sensitivity of light and dark adapted eyes of the honeybee is obvious and has often been shown (e.g. GOLDSMITH, 1963). Following the terminology of BURKHARDT (1960, 1961), there remain three mechanisms that could influence the excitation of the sense cells. Loss and recovery of sensitivity resulting from the momentary conditions of the photochemical reactions are well known to occur. Dynamic excitation changes and centrifugal control of excitation allow no final statements. In 1961, KUNZE described a mechanism as "an adaptation in the valuation of the stimuli, a dynamic process of temporal changements of a value dependent on the actual frequency of light alternations". Its fundamental processes are unknown. From electron microscope studies (VARELA, 1970) the possibility of centrifugal control of receptor neurons activity is given, but so far, no electrophysiological analyses of the neural connections have been reported.

The findings reported in this article are in agreement with the results of earlier behaviouristic investigations (see EHEIM and WEHNER, 1972; WEHNER, this volume) and especially with the electrophysiological results by LAUGHLIN and HORRIDGE (1971). This insures for the worker bee a bell-shaped intensity distribution curve respectively sensitivity distribution curve determining an acceptance angle $\triangle \varphi$ of about 2.6°.

The difference to the intensity distribution curve published by KUIPER (1962) and to the lens admittance function calculated by VARELA and WIITANEN (1970) must be affected by factors of great importance relative to the optical properties of the ommatidia, however not taken into account by the authors mentioned. Concerning the conditions in KUIPER's experiments changes in the optics of the eyes might be responsible resulting from the use of frozen material or owing to the long time needed for the distinct measurements. VARELA and WIITANEN got their results by means of ray optics considering refractive indices, radii of curvature and dimensions of the constituent parts of an ommatidium. In that indices of refraction and radii of curvature were varied in a wide range to evaluate possible errors, these cannot be the reason for the discrepancy. But VARELA and WIITANEN did not take into consideration diffraction effects that must be important, the way of the light being narrowed down by the corneal lenslet to about 30 μm. For the worker bee the role of diffraction was discussed by BARLOW (1952) and by LAUGHLIN and HORRIDGE (1972) with reference to the lens admittance function; EHEIM and WEHNER (1972) have shown that diffraction in addition to the refraction power of the corneal lenslets is also responsible for the overlap of the visual fields of neighbouring ommatidia.

References

AUTRUM, H.J., WIEDEMANN, I.: Versuche über den Strahlengang im Insektenauge (Appositionsauge). Z. Naturforsch.17 b, 480 - 482 (1962).
BARLOW, H.B.: The size of ommatidia in apposition eyes. J. exp. Biol. 29, 667 - 674 (1952).
BAUMGAERTNER, H.: Der Formensinn und die Sehschärfe der Bienen. Z. vergl. Physiol. 7, 56 - 143 (1928).
BRUNNERT, A., WEHNER, R.: Fine structure of light- and dark-adapted eyes of desert ants, Cataglyphis bicolor (Formicidae, Hymenoptera). J. Morph., in press.
BURKHARDT, D.: Die Eigenschaften und Funktionstypen der Sinnesorgane. Ergebn. Biol. 22, 226 - 267 (1960).
BURKHARDT, D.: Allgemeine Sinnesphysiologie und Elektrophysiologie der Receptoren. Fortschr. Zool. 13, 146 - 189 (1961).
BURKHARDT, D., STRECK, P.: Das Sehfeld einzelner Sehzellen - eine Richtigstellung. Z. vergl. Physiol. 51, 151 - 152 (1965).
EHEIM, W.P., WEHNER, R.: Die Sehfelder der zentralen Ommatidien in den Appositionsaugen von Apis mellifica und Cataglyphis bicolor (Apidae, Formicidae; Hymenoptera). Kybernetik 10, 168 - 179 (1972).

GOLDSMITH, T.H.: The course of light and dark adaptation in the compound eye of the honeybee. Comp. Biochem. Physiol. 10, 227 – 237 (1963).

HAMDORF, K., LANGER, H.: Veränderung der Lichtabsorption im Facettenauge bei Belichtung.
Z. vergl. Physiol. 51, 172 – 184 (1965).

KIRSCHFELD, K.: Die Projektion der optischen Umwelt auf das Raster der Rhabdomere im Komplexauge von Musca. Exp. Brain Res. 3, 148 – 270 (1967).

KUIPER, J.W.: The optics of the compound eye. Symp. Soc. exp. Biol. 16, 58 – 71 (1962).

KUNZE, P.: Untersuchung des Bewegungssehens fixiert fliegender Bienen. Z. vergl. Physiol. 44,
656 – 684 (1961).

LAUGHLIN, S.B., HORRIDGE, G.A.: Angular sensitivity of retinula cells in dark-adapted
worker bee. Z. vergl. Physiol. 74, 329 – 335 (1971).

SHAW, S.R.: Interreceptor coupling in ommatidia of drone honeybee and locust compound eyes.
Vision Res. 9, 999 – 1029 (1969).

VARELA, F.G.: Fine structure of the visual system of the honeybee (Apis mellifera). II. The
lamina. J. Ultrastruct. Res. 31, 178 – 194 (1970).

VARELA, F.G., WIITANEN, W.: The optics of the compound eye of the honeybee (Apis mellifera). J. gen. Physiol. 55, 336 – 358 (1970).

WEHNER, R., HERRLING, P.L., BRUNNERT, A., KLEIN, R.: Periphere Adaptation und zentral-
nervöse Umstimmung im optischen System von Cataglyphis bicolor (Formicidae, Hymenoptera). Rev. Suisse Zool. 79, 197 – 228 (1972).

WIEDEMANN, I.: Versuche über den Strahlengang im Insektenauge (Appositionsauge). Z. vergl.
Physiol. 49, 526 – 542 (1965).

4. Pigment Migration and the Pupil of the Dioptric Apparatus in Superposition Eyes

P. Kunze

Max-Planck-Institute of Biological Cybernetics, Tübingen, Germany

1. Introduction

The superposition eyes of many insect species show, when dark adapted, the phenomenon of eye glow (literature in HOEGLUND, 1966). It is changed by light induced pigment migration: initially a bright disc is visible in the eye which gradually becomes darker and contracts concentrically. This process was explained by EXNER (1891) on the basis of the superposition ray path. In the incident light microscope an additional observation can be made: the initially completely illuminated surface of each single facet is gradually reduced by a dark zone progressing from the edge to the centre; simultaneously its brightness decreases. This process was already observed by KIESEL (1894), however, with eyes which were immersed in a highly refracting medium in order to avoid reflexion and refraction by the corneal surface. In the following this observation is made with optically intact eyes. It allows conclusions on the nature of the dioptric apparatus of superposition eye ommatidia, and on the interference of pigment migration with the ray path which it produces.

2. Experiments and observations

With an experimental setup described elsewhere (KUNZE, 1972), partially illuminated superposition eyes with a reflecting proximal tapetum can be observed in a "reversed ray path": light from the inside of the eye passes through the dioptric systems. (Thus, disturbing corneal reflexions are avoided in the investigated eye areas.) The experiments of this paper were carried out with Ephestia kühniella (Lepidoptera) and Chrysopa vulgaris (Neuroptera).

In fig. 1 completely dark adapted eyes and partially light adapted eyes (the same eyes under the same conditions of observation) are shown. Light was entering the eyes through only 3 (Ephestia) and 4 – 5 facets (Chrysopa). The microscope was focussed on the cornea. The photographs were taken through a Leitz UM 20 objective (n.a. = 0.22). The photographic procedures were the same for the corresponding light and dark adapted states.

The eyes of both animals show the same basic phenomena. Dark adapted, their facets are brightly lit right to the edge. With progressing light adaptation characteristic changes appear which are caused by pigment migration: a dark zone progresses gradually from the edge to the centre of the facet. Simultaneously the still illuminated facet area becomes darker.

Here the question arises, how it is possible for screening pigment to shield the cornea if it is not situated in the cornea itself. As is well known, the screening pigment which, in the dark adapted eye, is concentrated between the crystalline cones, migrates with illumination into the area proximal to the crystalline cones. Apparently it was not the material pigment that was observed in the cornea, but rather its image – the dark zone – which was produced by the dioptric apparatus of each ommatidium. Hence, it should also be possible to observe in the cornea the image of an object which was introduced into the eye proximally to the crystalline cones.

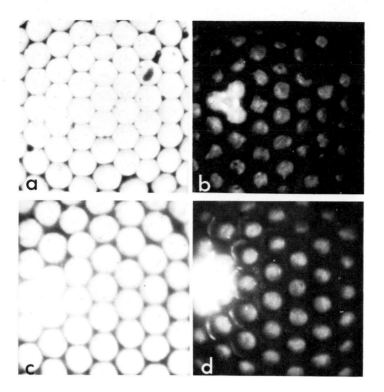

Fig. 1 a–d. Eye glow as observed in the corneal plane. Left: dark adapted eyes,
right: partially light adapted eyes. a, b: Ephestia kühniella, wild type eye;
c, d: eye of Chrysopa vulgaris

For this experiment eyes of the Ephestia mutant "transparent" were used. They are free of
screening pigment whose migration would interfere with the observation. A small metal needle
served as an object. Its diameter was 6 μm, conically tapered to a tip radius of 1 – 2 μm.
(Dr. R. HENGSTENBERG, Tübingen, kindly supplied a used microelectrode.) With the help of
a micromanipulator this needle was inserted into the eye 10 – 15 facets to one side of the observed ommatidium, approximately at right angles to its axis, and advanced. According to the
structural dimensions of the eye, the needle tip can thus be expected to reach a position immediately proximal to the crystalline cones of the ommatidium under observation. Under these
conditions the object appears in the plane of the cornea. It appears inverted. Therefore it must
be an image. When the object is moved, the movement within the cornea facet occurs in the
opposite direction; the image, however, appears successively in neighboring ommatidia which
follow in the objective movement direction.

In fig. 2 the needle tip can be seen in the upper facet of the middle row and also, less sharp and
somewhat shorter, in the centre facet. The ommatidia on the right were illuminated (corneal reflexions), and the light was projected by their dioptric systems onto the more proximal metal
needle.

Fig. 2. Inverse image of a needle tip in a corneal facet. The needle itself was placed into the interior of the eye, just proximal to the crystalline cones. Eye of Ephestia mutant transparent

3. Explanation of the observations

An explanation can be based on a model of the dioptric apparatus which is analogous to an astronomical telescope (Kepler telescope). It consists of two collecting lenses, the distance between which is the sum of their focal lengths for the telescopic system. This model was assumed by EXNER (1891) for the dioptric apparatus of superposition eye ommatidia (see also WINTHROP and WORTHINGTON, 1966).

In fig. 3 such a model is shown. The parallel ray bundle M passes through the system in an erect ray path. With no other stops in the ray path, the boundary of the object glass L_1 will serve as the entrance pupil and its image produced by the ocular L_2 as the exit pupil. Thus, the entire object glass is used for the passage of light. If, however, the ray path is narrowed in the plane of the exit pupil, the image of this pupil, which lies in the main plane of the object glass, will serve as the entrance pupil. Fig. 3 also demonstrates that an object in the plane of the exit pupil is inversely imaged in the plane of the entrance pupil.

Fig. 3. Ray path in a lens system of two collecting lenses L_1, L_2 (and their according foci F) at a distance equal to the sum of their focal length, and the position of the entrance pupil (Eintrittspupille) and the exit pupil (Austrittspupille) of this system

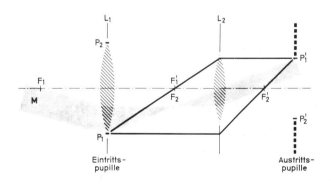

Assuming that the entrance pupil of the dioptric system was focussed in fig. 1, and that with pigment migration the exit pupil was restricted, the image of this exit pupil was observed. Accordingly, in fig. 2 the inverse image of the needle tip was observed in the cornea, with the needle itself placed approximately in the plane of the exit pupil of the particular dioptric system. Thus, the imaging properties of the curved corneal surface may be excluded for the observations documented in fig. 1 and 2, since the images are observed in the plane of the cornea itself. It seems fair to assume that the dioptric properties of the crystalline cones produced the observed images. KUNZE and HAUSEN (1971) gave evidence for imaging properties of the crystalline cones of Ephestia.

The dioptric apparatus of superposition eye ommatidia contains a true pupil which can be changed by pigment migration. Since the pigment migration depends on the light flux through the system, this pupil is functionally analoguous to the variable pupil in vertebrate eyes.

Acknowledgements.   I thank Miss I. GEISS and Dr. B. ROSSER who helped to prepare the manuscript, and E. FREIBERG for drawing fig. 3.

References

EXNER, S.: Die Physiologie der facettirten Augen von Krebsen und Insecten. Leipzig und Wien: Franz Deuticke (1891).

HOEGLUND, G.: Pigment migration, light screening and receptor sensitivity in the compound eye of nocturnal Lepidoptera. Acta Physiol. Scand. 69, Suppl. 282, 1 - 56 (1966).

KIESEL, A.: Untersuchungen zur Physiologie des facettirten Auges. Sber. Akad. Wiss. Wien, math.-nat. Kl. 103, 97 - 139 (1894).

KUNZE, P., and HAUSEN, K.: Inhomogeneous refractive index in the crystalline cone of a moth eye. Nature 231, 392 - 393 (1971).

KUNZE, P.: Comparative studies of arthropod superposition eyes. Z. vergl. Physiol. 76, 347 - 357 (1972).

WINTHROP, J.T., and WORTHINGTON, C.R.: Superposition image formation in insect eyes. Biophys. J. 6, Proc. Biophys. Soc. 124 (1966).

5. Eye Movements in the Housefly Musca domestica

R. Hengstenberg
Max-Planck-Institute of Biological Cybernetics, Tübingen, Germany

Abstract.   "Clock-spikes" in Musca are produced by a motoneurone which lies in the subesophageal ganglion. It innervates a small muscle, attached to the inner frontal margin of the retina.
In the frontal eye region, muscular activity causes the distal rhabdomere tips to move perpendicular to the ommatidial axes. Slow angular movements of the optical axes of the retinula cells
result herefrom. The spike rate can be slightly changed by a variety of gross visual stimuli. The
adequate visual stimulus as well as the functional significance of the eye muscle system with
respect to the behaviour of the unrestrained insects are still unknown.

"Clock-spikes" have been found in the optic lobes of the blowfly Calliphora by KUIPER and
LEUTSCHER-HAZELHOFF (1965, 1966). These impulses occur spontaneously at an extremely
regular rate of 50 - 60 imp/sec. They seemed to be independent of any stimulation. The same
kind of spike activity has been found in a number of Dipteran families like Syrphidae, Muscidae
and Drosophilidae. Fig. 1 shows a half second section from a 12 hrs continuous spike train, which
was recorded from the housefly Musca domestica. Under constant conditions, interval histograms
appear to be almost gaussian, where the standard deviation amounts to about .025 of the mean
value. Occasionally (< 0,2/min), the interval length increases spontaneously for about 10 intervals and then recovers within approximately 100 intervals to its initial value. This activity most
probably is also maintained in free living animals as long as they are alive.

These impulses can be related to structural elements by searching the brain with two independent
recording electrodes: one is fixed at a site of clock-spike activity, the other one is roved through
the brain until cross-correlated signals are obtained. This procedure revealed (HENGSTENBERG,
1971) that on both sides of the head there are two distinct sites of activity respectively: a central
region near the antennal glomerulus and a peripheral region at the ventral margin of the medulla.
Spikes originate in the central region and travel at 2 m/sec to the ipsilateral peripheral region.
Under constant conditions the impulses in the left and right side of the head are statistically independent from one another.

Fig. 1. "Clock-spikes" from
Musca; extracellular record

500 μV

100 ms

By marking the recording sites and serial sectioning of the experimental flies, the cells, producing clock-spikes could be identified. Fig. 2 shows the results: a motoneurone lies on either
side of the subesophageal ganglion, and produces action potentials regularely. These are centrifugally conducted along a thin peripheral nerve, which contains only a single axon of 6 μm
diameter. The motor fibre supplies a very small muscle, consisting of 14 - 20 tubular muscle
fibres, which run from the eye to the back of the head. The identity of nerve and muscle with
the generators of clock-spikes was verified by extra- and intracellular recording under microscopic observation (BURTT and PATTERSON, 1970; HENGSTENBERG, 1971).

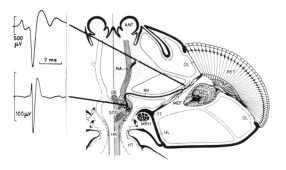

Fig. 2. Anatomical reconstruction of the eye muscle and its motoneurone. Two sites of "clock-spike" activity are indicated by averaged extracellular potential changes. MOT: eye muscle, NV: motor nerve

One end of the muscle is inserted at the tentorium at the back of the head. This region is relatively rigid and therefore considered to be static. The other end is fixed to the inner margin of the orbital ridge, which surrounds the retina like a cone mantle. According to its insertions the muscle is called M. orbito-tentorialis. The muscle transmits its force to the eye near the "equator" and at about $90°$ to the plane of the orbital ridge (fig. 2). Apparently the eye is thereby moved parallel to the corneal surface. No other muscles are attached to the eye skeleton, which suggest that the muscle operates against the elastic forces of the eye. Movements caused by the orbito-tentorial muscle are most likely restricted to the eye since no other movements in the head region could be correlated with clock-spikes. This was checked by recording clock-spikes while observing spontaneous movements of the proboscis, the maxillary palpi, the antennae, the pulsation of hemolymph, the peristaltic movements of the esophagus, and the respiratory movements of the tracheal system. The spike rate is likewise invariant against tilting and rotation of the head, relative to the thorax. Clock-spike dependent movements of the whole compound eyes can also be excluded by recording the light, reflexted from the edges of a tiny white dot, placed on the corneal surface.

Movements of the retina would affect the input to the receptors only if the distal ends of the rhabdomeres (which act as light guides) are displaced with respect to the corneal lenses. Depending on the direction, size and mode of such movements, different consequences might be expected. Movements in the focal plane of the dioptric system, for example, would cause angular shifts of the optical axes of the retinula cells. These shifts could be detected by a displacement of the pseudopupil. Fig. 3 shows a preliminary experiment, where clock-spikes are recorded simultaneously with the light, reflected from the edges of the ipsilateral deep pseudo-pupil (ventro part of the eye, frontal margin). The correlation between the two signals is evident and reverses its sign if the aperture of the photomultiplier tube is shifted from the anterior edge (Aa) of the pseudopupil to the posterior one (Ap). This relation becomes particularly clear, when the spike train shows one of the spontaneous frequency discontinuities, mentioned earlier. When averaging the photomultiplier signal phase-locked to the clock-spikes a modulation of the reflected light with the average spike frequency is observed which also reverses its sign between Ba and Bp. This modulation is much smaller in amplitude than the range of fluctuations in fig. 3A. It reflects an imperfect tetanus of the eye muscle.

One can conclude from these qualitative findings that the clock-spike frequency is transformed into a tetanic force by the muscle, which acts onto the eye skeleton and causes movements of the distal tips of the rhabdomeres, perpendicular to the optical axes of the retinula cells. Under constant conditions, these movements contain an approximately horizontal component which shifts the optical axes anteriorly with increasing spike frequency. A crude estimate of the angle through which the optical axes are shifted yields about $0,1 - 0,5°/10$ Hz near the average spike frequency for the horizontal component. This means that the actual movements would be larger if their direction did not coincide with the horizontal plane. Unfortunately, the direction of the movements and their actual size could not yet be precisely determined for technical reasons.

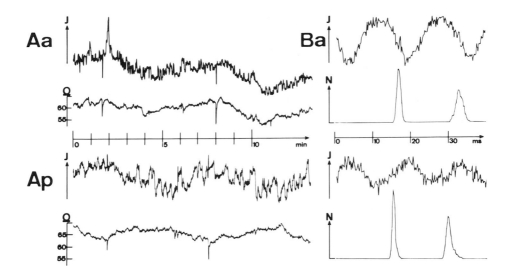

Fig. 3. Correlation between the clock-spike frequency Ω and the light intensity I, reflected from the anterior (Aa) and the posterior (Ap) edge of the deep pseudo-pupil. Ba, Bp: same records as in A, phase locked average of I, when the averager is started by clock-spikes. N: number of spikes per channel

So far the results suggest that the eye muscle is related in some way to the visual system. This idea is further supported by changes in spike frequency due to gross visual stimuli. Stepwise increased ambient light intensity produces a transient as well as a steady state decrease in spike frequency (BURTT and PATTERSON, 1970; HENGSTENBERG, unpublished results). The reverse holds for stepwise decreased light intensity. This influence is mediated via the compound eyes and is independent of the ocelli. Periodic light flashes cause a partial driving of the clock-spikes within a narrow frequency band, close to the undisturbed spike frequency, its integral or fractional multiples. The phase relation between flashes and spikes shifts through 180° when passing the resonance frequency. Small changes in spike frequency are also caused by moving striped patterns, where the mean brightness within the visual field is kept constant (Van BARNE-VELD, 1971; HENGSTENBERG, unpublished results). Here the frequency depends on the direction of pattern motion. It increases with horizontal front-to-back stimuli on the ipsilateral side, decreases with back-to-front stimuli and does not change when movement is about vertical. The directional specificity of this response strongly favours the idea of a specific visual input to the eye muscle system.

All of the observed changes in spike frequency are in the order of a few percent of the average. This might simply mean that none of the applied stimuli meets the specific requirements of the input to the eye muscle system. Unfortunately, there is no behavioral reaction of flies known which would require eye movements as described. Therefore, it is necessary to speculate about the possible significance of such movements, and to design critical experiments which would allow to test these ideas in the intact animals.

References

Van BARNEVELD, H.H.: Statistical analysis of repetitive neuronal activity. Clock-spikes in Calliphora erythrocephala. Proefschrift Rijksuniversiteit te Groningen (1971).

BURTT, E.T., PATTERSON, J.A.: Internal muscle in the eye of an insect. Nature 228, 183 - 184 (1970).

HENGSTENBERG, R.: Das Augenmuskelsystem der Fliege Musca domestica. I. Analyse der "Clock-spikes" und ihrer Quellen. Kybernetik 9, 56 - 77 (1971).

HENGSTENBERG, R.: unpublished results.

KUIPER, J.W., LEUTSCHER-HAZELHOFF, J.T.: High-precision repetitive firing in the insect optic lobe and a hypothesis for its function in object location. Nature 206, 1158 - 1160 (1965).

LEUTSCHER-HAZELHOFF, J.T., KUIPER, J.W.: Clock-spikes in the Calliphora optic lobe and a hypothesis for their function in object location. In C.G. BERNHARD (ed.) The Functional Organization of the Compound Eye. pp. 483 - 492, Oxford: Pergamon Press (1966).

III. Biochemistry of Visual Pigments

1. Photoreconversion of Invertebrate Visual Pigments

K. Hamdorf, R. Paulsen, J. Schwemer, and U. Taeuber
Institute of Animal Physiology, University of Bochum, Germany

Abstract. The reconversion of thermostable metarhodopsin to rhodopsin seems to be a basic principle of receptor function in invertebrates as studied in cephalopods and insects. The absorption of a quantum of uv light converts an uv–pigment molecule to its thermostable photoproduct, initiating simultaneously a generator potential; due to the high probability of the absorption of a quantum of blue light by photoproduct B, the uv–sensitive pigment molecule is regained quickly. By calculation it is estimated that at every moment about 90% of the whole pigment within the receptor has the uv–sensitive configuration. This implies that the sensitivity of the uv receptor is controlled by the spectral distribution of the sky.

It is well known that visual pigments are chromoproteids with 11–cis retinal$_1$ or 11–cis retinal$_2$ as chromophoric group. These visual pigments are located in the membranes of photoreceptor cells, the disc–membrane system of rods and cones in vertebrates and the microtubular membranes of the rhabdomeres in invertebrates. The primary event in vision involves the absorption of light quanta by the visual pigments: When a quantum of light is absorbed by a pigment molecule, the steric hindered 11–cis retinal changes into the all–trans configuration. Thermodynamic investigations by HUBBARD, BOWNDS and YOSHIZAWA (1965) showed that this first step of the bleaching sequence requires an activation energy of approximately 20 kcal/mole. The energy content of light rises from approximately 50 kcal for one mole of light quanta of 600 nm to approximately 50 kcal for one mole of light quanta of 600 nm to approximately 100 kcal for that of 300 nm. Therefore, within the spectral range each quantum has enough energy to start the bleaching process.

Fig. 1 shows the agreement of the absorption spectrum of the visual pigment with the sensitivity curve of the dark adapted eye of the cephalopod Eledone moschata over the whole spectral range. This demonstrates that the quantum yield for the isomerization process is independent from the energy content of the quanta. That the absolute quantum efficiency for a variety of visual pigments is approximately 1 has been shown by DARTNALL (1968).

The absorption of a light quantum by a rhodopsin molecule initiates a reaction sequence, the intermediates of which are well defined for the visual pigments of vertebrates. The first photoproducts after light absorption are the short-living intermediates prelumi– and lumirhodopsin (YOSHIZAWA and WALD, 1963). These are followed by metarhodopsin I, which has a longer lifetime and enters into a pH–dependent equilibrium with the ultraviolet absorbing metarhodopsin II (MATTHEWS, HUBBARD, BROWN and WALD, 1963). The pK point is about pH 7. The next stage in the decay sequence is metarhodopsin III, which decomposes into the all–trans retinal and the protein opsin. While previously, it was assumed that this splitting step triggers the excitation of the photoreceptor cell, nonetheless, in 1958 HUBBARD and ST. GEORGE showed that in cephalopods the reaction sequence stops at the metarhodopsin. Therefore, an

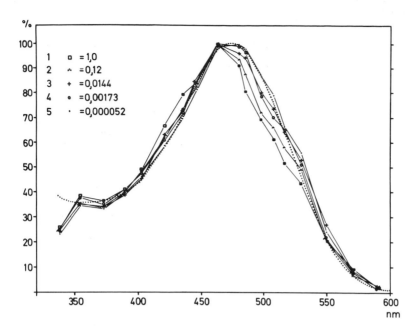

Fig. 1. Relative spectral sensitivity of the dark- and light-adapted retina of Eledone moschata. The curves 1 up to 5 are measured at different quantum levels from 1 to 10^{-5}. The dark-adapted eye shows a spectral sensitivity (curve 5) that fits very well the rhodopsin curve (dotted line). Light adaptation reduces the sensitivity in the range of metarhodopsin absorption (curve 1)

earlier step in the bleaching sequence has to be responsible for the excitation process.

In order to obtain further information concerning these primary processes and the function of photoreceptors, we studied the visual pigment system of the cephalopods. In the cephalopod E. moschata, irradiation of rhodopsin (λ_{max} 470 nm) leads to a thermostable, pH-dependent mixture of acid and alkaline metarhodopsin. The pK point of this equilibrium is at pH 6.4 (SCHWEMER, 1969). The thermostable photoproducts can be reisomerised by light to rhodopsin. Insofar as the intermediates of the reverse process are concerned, the exact number and kind are as yet unknown. Some aspects of this photochemical regeneration of rhodopsin can be illustrated as follows: Fig. 2 shows the absorption spectra of rhodopsin (A) and acid metarhodopsin (B). The point of intersection of the two absorption spectra at 480 nm - the specific absorption coefficients of rhodopsin and metarhodopsin are equal - is the isosbestic point of the rhodopsin - metarhodopsin system. With increasing wavelengths from the isosbestic point the absorption of metarhodopsin is higher than that of rhodopsin, whereas with decreasing wavelengths the reverse is true. Exposure of an acid (pH 5,0) solution of rhodopsin to light at the wavelength of the isosbestic point leads to an equilibrium of 50% rhodopsin and 50% metarhodopsin. Irradiation with light of shorter wavelengths than the isosbestic point, e.g. 440 nm, yields an equilibrium of 38% rhodopsin and 62% metarhodopsin. In contrast, irradiation with light of longer wavelengths than the isosbestic point, e.g. 550 nm, a ratio of 90% rhodopsin to 10% metarhodopsin is obtained (SCHWEMER, 1969). The mathematic description of photochemical equilibria leads to the equation depicted to the right of fig. 2. In this equilibrium, the quotient of the concentration of rhodopsin (A_{gl}) and the concentration of metarhodopsin (B_{gl}) is independent of light intensity, being only a

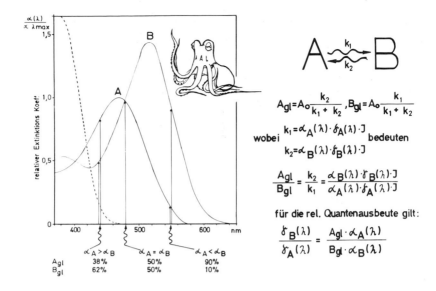

$$A \rightleftharpoons B$$

$$A_{gl} = A_o \frac{k_2}{k_1 + k_2} \quad , B_{gl} = A_o \frac{k_1}{k_1 + k_2}$$

$$wobei \quad \begin{aligned} k_1 &= \alpha_A(\lambda) \cdot \delta_A(\lambda) \cdot J \\ k_2 &= \alpha_B(\lambda) \cdot \delta_B(\lambda) \cdot J \end{aligned} \quad bedeuten$$

$$\frac{A_{gl}}{B_{gl}} = \frac{k_2}{k_1} = \frac{\alpha_B(\lambda) \cdot \delta_B(\lambda) \cdot J}{\alpha_A(\lambda) \cdot \delta_A(\lambda) \cdot J}$$

für die rel. Quantenausbeute gilt:

$$\frac{\delta_B(\lambda)}{\delta_A(\lambda)} = \frac{A_{gl} \cdot \alpha_A(\lambda)}{B_{gl} \cdot \alpha_B(\lambda)}$$

Fig. 2. The visual pigment system of Eledone moschata. A: Absorption spectrum of rhodopsin; B: Absorption spectrum of acid metarhodopsin. For further explanation of the equations on the right, see text

function of the wavelength of irradiation. The intensity of irradiation determines only the speed at which this equilibrium is obtained. Calculation of the relative quantum efficiency (δ_B / δ_A) from equilibrium measurements of the concentrations of rhodopsin and metarhodopsin, and the known absorption coefficients (α_A and α_B), yielded values of about 1 over the whole spectral range (SCHWEMER, 1969).

Regarding the dependence of the equilibrium rhodopsin to metarhodopsin on the wavelength of irradiation, it follows that it is possible to reconvert all metarhodopsin to rhodopsin by light of wavelengths greater than 590 nm. The reverse process, the complete conversion of rhodopsin to metarhodopsin, cannot be effected by any wavelength of irradiation. The lowest concentration of rhodopsin is obtained by light of that wavelength where the ratio of the absorption coefficients (α_B / α_A) has a minimum value.

The wavelength-dependent photochemical equilibria can be investigated by two different methods:

1. Difference spectra between rhodopsin solution and the monochromatic irradiated solution: From such difference spectra (fig. 3a, solid line) the isosbestic point (now the point of intersection of the difference curve with the abscissa) may be determined. The minimum and maximum of the difference curve corresponds to the wavelengths where the difference of the absorption coefficients is at its lowest or highest, respectively. The dashed curve in fig. 3a shows a difference spectrum for a microphotometric experiment on the isolated retina of E. moschata. The good fit of both curves indicates that the photochemical properties of the pigment system, in vitro and in vivo, are the same.

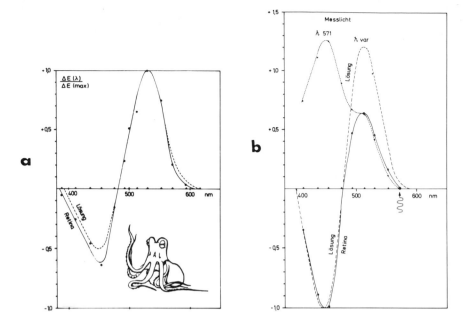

Fig. 3. a) The difference spectrum between a rhodopsin solution and an irradiated (440 nm) solution (dashed line) in comparison that of a dark-adapted and irridiated (440 nm) retina (solid line) of E. moschata.

b) The change in absorption during irradiation with different wavelengths (abscissa), each after irradiation with light of 571 nm for solution (dashed line) and the retina (solid line). The dotted curve shows the change in absorption measured at 571 nm after irradiation with the wavelengths indicated on the abscissa

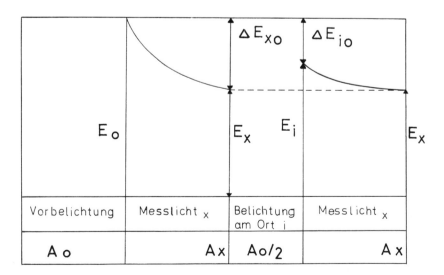

Fig. 4. Measuring program for determination of the absorption coefficients. For further explanations see text

2. The absorption coefficients of rhodopsin and metarhodopsin cannot be determined from the above difference spectra. However, they may be obtained in another way by microphotometric measurements (fig. 4): At first, the retina is irradiated with a wavelength that is only absorbed by the metarhodopsin, which is converted completely to rhodopsin("Vorbelichtung"). Then the retina is irradiated with light of given wavelength ("Messlicht X") at that the relative absorption coefficients is to determine. During this illumination the absorption at wavelength X will change to that of the new photo-equilibrium (ΔE_{xo}). Following irradiation with a wavelength corresponding to the isosbestic point yields rhodopsin and metarhodopsin in equal concentrations and allows the determination of the absorption difference ΔE_{io}. As a control, the change in absorption during irradiation with "Messlicht X" is measured. From the differences in absorption (ΔE_{xo} and ΔE_{io}) the absorption coefficients of rhodopsin and metarhodopsin at wavelength x can be determined by the following equations:

$$\alpha_X = \frac{\Delta E_{xo} \cdot \Delta E_{io}}{\Delta E_{xo} - \Delta E_{io}}$$

$$\beta_X = \frac{\Delta E_{io}(2\Delta E_{io} - \Delta E_{xo})}{\Delta E_{xo} - \Delta E_{io}}$$

Varying the wavelength x of the measuring light throughout the whole spectral range yields absorption spectra for rhodopsin and metarhodopsin.

The values of ΔE_{xo} in solution and in the retina are shown in fig. 3b. The difference between the curves for the solution and the retina in the spectral range of 480 nm to 570 nm is caused by difference in pH: While the solution was measured under acid conditions, the retina was measured at physiological pH. Under these physiological pH conditions, some of the acid meta-rhodopsin is converted into the alkaline form and in this way, the loss of absorption in the range around 530 nm can be explained. The dotted curve (λ 571) shows the relative metarhodopsin content after irradiation at different wavelengths. The highest concentration of metarhodopsin is formed after illumination at 445 nm.

It was of interest, therefore, to see whether the thermostable pigment system of the cephalopods is a special case or is also valid for other invertebrate systems. Since insect vision has been investigated to a great extent by electrophysiological methods, these animals seemed to be suitable objects for the study of visual pigment systems: In 1967, GOGALA demonstrated by extracellular recordings for the frontal eye of Neuroptera Ascalaphus macaronius that only a single type of receptor sensitive to ultraviolet light is present (GOGALA, 1967). By a special method we have successfully extracted the visual pigment from the dark adapted retinulae of these frontal eyes (HAMDORF, SCHWEMER and GOGALA, 1971). The solution exhibits an absorption spectrum with a peak in the near ultraviolet at 345 nm. Exposure of this extract, at physiological pH conditions, to ultraviolet light converts the visual pigment A to a thermostable photoproduct B, absorbing maximally at 480 nm (fig. 5). Whereas the most efficient wavelength for this conversion was 350 nm, all other ultraviolet wavelengths used in experiment convert smaller amounts of the ultraviolet pigment to photoproduct B. Furthermore, the amount of inter-mediate B which is formed, is the result of a photoequilibrium (HAMDORF, SCHWEMER and GOGALA, 1971). When the solution of photoproduct B is irradiated with light of wavelengths between 440 to 550 nm, B is completely reconverted to the uv pigment A. Fig. 5b shows a pair of difference spectra for this photo-interconversion. Furthermore, it can be demonstrated that photoproduct B is convertable to an intermediate C (absorbing maximally at 380 nm) by rising the pH of the solution (fig. 5a). The pK of this pH-dependent equilibrium is 9.3. The experiments described above were also carried out on isolated retinulae employing the microphotometric method.

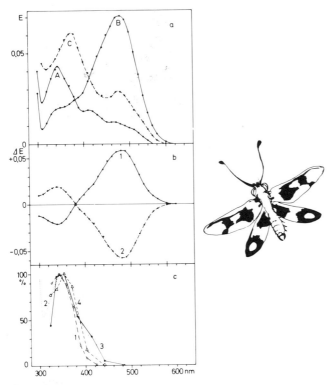

Fig. 5. The ultraviolet visual pigment of Ascalaphus macaronius.
a) Curve A: uv-sensitive pigment extracted from dark-adapted retinulae. The absorption in the range 420 nm to longer wavelengths is due to impurities; curve B: Acid photoproduct; curve C: The conversion of B to the alkaline intermediate C at pH 9.3.
b) Difference spectra between the photopigment A and the photoproduct B. Curve 1 is obtained by subtracting B from A; curve 2 shows the difference spectrum between B and the uv pigment regenerated by irradiating of B with 483 nm.
c) The relation between the spectral sensitivity of the frontal eye (AUTRUM and ZWEHL, 1964), the absorption of uv pigment (DARTNALL, 1968), and the amount of photoproduct B with the wavelengths of irradiation in solution (DOWLING, 1963) and the isolated retinula (GOGALA, 1967)

The results obtained by the various methods used in studying this pigment system are summarized in fig. 5c: From the good fit between the curves for the spectral sensitivity of the frontal eye (curve 1), the absorption spectrum of the uv pigment in solution (curve 2), the amount of photoproduct B formed by light of different ultraviolet wavelengths (curve 3) as well as the amount of B formed by light of different ultraviolet wavelengths in the isolated retinula (curve 4) we suggest that the same interconversions found in solution occur as well in the photoreceptor membrane.

This uv pigment system is comparable in every respect to the rhodopsin system of cephalopods, whereby the uv-visual pigment corresponds to rhodopsin, intermediate B and C to acid and alkaline metarhodopsin, respectively, in particular, because recent investigations demonstrated the uv pigment to be a chromoproteid similar to the known rhodopsins containing retinal$_1$ (PAULSEN et al., in prep.).

From electrophysiological measurements of the spectral sensitivities of other insects it is known that in addition to the ultraviolet receptor there exist two other types of receptors. This was first demonstrated by AUTRUM and ZWEHL (1964) in the honey bee: Intracellular recordings revealed the existence of a green, blue and ultraviolet receptor. A corresponding system with three different types of photoreceptors, the basis of a trichromatic colour vision, seems to be present in the moth Deilephila elpenor. Microphotometric measurements, however, unveiled the same uv pigment system as described above for A. macaronius. Applying the second technique (summarized in fig. 4) a second pigment was found which absorbs maximally in the green spectral range and gave rise to a photoproduct which absorbs maximally the blue range, the maximum of absorption similar to that of the uv intermediate. The photoproduct of the green absorbing pigment can also be reconverted by light (figs. 6 and 7). The third visual pigment system which was postulated from electrophysiological data and which should have a maximal absorption at about 450 nm could not be elucidated.

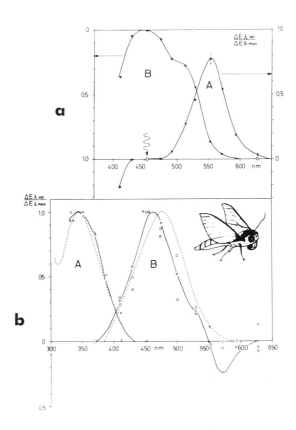

Fig. 6. The two photopigment systems of the moth Deilephila elpenor.
a) Difference spectra of the two components of the green-sensitive pigment at one measuring point: A: Change in absorption after irradiation with different wavelengths, when the retinula was previously irradiated with 453 nm; B: Change in absorption at 453 nm after prolonged irradiation with different wavelengths in green and yellow spectral range.
b) Difference spectra of the uv pigment and its photoproduct (solid curves). A: Change in absorption at 453 nm after prolonged irradiation with different wavelengths of the ultraviolet range; B: Change in absorption after irradiation with different wavelengths of the blue and green spectral range, when the retinula was previously irradiated with 371 nm. For comparison, the dotted curves show the uv pigment and its photoproduct of A. macaronius

Photoreconvertable pigment systems are apparently not restricted to cephalopods and insects but are found as well in crustacea (TAEUBER, in prep.).

Thus, the reconversion of thermostable metarhodopsins to rhodopsins by light seems to be a basic principle of receptor function in invertebrates. The electrophysiological experiments on E. moschata (fig. 1) demonstrate that only the transition of rhodopsin to metarhodopsin generates a receptor potential (HAMDORF, SCHWEMER and TAEUBER, 1968). The spectral sensitivity curve of the

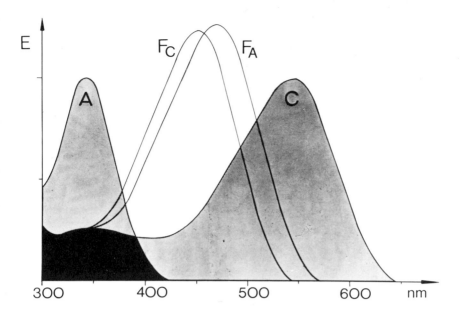

Fig. 7. The calculated absorption spectra of the two pigment systems of Deilephila elpenor.
A: the uv-visual pigment, F_A its photoproduct;
C: the green visual pigment, F_C its photoproduct

light-adapted eye of E. moschata shows a drop in sensitivity in the range above 480 nm. This decrease in sensitivity can be explained by a screening effect of the metarhodopsin which is present in a relatively high concentration in the light-adapted receptor. In the case of A. maca-ronius a screening effect by the photoproduct B cannot be observed because negligible absorp-tion of B in the ultraviolet range. That a correlation exists between the sensitivity of the photo-receptor and the concentration of rhodopsin is well known for vertebrate photoreceptors. In the rat, when 17% of rhodopsin is bleached, the sensitivity is reduced by one log unit (3). Each further bleaching step (of 17% rhodopsin) diminishes the sensitivity by another log unit. In E. moschata a decrease of only 6% rhodopsin is sufficient to reduce the sensitivity by one log unit. This means that the reduction of 40% rhodopsin depresses the sensitivity by 6 log units. While in the light-adapted situation the probability of a quantum absorption by a rhodopsin molecule is only slightly reduced, the sensitivity is greatly diminished. Furthermore, a low concentration of metarhodopsin decreases the efficiency of quantum absorption by a rhodopsin molecule.

To keep the sensitivity of a photoreceptor in a physiological range, it is necessary that the rhodopsin concentration is maintained at a high level. The rate of photoregeneration depends directly on the intensity of light, as shown in microphotometric measurements (fig. 8): In this experiment the retinula of A. macaronius was irradiated with intense ultraviolet light to produce a high concentration of photoproduct B. The transmission by the retinula was then measured at 475 nm for light of different intensities as a function of time. The lower figure shows that the half-time of regeneration is a linear function of the light intensity (SCHWEMER, GOGALA and HAMDORF, 1971). In the following electrophysiological experiment the effect of photoregene-ration on the sensitivity of the eye was tested (fig. 9). When the frontal eye of A. macaronius

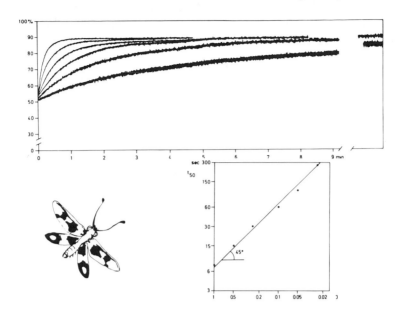

Fig. 8. Original registrations of the increase in transmission after light adaptation by prolonged monochromatic ultraviolet irradiation, measured at 475 nm. For the regeneration of the uv pigment, different intensities (rel. intensities 1; 0,5; 0,25; 0,1; 0,05; 0,025) of the measuring light were used. These curves demonstrate that the rate of photoregeneration is proportional to the intensity of the measuring light. This linear correlation is shown in the smaller figure, where the half-time of regeneration is plotted against intensity.

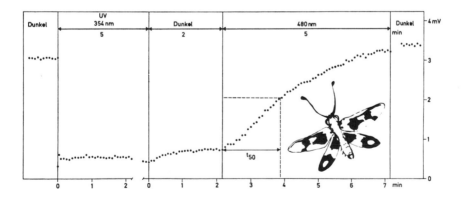

Fig. 9. The regeneration of the uv pigment of A. macaronius determined by extracellular recordings. For further explanation, see text

is irradiated by constant uv test flashes (0,2 cycles/sec) it responds with constant potentials with more than 3 mV. During exposure to prolonged uv-adapting light (5 min) the amplitude of the electroretinogram is reduced to approximately 20% of its original value. When the adapting light was turned off, the amplitude of the ERG increased slightly. After two minutes of dark adaptation, a blue adapting light (480 nm) was switched on and from that moment, the amplitude of the ERG increased rapidly to its original value (HAMDORF, GOGALA and SCHWEMER, 1971). In a further experiment was shown that the spectral efficiency of the electrophysically determined regeneration (the reciprocal time after the amplitude of the ERG had reached the height of 50% of the original value) fits well with the difference spectra received from solution and retinula (fig. 10). These experiments prove that the photoreconversion of intermediate B to the uv-sensitive pigment A increases the generator potential (HAMDORF, GOGALA and SCHWEMER, 1971).

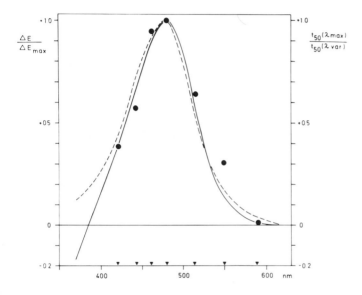

Fig. 10. The relative spectral efficiency of different wavelengths on the photoregeneration. Solid line: difference spectrum for uv pigment in solution, irradiated by uv light; dashed line: difference spectrum between a dark- and a light-adapted (uv-irradiated) retinula. The dots show the regenerative effect of 7 different wavelengths in an electrophysiological experiment

The biological significance of this visual cycle is supported by the interesting fact that the stray light of the sky has two maxima, one at 350 nm and the other at 450 nm. As seen from fig. 11, the absorption peaks of the uv pigment and its photoproduct B fits very well with the emission maxima of the sky (HAMDORF, GOGALA and SCHWEMER, 1971).

In summary, the absorption of a quantum of uv light converts an uv pigment molecule to its thermostable photoproduct, initiating simultaneously a generator potential; due to the high probability of the absorption of a quantum of blue light by photoproduct B, the uv-sensitive pigment molecule is regained quickly. By calculation it is estimated that at every moment about 90% of the whole pigment within the receptor has the uv-sensitive configuration. This implies that the sensitivity of the uv receptor is controlled by the spectral distribution of the sky.

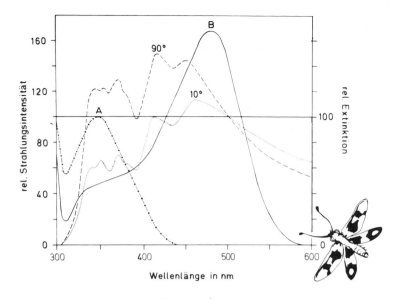

Fig. 11. The emission curves of the sky under 90° and 10° (azimuth). Curve A: absorption spectrum of the uv pigment of A. macaronius; Curve B: photoproduct B derived from the visual pigment A

In the photoreceptors of the cephalopods the chemical reconversion from metarhodopsin to rhodopsin is very slow: at physiological temperatures the receptor needs between 3 to 5 hours to regenerate the photopigment in the dark. The supposition is that in cephalopods the photo-chemical regeneration is of some importance.

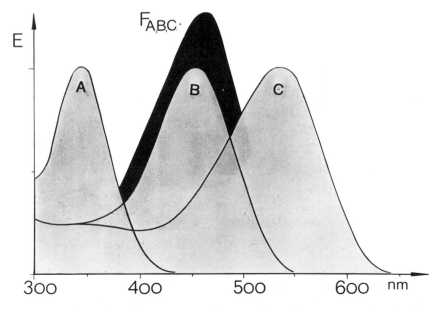

Fig. 12. A hypothetical trichromatic visual system. A and C represents the visual pigment of the uv- and the green-receptor (comparable to the systems found for Deilephila elpenor); B shows the pigment of the postulated blue-sensitive receptor. $F_{A, B, C}$: Absorption spectrum of all three metarhodopsins

In the moth D. elpenor, the blue light of the sky regenerates both the uv-sensitive pigment and the green sensitive pigment, since the corresponding photoproducts of each absorbs maximally in the blue spectral range. Thus, we propose that, as in the case of the honey bee, the third photoreceptor is present having a maximum absorption at 450 nm; furthermore, we assume that the photoproduct of this postulated receptor absorbs maximally in the same range as the photo-products of the green- and uv-sensitive pigments (fig. 12).

References

AUTRUM, H. and ZWEHL, V.v.: Die spektrale Empfindlichkeit einzelner Sehzellen des Bienen-auges. Z. vergl. Physiol. 48, 357 - 384 (1964).
DARTNALL, H.J.A.: The photosensitivities of visual pigments in the presence of hydroxylamine. Vision Res. 8, 339 - 358 (1968).
DOWLING, J.E.: Neural and photochemical mechanisms of visual adaptation in the rat. J. Gen. Physiol. 46, 1287 - 1301 (1963).
GOGALA, M.: Die spektrale Empfindlichkeit der Doppelaugen von Ascalaphus macaronius Scop. Z. vergl. Physiol. 57, 232 - 243 (1967).
HAMDORF, K., SCHWEMER, J. and TAEUBER, U.: Der Sehfarbstoff, die Absorption der Rezepto-ren und die spektrale Empfindlichkeit der Retina von Eledone moschata. Z. vergl. Physiol. 60, 375 - 415 (1968).
HAMDORF, K., GOGALA, M. and SCHWEMER, J.: Beschleunigung der "Dunkeladaptation" eines UV-Rezeptors durch sichtbare Strahlung. Z. vergl. Physiol. 75, 189 - 199 (1971).
HAMDORF, K., SCHWEMER, J. and GOGALA, M.: Insect visual pigment sensitive to ultra-violet light. Nature (Lond.) 231, 458 - 459 (1971).
HAMDORF, K., HOEGLUND, G. and LANGER, H.: Mikrophotometrische Untersuchungen an der Retinula des Nachtschmetterlings Deilephila elpenor (in press, 1972).
HUBBARD, R. and St.GEORGE, R.C.C.: The rhodopsin system of the squid. J. Gen. Physiol. 41, 501 - 528 (1958).
HUBBARD, R., BOWNDS, D. and YOSHIZAWA, T.: The chemistry of visual photoreception. Cold Spring Harbor Symp. on Quantitative Biology XXX, 301 - 315 (1965).
MATTHEWS, R.G., HUBBARD, R., BROWN, P.K. and WALD, G.: Tautomeric forms of metarho-dopsin. J. Gen. Physiol. 47, 215 - 240 (1963).
SCHWEMER, J.: Der Sehfarbstoff von Eledone moschata und seine Umsetzungen in der lebenden Netzhaut. Z. vergl. Physiol. 62, 121 - 152 (1969).
SCHWEMER, J., GOGALA, M. and HAMDORF, K.: Der UV-Sehfarbstoff der Insekten: Photo-chemie in vitro und in vivo. Z. vergl. Physiol. 75, 174 - 188 (1971).
YOSHIZAWA, T. and WALD, G.: Pre-lumirhodopsin and the bleaching of visual pigments. Nature (Lond.) 197, 1279 - 1286 (1963).

2. Metarhodopsin in Single Rhabdomeres of the Fly, Calliphora erythrocephala

H. Langer
Institute of Animal Physiology, University of Bochum, Germany

Abstract.   Stable metarhodopsin is the photoproduct of illuminated rhodopsin in the eye of Calliphora. Rhodopsin is resynthesized in darkness, evidently by chemical means dependent on energy supplied by metabolism within the retina. The thermostability of the product makes it probable that a photoreisomerization may also occur.

Recently, the eyes of a few insect species have been found to contain rhodopsins, the metarho-dopsins of which are thermostabile and can be reisomerized by light (Ascalaphus, GOGALA, HAMDORF and SCHWEMER, 1970; Deilephila, HAMDORF, HOEGLUND and LANGER, 1972). Rhodopsin has been demonstrated by microspectrophotometry in the rhabdomeres of Calliphora erythrocephala (LANGER, 1967; LANGER and THORELL, 1966). This could not be bleached in a simple way and accordingly tests for the presence of metarhodopsin were carried out on illu-minated rhabdomeres.

Measurements were made in a double-beam microspectrophotometer (according to CHANCE, PERRY, ÅKERMAN and THORELL, 1959) on slice preparations of fresh eyes from the chalky mu-tant of Calliphora erythrocephala, as described elsewhere (LANGER, 1965; LANGER and THO-RELL, 1966). In most of the experiments, the pH of the saline (JONES, 1956), in which the tis-sue was maintained, was held between 6.5 and 7.0. Some others were performed first at pH 6.0 and then at 9.0. Illumination of the specimen was provided by a tungsten lamp (6V, 5A), the beam of which was brought to the axis of the measuring beam by a mirror. The light contained predomi-nantly the yellow parts of the spectrum and any residual u.v. and far red components were re-moved with glass filters. Some additional experiments were undertaken with monochromatic light, but the intensities available were too small to obtain recognizable effects.

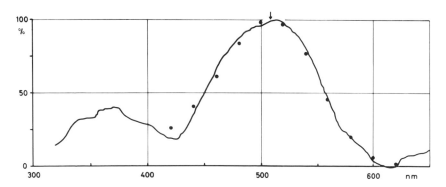

Fig. 1. Extinction curve (——) of rhabdomere no. 6 (of a "green receptor" cell). The difference between microspectrophotometric measuring curve and baseline has been normalized; 100% = $\Delta E_{max} \approx 0.1$. The points (●) represent relative extinction values for a theoretical rhodopsin with λ_{max} = 510 nm obtained from Dartnall's nomogram. Such a good agreement is only found in fresh and non-illuminated preparations

Extinction curves of single rhabdomeres are similar to those of rhodopsins in solution only if they are measured on fresh preparations made in dim red light. In these cases, the portion above 400 nm agrees well with figures for extinction calculated from Dartnall's nomogram (DARTNALL, 1953) for theoretical rhodopsins with an E_{max} of 510 to 530 nm (fig. 1). For older preparations, the main maximum broadens as a result of increasing extinction at shorter wavelengths. Frequently a shoulder in the curve is evident, and occasionally an adjacent peak is found at 480 to 500 nm. The extinction often increases in the near u.v. as well, at around 380 nm. Short illumination intensifies all these changes in the shape of the curves. A bleaching of the type recorded in microspectrophotometric measurements on vertebrate rods and cones (LIEBMAN, 1962; LIEBMAN and ENTINE, 1968; MARKS, 1965) can not be observed in these preparations. Long-lasting illumination, the intensity of which was raised by opening the iris of the condenser, caused an overall decrease in the absorption, probably by destroying the rhabdomere structure and possibly the visual pigment molecules themselves (fig. 2).

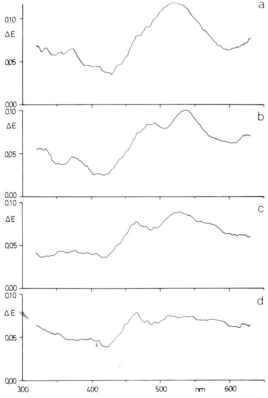

Fig. 2. Series of extinction curves of rhabdomere no. 5 following periods of illumination, redrawn from the originals (the baseline is somewhat oblique to the abscissa):
a) fresh preparation before beginning of illumination
b) after exposure to the amount of light required for the measurement of the extinction in the other rhabdomeres of the ommatidium
c) after 1 minute of illumination
d) after additional 3 minutes of illumination of higher intensity.
The rhodopsin maximum at about 530 nm (curve a) decreases and a maximum of the photoproduct at about 490 nm arises with duration of illumination (curves b and c). The state in curve d) is only obtained under non-physiological conditions

If a short-time illuminated rhabdomere is kept in total darkness for 10 to 30 minutes, the shoulder or adjacent peak becomes smaller, and concomitantly the main maximum at about 520 nm becomes more prominent. The results of a complete experiment of this type are shown in the left hand part of figure 3. After having recorded an extinction curve (a) of a rhabdomere in a fresh preparation, it was kept in darkness for about 20 minutes and the measurement was repeated. Except for an over-all decrease in extinction, no change is evident after this period (b). Then an illumination of one minute was given, and a new curve was immediately obtained. This showed an increase in extinction at somewhat less than 500 nm (c). An additional dark period of 20 minutes yielded a curve with the maximum once more at 520 nm (d). Finally, illumination for three minutes led to a curve (e) with two maxima, the larger of which was located at 490 nm.

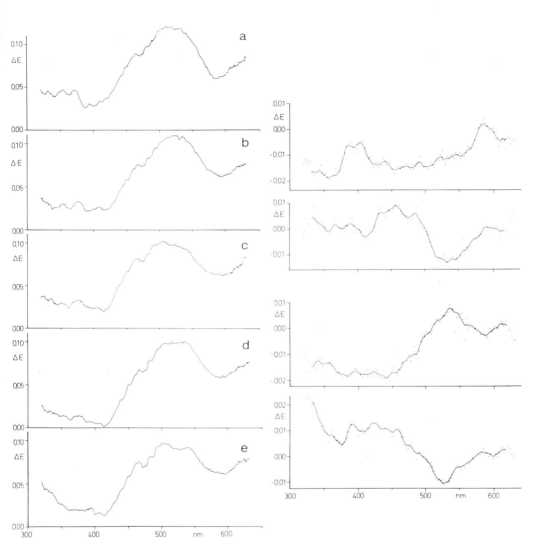

Fig. 3. Left: Series of extinction curves of rhabdomere no. 6 following periods of illumination and darkness (originals as in fig. 2):
a) preparation at the beginning of the experiment
b) after 20 minutes of darkness
c) after 1 minute of illumination
d) after further 20 minutes of darkness
e) after additional 3 minutes of illumination.
Right: Difference plots between pairs of extinction curves taken successively (calculated

from digitalized values). Storage in darkness gives only a slight overall decrease in
extinction. Illumination causes a decrease in the green and an increase in the blue
regions of the spectrum, which is at least partially regained during darkness. The difference curves c/d and d/e are nearly mirror images. This indicates the existence of a
thermostabile metarhodopsin

Since the changes were not very pronounced, differences between pairs of curves taken succesi-
vely were calculated after digitalizing the curves at intervals of 2.5 nm. The calculated trend
lines are plotted, the ordinates amplified by a factor of three, in the right hand part of figure 3.
The difference plots between pairs of curves following illumination and darkness show extreme
values at about 530 nm, and are nearly mirror images. The first illumination produced a difference
plot with a clear maximum at about 460 nm which, however, does not appear in the following
difference plots. It may be masked by alterations in the u.v. which are not yet understood. In
the portion of the spectrum above 400 nm the plots point to a light-dependent conversion of
rhodopsin to a thermostabile photoproduct with a maximum between 460 and 500 nm, and its
subsequent restitution during periods of darkness.

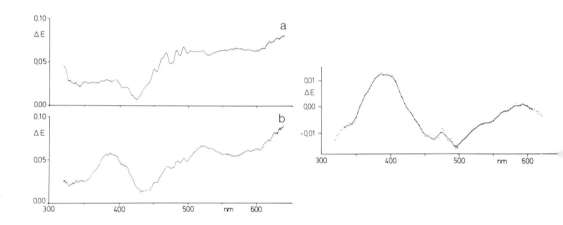

Fig. 4. Left: Extinction curves of rhabdomere no. 3 after long-lasting illumination
a) in solution of pH 6, and
b) in solution of pH 9, about 5 minutes later (without further illumination).
Right: Mean of six difference plots between the extinction curves a) and b) calculated
from digitalized values. The pH-change causes a decrease in the maximum at about
490 nm, and the appearence of a new maximum at about 380 nm. This change is characteristic of the two forms of metarhodopsin

The photoproduct was expected to be acid metarhodopsin. To prove this assumption, a series of
experiments was carried out, an example from which appears in fig. 4. A rhabdomere in a preparation mounted in saline of pH 6 was illuminated with bright light in order to obtain a maximal
amount of the product. After the extinction spectrum was measured, the medium was exchanged,
without moving the specimen, for a solution of pH 9. The curve obtained subsequently from the
same rhabdomere is clearly different from the preceding one; the maximum at about 490 nm
has levelled out and a new one at about 380 nm has arisen. The difference plot between these
curves (calculated as for fig. 3) corresponds in the range from less than 350 to more than 500 nm
to that known for acid and alkaline metarhodopsin in solution (SCHWEMER, 1969; SCHWEMER,
HAMDORF and GOGALA, 1971).

It is therefore believed that stable metarhodopsin is the photoproduct of illuminated rhodopsin in the eye of Calliphora. The experiments demonstrate that rhodopsin is resynthesized in darkness, evidently by chemical means dependent on energy supplied by metabolism within the retina. The thermostability of the product makes it probable that a photoreisomerization may also occur, as has been shown for other species of insects (HAMDORF, HOEGLUND and LANGER, 1972; SCHWEMER, HAMDORF and GOGALA, 1971). Further investigations will be carried out to test this hypothesis for the eye of the fly.

References

CHANCE, B., PERRY, R., ÅKERMAN, L. and THORELL, B.: Highly sensitive recording micro-spectrophotometer. Rev. Sci. Instr. 30, 735 - 741 (1959).

DARTNALL, H.J.A.: The interpretation of spectral sensitivity curves. Brit. Med. Bull. 9, 24 - 30 (1953).

GOGALA, M., HAMDORF, K. and SCHWEMER, J.: UV-Sehfarbstoff bei Insekten. Z. vergl. Physiol. 70, 410 - 413 (1970).

HAMDORF, K., HOEGLUND, G. and LANGER, H.: Mikrospektrometrische Untersuchungen an der Retinula des Nachtschmetterlings Deilephila elpenor. Verh. Dtsch. Zool. Ges., Helgoland 1971, 65, 276 - 281 (1972).

JONES, B.M.: Endocrine activity during insect embryogenesis. Function of the ventral head glands in locust embryos (Locustana pardalina and Locusta migratoria, Orthoptera). J. exp. Biol. 33, 174 - 185 (1956).

LANGER, H.: Nachweis dichroitischer Absorption des Sehfarbstoffes in den Rhabdomeren des Insektenauges. Z. vergl. Physiol. 51, 258 - 263 (1965).

LANGER, H.: Grundlagen der Wahrnehmung von Wellenlänge und Schwingungsebene des Lichtes. Verh. Dtsch. Zool. Ges., Göttingen 1966. Zool. Anz., Suppl. 30, 195 - 233 (1967).

LIEBMAN, P.A.: In situ microspectrophotometric studies on the pigments of single retinal rods. Biophys. J. 2, 161 - 178 (1962).

LIEBMAN, P. and ENTINE, G.: Visual pigments of frog and tadpole (Rana pipiens). Vision Res. 8, 761 - 775 (1968).

MARKS, W.B.: Visual pigments of single goldfish cones. J. Physiol. (Lond.) 178, 14 - 32 (1965).

SCHWEMER, J.: Der Sehfarbstoff von Eledone moschata und seine Umsetzung in der lebenden Netzhaut. Z. vergl. Physiol. 62, 121 - 152 (1969).

SCHWEMER, J., HAMDORF, K. and GOGALA, M.: Der UV-Sehfarbstoff der Insekten: Photochemie in vitro und in vivo. Z. vergl. Physiol. 75, 174 - 188 (1971).

IV. Intensity-dependent Reactions

1. The Discrimination of Light Intensities in the Honey Bee

T. Labhart
Department of Zoology, University of Zurich, Switzerland

Abstract.   The phototactic attractivity of two luminous screens of different brightness was examined and the discrimination threshold over a range of five logarithmic units has been determined. $\triangle$ I/I was found to be 0.135 at the highest and 0.60 at the lowest intensity used.

The white light intensity discrimination of bees is tested by means of the photopositive reaction.

1. Apparatus (fig. 1)

The light of a xenon high-pressure lamp (XBO 450 W. Osram) collected by two quartz lenses (Herasil, Heraeus Schott) and reflected by four surface mirrors (Alflex A, Balzers AG) illumi- nates two circular groundglass disks (UV-transmitting plexiglass 218, Röhm & Haas) of 40 mm diameter, each at the end of a Y-tubus. The Y-tube, the branches of which are 22 cm long and enclose an angle of 120°, may be rotated within a plexiglass cylinder. To avoid reflexions in the Y-tubus constructed of plexiglass, the inner surface is painted with dull black colour except the coverglass, where blinds are mounted.

From an opening in the outside wall a rectangular plexiglass channel (pretraining channel) runs beyond the Y-tubus system and ends in a cylindrical box (pretraining box). A hole in the bottom of the box gives access to a little pot with sugar solution automatically refilled by a level re- gulator.

The intensity of each of the light beams can be varied by neutral density filters (made of quartz, Balzers AG) placed into two filterholders rotably mounted around the xenon lamp.

2. Light measurements

For the measurement of relative intensities a selenium photocell (11 x 9 mm, Evans Electrosele- nium LTD) was used together with a galvanometer (Micrograph BD 5, Kipp & Zonen). The spectral sensitivity of the photocell was determined by comparing its electrical response to monochromatic lights from a monochromator (Jarrel Ash 82–400) with the response of a quantum converter.

For light measurement the photocell is placed in the middle of the Y-tubus, the point of decision (see below). By means of 10 interference filters (Filtraflex-B-40, Filtraflex-R-UV, Balzers AG), the exact transmission curves of which have been mapped previously by a spectrophotometer (Spectronic 505, Bausch & Lomb), the relative spectral composition of the light from 300 to 634 nm wavelength has been measured (fig. 2).

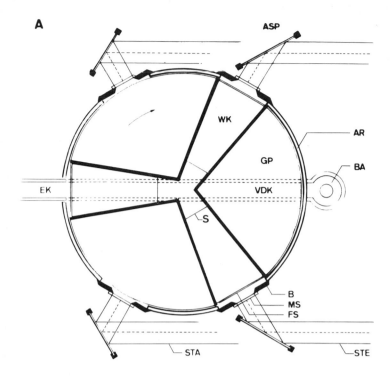

A

ASP

WK

AR

GP

BA

EK

VDK

S

B
MS
FS

STA

STE

Fig. 1. Binary choice apparatus: AR plexiglass cylinder; ASP surface mirrors; B diaphragm, BA pretraining box; EK entrance channel; FS front glass; GP basal plate, carries Y-tubus; MS groundglass disk; S slide; STA, STE light beams for different mirror positions, STE used here; VDK pretraining channel; WK Y-tubus

The absolute intensity was determined by a compensated thermophile (CA 1, Kipp & Zonen) and the galvanometer by the following procedure: The absolute intensity at 551 nm, which was calculated after measuring of the light passing the interference filter with λ_{max} = 551 nm was compared with the relative value measured by the photocell. By means of the calibrating factor

$$A = \frac{\text{absol. int. at 551 nm } [W/m^2 \text{ nm}]}{\text{relat. int. at 551 nm}}$$

the absolute intensity could be calculated for each wavelength examined for the determination of the relative spectral intensity. The integral

$$\int_{\lambda = 300 \text{ nm}}^{\lambda = 634 \text{ nm}} \text{absol. int. Id } \lambda$$

gave the absolute intensity of this range. It comes to 3.6 W/m^2.

The spectral transmission of the neutral density filters was measured by the spectrophotometer and found to be not exactly constant over the whole spectrum. For the calculation of the different intensities this and the relative spectral sensitivity of the bees (HELVERSEN, in press) were taken into account.

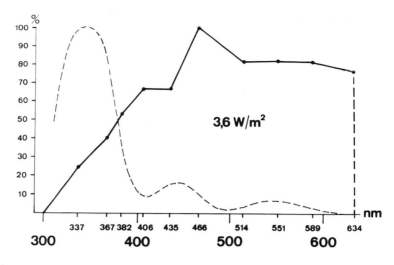

Fig. 2. Relative spectral composition of the light at the point of decision •——•; relative spectral sensitivity of bees (by HELVERSEN) ------

3. Test procedure

Bees are trained to enter the opening in the outside wall and to crawl through the pretraining channel into the darkened pretraining box, where they can suck some 0.15 nm sugar solution. After this pretraining period of at least two hours, single bees are allowed to enter the Y-tubus tilting up the movable entrance channel, which now fits to one arm of the Y-tubus (fig. 1). After lowering of the vertically sliding front glass (UV-transmitting plexiglass) and the slide (fig. 1) the bee is enclosed in one side of the Y-tubus. Now the slide is lifted and the bee crawls against the point of decision, where it decides between the two lights according to its photopositive reaction. After arriving at the end of one arm, the slide is closed, the lights are turned off and the Y-tubus is turned back into the testing position. Each bee is tested 10 times and then sent to Allah.

Five experimental series were carried out. For each series an intensity Ic was kept constant and compared with a variable lower intensity Iv. The reaction of the bees to 2 to 4 different Iv in each series was examined. For each Iv 50 choices were registrated, thus 5 bees were tested.

Examining infinitely many Iv, one would get the lower branch (because Iv <Ic) of the s-shaped reaction frequency function. For our investigations only Iv giving points in the middle, linear part of the curve were chosen.

To avoid any right-left inhomogenities of the apparatus the position of Ic and Iv is changed during the experiment following the pattern: l, r, l, l, r, l, r, r, l, r.

4. Results

Reaction frequency functions were determined over an intensity range of five powers of ten. The results are shown in fig. 3. For the determination of the discrimination thresholds the following method was chosen: the points of intersection of the straight lines with the value of the confidence limit for %lv = 50% (n = 50, p = 0.05) gives a threshold intensity (I th), i.e. the intensity, which is just discriminated from Ic. The ratio

$$\frac{I\,c_i - I\,th_i}{I\,c_i} = \frac{\Delta I_i}{I\,c_i}$$

is an index for the ability of discrimination at $I\,c_i$. Fig. 4 shows that the discrimination threshold increases only slightly from high towards medium intensities, but towards low intensities the curve ascends quite steeply. At the highest intensity examined $\Delta I/I = 0.135$, at the lowest $\Delta I/I = 0.603$, i.e. intensity differences of about 13% resp. 60% are discriminated.

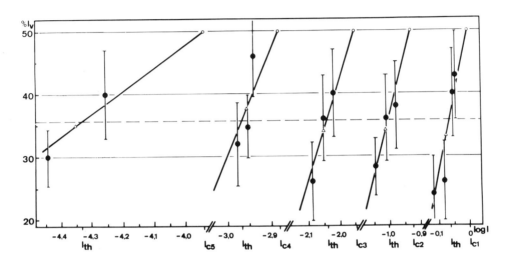

Fig. 3. Reaction frequency functions: I_{cl-5} = constant intensities; • test values with s.d., △ their center of gravity; ○ 50% point i.e. Ic = Iv

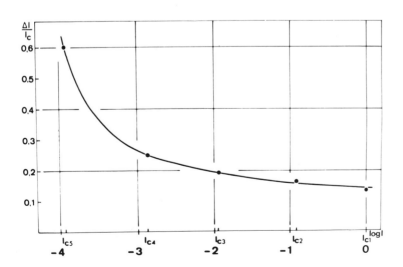

Fig. 4. Threshold curve of intensity discrimination

5. Discussion

The first work on light-intensity discrimination in bees was done by BERTHOLF (1931). He also used the phototaxis method. His $\triangle$ I/I-value is quite high, about 0.85. The absolute intensity is not given; it must be assumed, however, to be quite low.

Observing the optomotoric reaction, WOLF (1933) gives a threshold curve ranging over 4.5 powers of ten of intensity. At the highest intensity I $c \triangle$ I/I = 0.185.

KUNZE (1961) used the same method but a more refined for technique measuring the mechanical torque of the bees. His threshold value as found by extrapolation is about 0.25 = $\triangle$ I/I.

Neither author gives any useful values of absolute intensities. But the ability of discrimination at the highest intensities used by WOLF is nearly constant over a range of three powers of ten; in the investigations of KUNZE over about half a power of ten. This suggests it may fall in that intensity range where Weber's law is valid. Thus their values seem to reveal the highest possible ability of discrimination for their methods.

HOERMANN (1934) trained bees to different grey cardboards and gives values as low as 0.03. But she failed to test spontaneous preferences before training them. She obviously assumed each cardboard to be equally attractive for an untrained bee.

Nor do any of the cited authors give any useful values for absolute intensities, nor any information about the spectral composition of the light, the second factor being especially important because of the big differences in sensitivity at different wavelengths.

Moreover, neither the spectral transmission of the neutral density filters, nor the spectral remission of the grey cardboards used for the variation of intensities or contrasts have been measured. These values often change over the spectrum. Thus, a certain brightness difference measured by conventional photocells or thermopiles which do not possess the same spectral sensitivity as the eyes of insects may not correspond to the real, subjective brightness difference for the bee.

In the present work we have tried to define these parameters exactly. It is noteworthy, however, that the results of KUNZE and WOLF are not very different from ours.

In flies, values of $\triangle$ I/I lower than 0.02 have been found by FERMI and REICHARDT (1963). Further work in training bees may reveal whether bees are able to distinguish between such closely similar intensities as flies do, (because the motivation for discrimination under such circumstances is much stronger than in spontaneous behaviour), or whether intensity discrimination is less precise in bees than flies.

Acknowledgement. The experimental work was supported by grant no. 3.315.70 awarded to RUDIGER WEHNER by the Fonds National Suisse de la Recherche Scientifique.

References

BERTHOLF, L.M.: Reactions of the honeybee to light. J. Agric. Research 42, 379-419 (1931).
FERMI, G., REICHARDT, W.: Optomotorische Reaktionen der Fliege Musca domestica. Kybernetik 2, 15-28 (1963).
HOERMANN, H.: Ueber den Helligkeitssinn der Bienen. Z. vergl. Physiol. 21, 188-219 (1934).
KUNZE, P.: Untersuchungen des Bewegungssehens fixiert fliegender Bienen. Z. vergl. Physiol. 44, 656-684 (1961).
WOLF, E.: The visual intensity discrimination of the bee. J. Gen. Physiol. 16, 407-422 (1933).

2. Photopositive Reactions of Honey Bees to Circular Areas of Varying Sizes and Light Intensities

M. Frischknecht
Department of Zoology, University of Zurich, Switzerland

Abstract.   Spontaneous phototactic decisions towards bright, uv-light emitting circular ground glass screens are tested. The diameter of one circular area is kept constant, the other one is vari ed, whereas light intensity at the decision point is constant. If the circular areas exceed 9.2°, the larger circular area was preferred significantly. If the light intensity decreased by two powers of ten, the preference shifted towards larger circular areas. Therefore RICCO's law is only valid for a limited range of circular areas ($r \leq 9.2°$).

1. Methods

The experimental sequence and the stimulus apparatus are fundamentally in accordance with those employed by LABHART (this volume). Also, the measurement of light was based on the same criteria and executed with corresponding apparatus.

However, several differences in the test apparatus (figs. 1 and 2) were stipulated by the nature of the problem. Since circles of varying diameters are to be tested one against the other, the axis of the circular areas must be on the same level as the axis of the bee's eye. This requirement postulates an inclination of the basis of the Y-tube towards the circular areas and the position of the decision point 5 mm below the axis. The dimensions of the Y-tube were kept as small as to remain technically significant, in order to maintain the area of the illuminated visual units under variation of the circular areas as large as possible.

Diaphragms were chosen to represent the varying circular areas. In order to insure a constant light intensity at the decision point, no matter which circle was used, a set of neutral density filters was photometrically coordinated with the circles in such a manner that the measurements of the light intensities remained constant. The following constant light intensities were employed:

(1) $1.35 \cdot 10^5$ erg cm^{-2} min^{-1},
(2) $1.20 \cdot 10^3$ erg cm^{-2} min^{-1}.

The Y-tube was pivoted so that the initial test position could be repeated 10 times for each ins ect For every readjustment to the initial position, the position of the circular areas was changed according to the system utilized by LABHART, so.as to avoid training to the left or right-hand side. Only spontaneous decisions were followed through. The bees were fed on their way in and tested on the way back. A run was considered valid if the bee followed the bottom of the Y-tube and if it passed a certain mark in the direction towards the pattern.

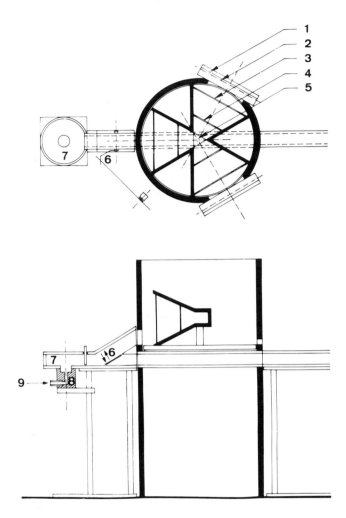

Fig. 1. Apparatus as seen from above (upper graph) and in lateral view (lower graph).
1, light screen; 2, diaphragm; 3, uv permeable PLEXIGLASS screen; 4, shutter; 5, center
of the circular arena = point, where the decisions are made by the bees; 6, apparatus
guiding the bees in the Y-tube; 7, 8, 9, feeding place

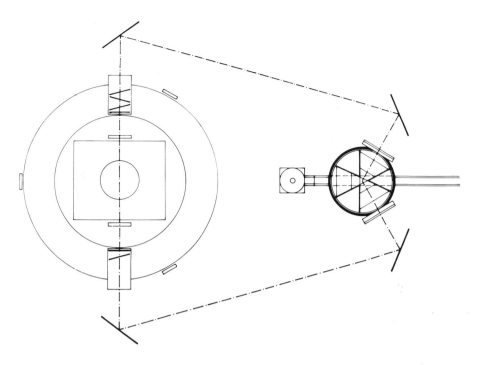

Fig. 2. Total experimental arrangement with the light source, neutral density filters and uv-reflecting mirrors on the left-hand   side (stimulus presentation) and the test appa-ratus on the right-hand   side. The latter one is shown in more details in fig. 1

2. Results (fig. 3)

Throughout series K_c I - K_c IV, the frequency of decisions towards circles up to a diameter of $9.2°$ of the varying comparison circle are all within the confidence limits of equivalent choices (p=o.95, n=50). As soon as the diameter of the comparison circle increases to $12.8°$, a significant preference for the larger circle occurs. The preference increases according to the further increase in the diameter of the comparison circle.

By installing a plane parallel to the decision point, it may be shown that the preference is not the consequence of the fact that the animal sees the larger circle before the smaller when it is running towards the decision point.

Experiments in which the light intensity at the decision point was decreased by two powers of ten led to shift in preference in the direction of the larger diameter, which however did not set in until $19.6°$.

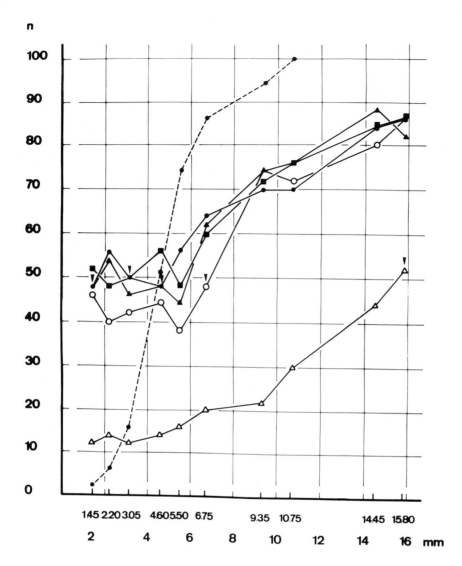

Fig. 3. Number of runs to the screens ranging in size in competition to a screen of constant size (marked by an arrow in each series). Solid curves: the total light intensity measured at the decision point remained constant ($1.35 \cdot 10^5$ erg cm^{-2} min^{-1}) in all series. Stippled curve: the light intensity per unit area was kept constant

3. Discussion

The significant preference for the larger circle at constant light intensities at the decision point raises the following problem: How many ommatidia are necessary to comprehend the circular areas? The results of PORTILLO (1936) on the determination of the interommatidial inclinations of the bee's eye transcribed to our specific experimental situation show that the horizontal axis of the circle lies at the level of the 86th ommatidium (measured from the dorsal boundary of the eye), the vertical axis at the level of the 12th ommatidium (measured from the frontal boundary of the eye). We ascertained a medium interommatidial inclination of 2.8° horizontally and 1.2° vertically. A definite number of ommatidia, through which the circles are seen, can therefore be coordinated to each circle in the horizontal and the vertical direction. We are now able to ask the following question: How does the bee assess the incidence of light energy?

Were the total amount of excitation formed by all ommatidia on which the circles are projected (1), all circles should cause the same effect on phototaxis. On the other hand, when the mean light intensity offered to one ommatidium would be processes (2), increasing circles should lead to a decrease in the strength of the phototactic reactions. Both possibilities, however, only hold for the ommatidia not maximally stimulated by the lights used in our experiments. If the latter would be true, the reaction frequencies for increasing circles would either increase also (1) or remain constant (2).

None of these relations fits the experimental results satisfactorily (fig. 3). So it must be supposed that the reaction frequencies can be completely correlated neither to the total light intensity nor to the mean light intensity offered to one ommatidium. Additionally, the size of the area itself must be regarded as an independent parameter for the phototactical assessment of circular areas. RICCO's law, however, does not hold for areas at least larger than 4 – 5 (horizontal) x 8 – 9 (vertical) ommatidia, respectively 6 – 7 x 12 – 13 ommatidia at the lower (10^{-2}) intensity level (compare CREUTZFELD, 1970, for the ganglion cells in the cat's retina).

If one wants to refer to units with specially sized receptive fields, one must consider the work of KAISER and BISHOP (1970), who only found units extending with their receptive fields over the whole eye.

Acknowledgements.   The work was supported by grant No. 3.315.70 awarded to RUDIGER WEHNER by the Fonds National Suisse de la Recherche Scientifique.

References

CREUTZFELDT, O.D.: Some principles of synaptic organization in the visual system. The Neuro-
     sciences. Second Study Program, ed. by F.O. SCHMITT. New York: Rockefeller Univ. Press,
     pp. 630 – 647 (1970).
KAISER, W., BISHOP, L.G.: Directionally selective motion detecting units in the optic lobe of
     the honey bee. Z. vergl. Physiol. 67, 403 – 417 (1970).
PORTILLO, J. del: Beziehung zwischen Oeffnungswurzel, Krümmung und Gestalt der Insekten-
     augen und ihrer funktionellen Aufgaben. Z. vergl. Physiol. 23, 100 – 145 (1936).

3. Screening Pigment and Visual Field of Single Retinula Cells of Calliphora

P. Streck
Department of Biology, Laboratory II, University of Regensburg, Germany

Abstract. Under identical illumination the receptor potentials recorded intracellularly from white-apricot and chalky are higher than from the red-eyed wild-type of Calliphora erythrocephala. The lack of screening pigment in mutants generates a light flow in the optical axis of a visual cell 10 x greater than in wild-type. The relative positions of the bell-shaped directional efficiency curves are in good agreement at 360, 495 and 625 nm with the extinction of the screening pigment measured by microspectrophotometer. The directional sensitivity curves are almost identical in breadth at 50% of the maximum and differ slightly in the height of their bases. That the mutants have poorer contrast perception than the wild-type flies is thus confirmed. This effect seems to be more prominent at lower intensities: the sensitivity curves of mutants are broader and the bases much higher than in wild-type.

Normally the ommatidia of apposition eyes are optically isolated from each other by screening pigment which prevents light from straying from one ommatidium to another. The transmission of the screening pigment was intensively studied in normal red-eyed flies and in mutants where the pigment was totally or partially lacking (LANGER, 1967). My purpose was to determine how the transmission would be reflected in the amplitude of the intracellularly recorded receptor potentials. The red-eyed wild-type of the fly Calliphora erythrocephala was compared with the mutants, white-apricot and chalky. An evenly illuminated disc including a visual angle of 2.5° could be moved around the eye placed in the center. The disc was centered about the optical axis of the visual cell which was being tested. The intensity of the three investigated wavelengths 360, 495, and 625 nm was adjusted to produce responses of equal amplitude in wild-type. These intensities were maintained for the mutants. Both the dynamic and the static phases of the intracellular potential were examined, i.e. the amplitude at 20 msec after stimulus onset and at stimulus end (= 100 msec). Aside from the absolutely lower amplitude of the static phase, the behaviour of both phases was basically the same. Therefore I shall limit myself to the behaviour of the dynamic phase (for more details see STRECK, 1972).

Two groups of curves become manifest: one for the red-eyed flies and another for the mutants (fig. 1). The difference in the groups could conceivably result from (1) unfavorable recording conditions in the wild-type, (2) race-specific sensitivity of the visual cells themselves, (3) the effect of screening pigments on the light flow along the visual axis. The first possibility is excluded by greatly increasing the intensity, for the potential of all three races reaches an amplitude of 50 mV. The form of the characteristic curves also render race-specific sensitivity improbable. They appear to be segments of a single curve, sigmoid in its entirety. With the light intensity available the upper half of the "S" was described by the response of the unshielded mutants and the lower half by that of the shielded red-eyed wild-type. Assuming that the concentration of visual pigment is lower in the wild-type rhabdomeres, these receptors should reach saturation earlier. But at maximum intensity the curves show no sign of flattening. The screening pigment therefore must somehow influence the amplitude of the receptor potential. But how? It should be recalled that the light radiating from the milky disc falls not only on the dioptric apparatus of the ommatidium being tested but on the surrounding ommatidia as well. The screening pigment would prevent any light from entering from the side. But in the mutants this pigment is lacking in both primary and secondary pigment cells. Light scattering off particles could also enter the rhabdomeres. A visual cell would thus receive additional light. Light intensity of the 2.5° disc must be reduced to 1/10 its value to produce the same amplitude in the mutants as in the wild-type, i.e. the light flow in the receptors of the mutants should be 10 x greater.

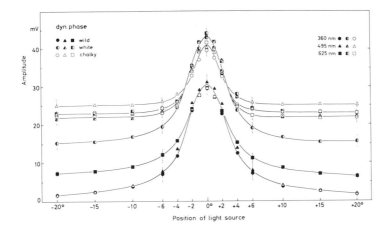

Fig. 1. Amplitude of the intracellularly recorded receptor potential as a function of the position of light source (directional efficiency), measured on 3 races of Calliphora at 3 wavelengths. The optical axis of the visual cell is at 0°. Mean curves for 8 experiments, bars indicate 2 x S.E

The transmission maxima of the 360 and 495 nm filters lie close to the spectral sensitivity maxima. At 625 nm, however, sensitivity is very slight (BURKHARDT, 1962). According to LANGER (1967) the screening pigments of wild-type absorb all wavelengths almost equally from UV to 600 nm; then the absorption drops off sharply. The pigments of the mutant white-apricot absorb from UV only up to 470 nm before allowing longer wavelengths to pass. It should be noted that absolute extinction in white-apricot is about 4 log units lower than in wild-type. Extinction in the colourless granula of chalky is so slight that it shows no dependence on wavelength at all. Under lateral illumination the visual cells themselves can serve as instruments to measure the relative absorption of the screening pigment. The course of the directional efficiency curves correspond perfectly with the extinction curves measured with the micro-spectrophotometer. Whereas the 360 and 495 nm curves of wild-type coincide and approach 0 at ± 20° displacement from the optical axis, the curve for red light lies above them, i.e. the visual cell receives red light even at oblique illumination. Thus single-cell recordings afford another proof that the secondary maximum in red found in spectral ERG efficiency curves results from light straying into the rhabdomeres from the side (AUTRUM, AUTRUM and HOFFMANN, 1961; BURKHARDT, 1962; GOLDSMITH, 1965). The UV-light is largely absorbed in white-apricot. Consequently the number of cells contributing to the ERG is smaller than in chalky (LANGER and HOFFMANN, 1966). The directional efficiency curves confirm this interpretation, for the UV curve of white-apricot runs between the wild-type and the other mutant curves. Chalky eyes on the other hand are so transparent that beyond ± 4° displacement the amplitude of the receptor potentials remains at a constant response level independently of the direction of the light.

The receptor is a non-linear transducer with a sigmoid logarithmic characteristic curve. The directional efficiency curves display corresponding distortion. If one calculates the sensitivity

curves from the characteristic and the efficiency curves, the differences are considerably reduced because the relative intensity is plotted linearly on the ordinate (fig. 2).

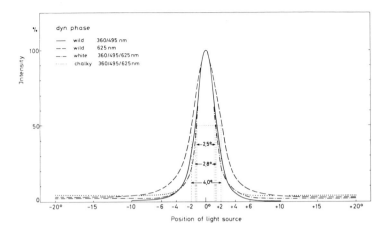

Fig. 2. Incidence of light into a visual cell as a function of the position of light source
(directional sensitivity), calculated from the efficiency and the characteristic curves.
The breadth of the curves at 50% intensity is marked by dashed lines

These curves show the incidence of light into a single visual cell as a function of the position
of the light source. They can be characterized by the height of their bases and the breadth at
50% of the maximum. The bases of the wild-type and the mutant curves do not differ as much as
in the efficiency curves, but it must be taken into consideration that the receptor is a system
where sensitivity extends over several log units. At a $\pm 10°$ displacement of the light source, the
UV-495 nm curve of wild-type drops below 0.1%; whereas the other curves keep a level of 2.5
to 4%. The height of these bases is of great significance for contrast perception. The other important factor, the breadth of the curves at 50% of the maximum, is nearly the same in mutants
and in wild-type (2.5 to 2.8°). The red-light curve of wild-type has a larger value (4°). The
mutants have generally poorer contrast perception on account of the higher bases. This conclusion is also borne out by optomotor responses and ERG recordings of Drosophila (GOETZ,
1964; HENGSTENBERG and GOETZ, 1967).

Experiments described so far were performed under the same stimulus conditions. Stimuli along
the optical axis produced lower responses in wild-type than in mutants. Some experiments carried
out at other intensities indicate that directional sensitivity of mutants may depend on intensity.
By diminishing the diameter of the light source from 2.5 to 0.7° the amplitude of the receptor
potential of the mutants could be reduced to that of wild-type. The result of such an experiment
with white-apricot is shown in fig. 3. Sensitivity curves yielded by a larger disc are narrower
and have lower bases. If disc size rather than intensity were decisive, the larger light source
should produce broader sensitivity curves. Wild-type on the other hand does not appear to be
affected by the size of the disc nor, consequently, by intensity. Since only three experiments
of this type were performed, however, some reservation is in order. Nevertheless, they fit in
well with AUTRUM's (1961) results from the optomotor reaction of white-apricot. With sufficiently intense illumination these mutants were able to resolve stripes almost as narrow as the
wild type could. But at lower illumination white-apricot ceased to turn with the pattern even
when the stripes were broader. Since all measurements were performed on one single cell, the
findings cannot be the result of switching from a system of small aperture to one of larger

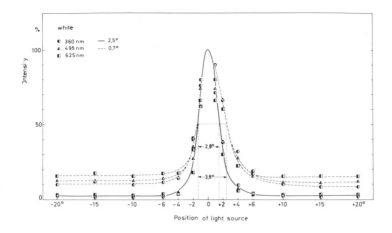

Fig. 3. Incidence of light into a visual cell of mutant white–apricot as a function of the position of light source (directional sensitivity). Single measurement with a light source of 0.7° and 2.5° visual angle. The breadth of the curves at 50% intensity is marked by dashed lines

aperture when the intensity is lowered (see: ECKERT, 1970; KIRSCHFELD, 1971). Whether the diameter of the light source or the intensity is decisive can only be determined by further experiment.

References

AUTRUM, H.: Die Sehschärfe pigmentfreier Fazettenaugen von Calliphora erythrocephala. Biol. Zbl. 80, 1 - 4 (1961).
AUTRUM, H., AUTRUM, J., HOFFMANN, Ch.: Komponenten im Retinogramm von Calliphora und ihre Abhängigkeit von der Spektralfarbe. Biol. Zbl. 80, 513 - 547 (1961).
BURKHARDT, D.: Spectral sensitivity and other response characteristics of single visual cells in the arthropod eye. Symp. Soc. exp. Biol. 16, 86 - 109 (1962).
ECKERT, H.: Diss. Freie Univ. Berlin 1970 (zit. n. KIRSCHFELD, 1971).
GOLDSMITH, T.H.: Do have flies a red receptor? J. gen. Physiol. 49, 265 - 287 (1965).
GOETZ, K.G.: Optomotorische Untersuchung des visuellen Systems einiger Augenmutanten der Fruchtfliege Drosophila. Kybernetik 2, 77 - 92 (1964).
HENGSTENBERG, R., GOETZ, K.G.: Der Einfluss des Schirmpigmentgehalts auf die Helligkeits- und Kontrastwahrnehmung bei Drosophila–Augenmutanten. Kybernetik 3, 276 - 285 (1967).
KIRSCHFELD, K.: Aufnahme und Verarbeitung optischer Daten im Komplexauge von Insekten. Naturwissenschaften 58, 201 - 209 (1971).
LANGER,H.: Ueber die Pigmentgranula im Facettenauge von Calliphora erythrocephala. Z. vergl. Physiol. 55, 354 - 377 (1967).
LANGER, H., HOFFMANN, Chr.: Elektro- und stoffwechselphysiologische Untersuchungen über den Einfluss von Ommochromen und Pteridinen auf die Funktion des Facettenauges von Calliphora erythrocephala. J. Ins. Physiol. 12, 357 - 387 (1966).

STRECK, P.: Der Einfluss des Schirmpigmentes auf das Sehfeld einzelner Sehzellen der Fliege
    Calliphora erythrocephala. Z. vergl. Physiol. 76, 372 - 402 (1972).
WASHIZU, Y., BURKHARDT, D., STRECK, P.: Visual field of single retinula cells and inter-
    ommatidial inclination in the compound eye of the blowfly Calliphora erythrocephala.
    Z. vergl. Physiol. 48, 413 - 418 (1964).

4. Circadian Sensitivity Changes in the Median Eyes of the North African Scorpion, Androctonus australis

G. Fleissner
Department of Zoology, University of Frankfurt/M., Germany

Abstract. Circadian sensitivity changes (up to 4 log units) were observed in the median eyes of the scorpion, Androctonus australis L. under constant light conditions. The time course of the sensitivity change (about 12 h low sensitivity, 9 h high sensitivity, 1.5 h transition time respectively) is endogenously controlled and stays unaltered in its general shape, even if the state of high sensitivity is interrupted by light adaptation. The phase position of this circadian periodicity can be synchronized exogenously by light–darkness cycles. Some observations have shown, that the two median eyes can influence their adaptation state mutually. Histological investigations have made it probable that these circadian sensitivity changes base on pigment migration within the visual cells.

1. Introduction

The light-dependent excitations in the median eyes of the night–active scorpion (VACHON, 1953; WUTTKE, 1966) are superposed by sensitivity changes depending on day-time (FLEISSNER, 1971). This phenomenon seems to be a general feature of the visual system of arthropods, which are active in darkness (WELSH, 1930; JAHN and CRESCITELLI, 1940; HOEGLUND, 1966; WADA and SCHNEIDER, 1968; ARECHIGA and WIERSMA, 1969). Since the eyes of scorpions are built very primitive, the detailed knowledge of these events and their physiological basis in scorpion eyes may elucidate also some of the problems in more complicated visual systems of other arthropods.

2. Methods

All experiments were done on Androctonus australis L. For longtime preparations the animals were mounted on a special device by wax–colophony, and survived without feeding in this position for many months. Thin, pointed platinum wires serve as electrodes. The different electrode was placed laterally in a little hole in the lense of the median eye, the indifferent one at a distance of about 2 mm in the tissue of the prosoma. Before the experiments animals were kept in 12 h light–12 h darkness cycle and during the experiments in continuous darkness. In the following "continuous darkness" will mean, darkness is interrupted by a flash light stimulus every half hour. The amplitudes of the ERG were registrated automatically (FLEISSNER, 1971).

3. Results

Identical flash lights evoke in the retina of dark-adapted median eyes of scorpions potentials, the shapes of which differ depending on day–time. Fig. 1 shows two typical electroretinograms (ERG); other forms observed can be held for transition forms. The ERG on the left side has been recorded during the day, the ERG on the right side during the night.

The ERG observed during the day resembles the generator potential, as it has been recorded intracellularly from single visual cells of the scorpion (FLEISSNER, 1971).

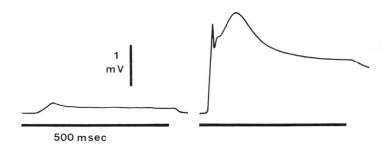

500 msec

Fig. 1. Electroretinograms of a median eye in "continuous darkness" recorded during day-state of adaptation (left) and night-state (right), induced by the same stimulus. Corneanegativity upwards. Horizontal bar: time calibration; vertical bar: amplitude calibration

The ERG recorded under the same conditions but during the night shows much greater amplitudes of the fast on-effect and the following waves. Besides some additional peaks emerge, which appear during the day only after high stimulation intensities.

If you try to get at night an ERG of the same amplitude as you get by day after stimulation intensity 1, you have to lower the intensity of stimulation light down to 10^{-4}. That means the median eye of scorpion is at night up to 10^4 times more sensitive than by day. The sensitivity can increase, so that stimulation intensities of less than 10^{-2} lx evoke distinct light reactions (FLEISS-NER, 1968).

a) Time course of sensitivity changes. In fig. 2 the changing amplitude of the ERGs evoked by identical test flashes interrupting "continuous darkness" every half-hour are plotted against time.

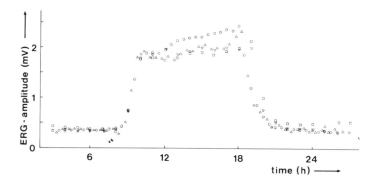

Fig. 2. The amplitudes of the ERG on-effect during "continuous darkness" plotted against time. Test flashes (150 msec, 10^4 lx) repeated every half hour. (o): Mean value of 8 different test days of the same animal; ($\triangle \square \triangledown$): Single test days of different animals

During a 11 - 12 h lasting period ERGs of constant low amplitude are recorded. Then suddenly the ERG amplitude increases up to a ten times higher value and stays so for 8 - 9 h. Then it decreases nearly as fast as it has increased, till the low value is reached again. The period of the low ERG amplitudes coincides with the light period of the 12 h light: 12 h darkness cycle the animal lived in before the experiments started. The period of the high ERG amplitudes coincides with the dark period. Therefore these two adaptation states of the eye will be distinguished in the following as physiological day-eye and physiological night-eye, or in short form day-eye and night-eye.

b) Control of sensitivity changes. Some observations will serve as first hints to the controlling of the spontaneous sensitivity change.

(1) Testing a scorpion under constant light conditions - "continuous darkness" or continuous light - you can realize that the phase position of the periodical sensitivity change shifts compared to the "standard day", though its typical time course stays unaltered. This shift can exceed one hour per day and can be observed as long as the animal lives, sometimes more than half a year. The period can be longer than 24h (fig. 3) as well as shorter than 24 h depending upon light conditions (FLEISSNER, 1971). Such a phase shift under constant conditions is considered as the most certain indication to an endogenous control.

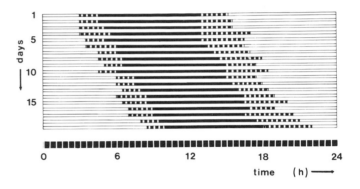

Fig. 3. Relation between the time course of the sensitivity change and the standard day. The animal lives in continuous darkness and is tested every half hour by a constant flash. Horizontal bars: successive days one below the other. (▭): day state of adaptation; (◼▭): transition state; (▬): night state

(2) The phase-position of this periodicity can be synchronized by light, e.g. during a 12 h light: 12 h darkness cycle it can be displaced within a few days in 180° corresponding to the light-program offered.

(3) Simultaneous ERG recordings of both median eyes have proved the spontaneous periodicity of sensitivity change to be synchronous in both eyes.

(4) The periodicity of the circadian adaptation is not influenced by a short time superposition of light-dependent decrease of sensitivity. Investigations concerning this topic have just started. In these experiments both eyes were tested periodically as described above. One eye is kept in "continuous darkness", while the other eye is conditioned in the midpoint of its nightstate by an additional 2.5 h-lasting light-program. Two observations have turned out to be important (fig.4). The conditioned eye lowers its sensitivity immediately after onset of the additional light. After the end of the additional light the stimulus effiency increases to the amplitude of the standard night ERG of this eye. The contralateral median eye, though carefully protected against the additional light, lowers its sensitivity, too, but with a latency of about one hour and returns to ERGs with a normal night amplitude past the conditioning light. Nevertheless the night state of adaptation finishes in both eyes at the same time as during test period without the additional light.

These observations yield the following preliminary results: The time course of circadian sensitivity change is controlled by a clock system not primarily depending on external light effects.

But this endogenous timer can be synchronized by exogenous stimuli. The endogenously controlled time course of sensitivity change can be additionally modified by direct light stimulus
as well as by stimulating the contralateral eye alone. That means the two median eyes communicate, whether by direct connections or via CNS, whether humorally or via nervous pathways,
is not yet known.

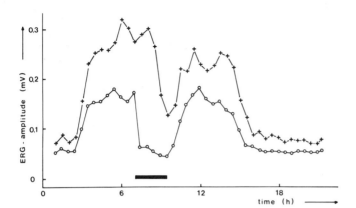

Fig. 4. The time course of
two simultaneously recorded
ERGs of contralateral median
eyes. "Continuous darkness"
is interrupted in one eye
(O) by a 2,5 h-lasting
conditioning light. Horizontal bar: additional light
stimulus. (O): ERG amplitude of the additionally
stimulated median eye;
(+): ERG amplitude of the
contralateral median eye
in "continuous darkness"

(c) Mechanisms of the circadian sensitivity change.   It may be that the efficiency of light
stimuli is altered by changing the sensitivity of the visual cells and that of other elements
synthesizing the ERG. It may be, too, that there exists a mechanism altering the relation between input quantity of light and utilized stimulus in changing the stimulus-conducting system.

Alterations of the sensitivity of the retinula cells may be ascertained in experiments where
either the utilized stimulus does not change, or its change can be estimated quantitatively. These
experiments are not yet accomplished.

The second possibility, that the quantity of the utilized stimulus is changed, has been verified
in comparatively simple experiments.

Other authors have observed light-dependent pigment migrations in the eyes of scorpions
(SCHEURING, 1913; MACHAN, 1966, 1968). Besides some arthropods control the circadian
light flux into their eyes by displacing screening pigments (summarized in STUDNITZ, 1952;
GLUUD, 1964), therefore the eyes of the scorpion were examined histologically in this regard.

The median eye of scorpion is constructed as simply as an ocellus (fig. 5). The dioptric apparatus
of the eye consists of a common great lens and a vitreous body. The retina contains in the main
retinula cells, five of which build a retinula with a fused rhabdom. The visual cells - radially
situated in the cup-formed retina - contact with their distal ends the praeretinal membrane,
which is the border between vitreous body and retina. The axons of the visual cells compose the
optic nerve.

Concerning the sensitivity change there are two important facts: 1. Visual cells are rich of
pigment granules. 2. The rhabdoms end 20 - 30 μm in front of the praeretinal membrane. A
layer of densely packed pigment granules serves as shielding against incident light.

During dark adaptation this pigment migrates within the visual cells backwards, thus making
the rhabdoms accessible for environmental light (SCHEURING, 1913; MACHAN, 1966, 1968).
Fig. 6 shows two histological preparations of eyes of scorpions which lived in the predescribed
standard light-darkness cycle. The section on the left comes from a day-eye fixated during the
light period, the section on the right from a night-eye fixated in darkness.

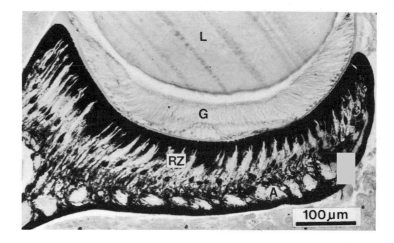

Fig. 5. Vertical section through a median eye parallel to the visual cells. L: lens ;
G: vitreous body; RZ: retinula cells; A: bundles of axons of the visual cells

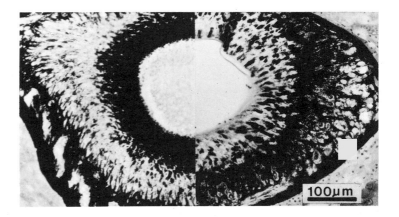

Fig. 6. Horizontal section through a light-adapted day-eye (left) and a dark-adapted
night-eye (right)

In this section the eye cup is cut horizontally, so that you see the direction of the incident
light into the eye from the middle part – through the lens and the vitreous body – centrifugally
to the inner layer of the retinal cells. In the day-eye you can observe that the stimulus light
has to penetrate a dense ring of screening pigment till it reaches the tips of the rhabdoms. In
the night-eye the rhabdoms are free accessible as the pigment has retracted towards the proximal part of the retina.

These eyes came from animals living in standard conditions, therefore the question rises whether
such a pigment migration occurs in constant darkness, too, and thus accounts for the circadian
sensitivity change.

Therefore median eyes which came from an animal living in constant darkness for several weeks were fixated during the electrophysiological day-state and night-state of adaptation, respectively. These eyes show the same differences between the two states. The result is not astonishing in respect to the night-eye as the animals lived in "continuous darkness". But it is rather astonishing that in the day-eye the pigment has migrated like during a light adaptation in spite of dark adaptation already lasting for several weeks (fig. 7).

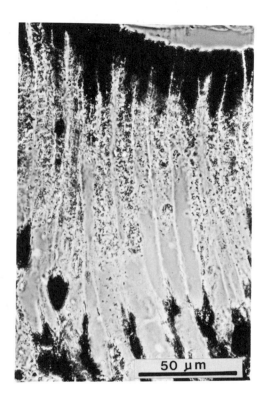

Fig. 7. Vertical section through a dark adapted day-eye

Though not yet quantitatively tested, these results make it probable that the pigment migration alone is sufficient for explaining the circadian sensitivity change.

Therefore the analysis of the mechanisms underlying the circadian sensitivity change resembles the problems arising in physiology of pigment migration in chromatophores.

Acknowledgements. The author wishes to thank Prof. Dr. D. BURKHARDT, Regensburg, for discussion and many valuable suggestions, and Dr. C. and Dr. K.E. LINSENMAIR, Regensburg, for the scorpions, captured in North Africa. The investigations were supported by a grant from the Deutsche Forschungsgemeinschaft.

References

ARECHIGA, H., WIERSMA, C.A.G.: Circadian rhythm of responsiveness in crayfish visual units. J. Neurobiol. 1, 71 - 85 (1969).

FLEISSNER, G.: Untersuchungen zur Sehphysiologie der Skorpione. Verh. Dt. Zool. Ges. Innsbruck (1968).

FLEISSNER, G.: Ueber die Sehphysiologie von Skorpionen. Belichtungspotentiale in den Medianaugen von Androctonus australis L. und ihre tagesrhythmischen Veränderungen. Inaug.-Diss. Nat. Fak. Univ. Ffm. (1971).

GLUUD, R.: Pigmentwanderung und Retinomotorik in Komplexaugen von Insekten. Zulassungsarb. H. Lehramt, Univ. München (1964).

HOEGLUND, G.: Pigment migration, light screening, and receptor sensitivity in the compound eye of nocturnal lepidoptera. Acta physiol. scand. 69, (Suppl. 282), 5 - 56 (1966).

JAHN, P.L., CRESCITELLI, F.: Diurnal changes in the electrical responses of the compound eye. Biol. Bull. 78, 42 - 52 (1940).

MACHAN, L.: Studies on structure, electroretinogram, and spectral sensitivity of the lateral and median eyes of the scorpion. Doctoral Diss. Univ. Wisconsin, Madison, Wis., U.S.A. (1966).

MACHAN, L.: The effect of prolonged dark adaptation on sensitivity and correlation of shielding pigment position in the median and lateral eyes of the scorpion. Comp. Biochem. Physiol. 26, 365 - 368 (1968).

SCHEURING, L.: Die Augen der Arachnoideen: I. Die Augen der Skorpioniden. Zool. Jb. Anat. Ont. 33, 553 - 588 (1913).

STUDNITZ, G. v.: Physiologie des Sehens. Retinale Primärprozesse. In: Probleme der Biologie 3, Akad. Verlagsges., Geest und Portig, Leipzig (1952).

VACHON, M.: The biology of scorpions. Endeavour 12, 80 - 89 (1953).

WADA, S., SCHNEIDER, G.: Circadianer Rhythmus der Pupillenweite im Ommatidium von Tenebrio molitor. Z. vergl. Physiol. 58, 395 - 397 (1968).

WELSH, I.H.: Diurnal rhythm of the distal pigment cells in the eyes of certain crustaceans. Prog. nat. Akad. Sci. U.S.A. 16, 386 - 395 (1930).

WUTTKE, W.: Untersuchungen zur Aktivitätsperiodik bei Euscorpius carpathicus. Z. vergl. Physiol. 53, 405 - 448 (1966).

5. Dark Activation of the Stationary Flight of the Fruitfly Drosophila

E. Buchner

Max-Planck-Institute of Biological Cybernetics, Tübingen, Germany

Abstract.   The fruitfly Drosophila and the housefly Musca develop , under conditions of stationary flight in still air, greater thrust in complete darkness than at constant illumination. This
effect of "dark activation" has been demonstrated in earlier experiments with Musca by L. MAF-
FEI. The present investigations establish similar reactions in Drosophila. The thrust reduction as
a function of luminance resembles the transducer characteristics of light receptors. Dark activation
can be elicited throughout the visual field. The effect is mainly mediated by the directly stimulated visual element of the compound eye, provided that the source luminance is low enough to
prevent stray light effects in neighbouring elements. Covering the ocelli has little, if any, effect
on the dark activation. An influence of position and distribution of the light stimulus on the dark
activation is expected, respectively, from the discontinuities of the receptive fields and from
possible lateral interactions between neighbouring visual elements. However, an estimate of the
expected variations shows that a considerable experimental effort would be necessary to distinguish these variations from statistical fluctuations of the dark activation.

In behavioral input-output measurements on the visual system of insects the optomotor response
has played a predominant role. However, since it depends on time-dependent stimulation, it is
not appropriate in investigations of (1) the level of excitation in single visual elements under
time-independent stimulation, (2) the summation of the excitation in different visual elements
under time-independent stimulation, and (3) the interaction between neighbouring visual elements under time-independent stimulation. These questions are of interest in the context of perception and evaluation of stationary patterns. Among the behavioral reactions that are elicited
by a constant light stimulus, the "dark activation" is distinguished by a well-defined stimulus
(open loop stimulation) and the possibility to readily quantify the response (thrust compensator).
The dark activation denotes a negative correlation between source luminance and thrust developed
by the fly under conditions of stationary flight in still air. The effect was first found by MAFFEI
in experiments with Musca. The present work investigates the dark activation with Drosophila and
tries to determine the applicability of the reaction to the treatment of the previous questions.

1. Methods

Due to the gyroscopic effect of the halteres, Dipterans are able to fly in complete darkness. Once
set into flight, Drosophila and Musca will in general continue until exhaustion unless the legs
make contact with an object. Stimulation of the Johnstone organ is not necessary. Thus the experiments can be carried out with the flies stationary in space. This greatly simplifies the well-defined stimulation of single visual elements of the compound eye. For the thrust measurements under
these conditions a compensation technique (GOETZ, 1968) was used, which keeps the displacement
of the fly smaller than 5 μ m.

The dark activation (DA in fig. 1) is given by DA = R = thrust without light stimulus - thrust with
light stimulus. Since there are large statistical fluctuations superimposed on this behavioral reaction, it is advantageous to integrate over "on" and "off" periods and then average over several
identical measurements:

The light sources in fig. 1 are of the LAMBERT type and subtend an arc of 0.75° (unless stated otherwise), so that, for the limited resolution of the eyes of Drosophila and Musca, they can be considered to be point sources. The maximum luminance of the light sources J_o was measured to be about 730 cd/m².

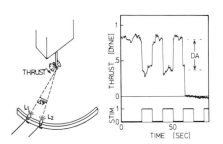

Fig. 1. (a) Experimental arrangement for thrust measurement under open loop stimulation. The fly is fixed to a position-compensated lever and flies stationarily in still air. (b) Dark activation (DA) of a fruitfly whose ocelli are covered, using a strong stimulus (ca. 30 facets stimulated, luminance >2000 cd/m²). The recording shows the thrust reduction when the light is turned on, the statistical fluctuations at constant stimulus and, for comparison, the noise and the drift of the apparatus when the insect stopped flying ($t > 60$ sec)

2. General properties of the dark activation

When a single light source is moved within the visual field of the compound eye of Drosophila, the characteristics of the dark activation can be summarized as follows: (1) The effect is positive at all stimulus positions. (2) It is most pronounced in the upper frontal eye region. (3) Covering the ocelli has no significant influence on the effect.

In a second set of experiments the dark activation of Drosophila was measured as a function of source luminance. Fig. 2a shows that the results strongly resemble the standard response-log J curves of light receptors (SCHOLES, 1969). The solid curve represents a tanh log J function that is theoretically predicted for light receptors (LIPETZ, 1969), the dashed curve shows a power

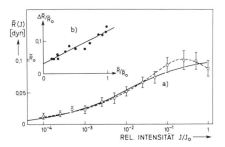

Fig. 2. (a) Dark activation of Drosophila as a function of light intensity. The curves (———) and (-----) represent two analytical functions fitted to the data. Maximum luminance of the light source was 730 ± 140 cd/m². (b) The (normalised) standard deviation of the mean values in (a) show a linear dependence on the (normalised) reaction

function fitted to the data. The dark activation reaches saturation at $x = 1$, as shown by experiments up to $x = 130$, so the power function certainly is not valid outside the interval $0 = \leqq$

$x \leq 1$. It is, however, more appropriate for calculations that are discussed later in this paper. The maximum value of the reaction amounts to about 20% of the total thrust developed in darkness. The size of the errors in fig. 2a can be approximated by a straight line, as shown in fig. 2b.

Thirdly, it was shown that the processing of the outputs of different visual elements can be described by a (nonlinear) superposition: The reaction increased by 30 to 50% when a second identical light source was turned on.

This latter property gave rise to the following question: Considering the flat slope of the response-log J curve in fig. 2a, is it conceivable that the sum of the light flux scattered into facets which are not directly illuminated contributes more to the dark activation than the light in the visual element whose optical axis points into the direction of light incidence? It will be seen that this question is of fundamental importance for all further discussions in this paper. The experiment shown in fig. 3 demonstrates that the influence of stray light becomes considerable only at high source luminance.

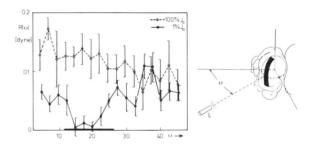

Fig. 3. Stray light test: About 8 adjacent vertical rows of facets are covered by a 120 μm broad stripe of blackened aluminium foil. When a light source of luminance $J = 1\% J_o$ is moved horizontally across the eye, the dark activation breaks down in the blinded part of the visual field. With $J = 100\% J_o$ the light flux scattered into unblinded facets saturates the dark activation, independent of the position of the source. The bar on the φ –axis denotes the range in which the "reduced surface pseudopupil" totally disappeared behind the black stripe

3. Investigations of the visual system using the dark activation

The present picture of structure and organisation of the visual system of flies has been extensively discussed in this volume. In the following the term "multiterminal visual element" represents 6 + 2 light receptors having parallel optical axes. To what extend the receptor systems "1 – 6" and "7 + 8" are responsible for the dark activation is unknown and of no concern for the following considerations.

For the compound eye, the light distribution generated by a given stimulus in the different visual elements depends on two optical parameters of the eye: The density of visual elements, given by the angle of divergence $\triangle \varphi$ between the axes of neighbouring visual elements, and by the size of the visual fields of the elements, given by the half-width $\triangle \varrho$ of their (approximally gaussian) sensitivity distribution. The overlap of the visual field of neighbouring elements is given by the ratio $\triangle \varrho / \triangle \varphi$. If the angle φ of incidence of parallel light is varied with respect to the axis of a visual element, not only the light flux in this visual element will vary, but also – especially at low degree of overlap – the sum of the light fluxes in all receptors. This is demonstrated in fig. 4a for different overlap ratios, with the assumption of a sufficiently large two-dimensional field of visual elements in hexagonal order.

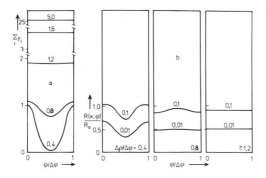

Fig. 4. (a) The total light flux in a large two–dimensional field of visual elements in hexagonal order with separation constant $\triangle \varphi$, under the assumption that each visual element has a gaussian sensitivity distribution of half-width $\triangle \varrho$. φ denotes the angle between the axis of a visual element and the direction of incidence of parallel light from a point source that is moved between two neighbouring elements. The parameter is the overlap ratio $\triangle \varrho / \triangle \varphi$. (b) Theoretical prediction of the dark activation as a function of φ. Since F (y) depends on x = J/J_o the modulation of the dark activation, when φ is varied, will depend on x. Therefore the calculations were carried out for x = 0.1 and x = 0.01. The overlap parameter was set to 0.4, 0.8 and 1.2, respectively

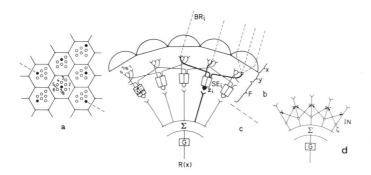

Fig. 5. (a) Horizontal section through 7 ommatidia at the level of the rhabdomere tips (not to scale). The filled circles denote rhabdomeres of receptors of equal viewing direction. (b) Vertical section along the dashed line in (a) with the optics (transfer characteristics y (x, φ)) and the receptive system of the "multiterminal visual element" (SE) (transducer characteristics F (y)). (c) Hypothetical block diagram for the transduction of the activities in the visual elements into thrust changes of the motor system (transducer characteristics G ($\sum z$)). (d) Block diagram as in (c) with backward inhibition between neighbouring visual elements

Now, it follows from the properties in section 2, that the dark activation to a first approximation depends on the total light flux in the eye. Therefore one would expect that the reaction should show a qualitatively similar modulation when the position of a point source is varied with respect to the axes of the visual elements.

With x = J/J_o being the incident light intensity, quantitatively, however, the modulation of the dark activation will not only depend on the sum over i of the light flux y_i in the i-th visual element (cf. fig. 5) but also on the transducer characteristics of the receptors F (y_i). Moreover it will depend on the actual kind of summation of activity in the perceptive system, and finally on the transducer characteristics of the motor system G ($\sum_i z_i$). Except for the total light flux in the eye, which can be calculated assuming a gaussian sensitivity distribution $y(x, \varphi)$ for a single visual element, none of these factors are known. However, one can improve the qualitative picture of fig. 4a by calculating the modulation using the following simplifying assumptions (cf. fig. 5): (1) The summation Σ is linear. (2) The transducer function G is a constant. Under these assumptions the theoretical dark activation as a function of light intensity x and position φ will be

$$R(x, \varphi) = R_o \sum_i F(y_i(x, \varphi))$$

This function closely resembles the measured function $\bar{R}(x)$ of fig. 2a, which represents an average of the reaction over φ, since φ was not kept constant in the experiments to fig. 2. With these assumptions it is possible to calculate the modulation of the dark activation due to the inhomogeneity of the visual system (BUCHNER and GOETZ). The results in fig. 4b show that for overlap ratios $\Delta \varrho /\Delta \varphi \gtrsim 0.8$ hardly any modulation can be expected. From optomotor experiments with Drosophila one knows $\Delta \varrho /\Delta \varphi$ to be about 0.8 (GOETZ, 1964). Considering the rather large fluctuations of the dark activation and the stray light effect discussed above, it seems impossible to do measurements precise enough to yield useful information about the overlap parameter $\Delta \varrho /\Delta \varphi$.

A similar qualitative estimation was carried out concerning the question of whether a lateral interaction between neighbouring visual elements could be detected with dark activation experiments using two point sources as in fig. 1. A backward inhibition between neighbouring visual elements as shown in fig. 5d would result in a decrease of the reaction if the two light sources stimulated neighbouring elements, that is, if the separation angle $\Delta \psi$ was on the order of the divergence angle $\Delta \varphi$. Fig. 6 shows the expected reaction as a function of $\Delta \psi$ for different inhibition coefficients, if the oversimplified assumption of a linear relationship between light flux and dark activation is used (cf. however fig. 2a). Actually, the two-source experiments by MAFFEI show a dependence of the dark activation on the separation angle $\Delta \psi$ that resembles the curves in fig. 6 for non-zero inhibition coefficient. However, calculations with more realistic assumptions similar to those in the discussion on the influence of the overlap parameter $\Delta \varrho /\Delta \varphi$ predict a minimum of the dark activation at $\Delta \psi = 0$. Furthermore they show that, even with assumptions leading to maximum inhibition, the influence of this interaction on the dark activation would be smaller than the standard deviations of the means in fig. 2a.

Fig. 6. Theoretical prediction of the dark activation as a function of the separation $\Delta \psi$ of two light sources as in fig. 1 under the oversimplified assumption of a linear relationship between light intensity and dark activation (cf. however fig. 2a). Region "C" shows the effects of lateral inhibition between neighbouring visual elements for different inhibition coefficients

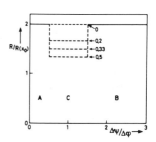

Extensive experiments with Drosophila and Musca show no significant effects that could be interpreted as lateral interaction. An explanation of this discrepancy with the experiments by MAFFEI cannot be given.

5. Discussion

It is of interest to compare the dark activation with the various behavioural light reactions described in the literature. In the theory of "stimulatory organs" (BUDDENBROCK, 1952), in the landing response (BRAITENBERG and TADDEI, 1966) and in free flight experiments with Calliphora

(SCHNEIDER, 1965; DIGBY, 1958) the sign of the light effect is opposite to that of the dark activation. Therefore speculations about the physiological meaning of the effect are problematic. The attempt to utilize the effect in the analysis of the visual system failed for three reasons: (1) The flat slope of the reaction-log J curve in fig. 2a predicts small effects. (2) These small effects cannot be detected due to large statistical fluctuations on the dark activation. (3) The stray light effect prevents the discrete stimulation of single visual elements at high light intensities. The dark activation may however be useful for the investigation of defects in the visual system of mutant flies.

Acknowledgement. I wish to thank Dr. K. G. GOETZ for his invaluable advice and support throughout this work.

References

BUCHNER, E., GOETZ, K.G.: Dunkelanregung des stationären Flugs der Fruchtfliege Drosophila (in press).

BRAITENBERG, V.v., TADDEI FERRETTI, C.: Landing reactions of Musca domestica induced by visual stimuli. Naturwiss. 53, 155 (1966).

BUDDENBROCK, W.v.: Vergl. Physiol. Bd. I, Basel: Verlag Birkhäuser (1952).

DIGBY, P.S.: Flight activity in the blowfly Calliphora erythrocephala in relation to light and radiant heat with special reference to adaptation. J. exp. Biol. 35, 1 (1958).

GOETZ, K.G.: Optomotorische Untersuchungen des visuellen Systems einiger Augenmutanten der Fruchtfliege Drosophila. Kybernetik 2, 77 (1964).

GOETZ, K.G.: Flight control in Drosophila by visual perception of motion. Kybernetik 4, 199 (1968).

LIPETZ, L.E.: The transfer functions of sensory intensity in the nervous system. Vision Res. 9, 1205 (1969).

MAFFEI, L.: Unpublished experiments (1969).

SCHNEIDER, P.: Vergleichende Untersuchungen zur Steuerung der Fluggeschwindigkeit bei Calliphora vicina. Z. wiss. Zool. 173, 114 (1965).

SCHOLES, J.: The electrical responses of the retinal receptors and the lamina in the visual system of the fly Musca. Kybernetik 6, 149 (1969).

V. Wavelength–dependent Reactions

1. Electrophysiological Studies on the Eyes of Diptera, Mecoptera and Hymenoptera

D. Burkhardt and I. de la Motte
Department of Biology, Laboratory VI, University of Regensburg, Germany

Abstract. The ERG of insect eyes – recorded with conventional techniques – was used for comparative studies of light sensitivity (Hymenoptera) and spectral response curves (Diptera, Mecoptera). Light sensitivity, range of adaption, and latency in the compound eye of tropical nocturnal Hymenoptera show markedly higher values than in related diurnal species. The off response in the ocelli is of extremely low threshold. In a certain range the response is increased with decreasing stimulus intensity as well with decreasing stimulus duration. Obviously the ocelli serve as light detectors which react to cessation of weak and short stimuli with enhanced signals to the CNS. Investigations of spectral response curves in Diptera so far yield two regions of maximal efficiency: the short-wavelength region with peaks varying between 350 and 460 nm, and the green region with peaks around 490 to 530 nm. The maximum in the red sometimes observed is due to lack of pigment screening for longer-wavelength irradiation. It is especially prominent in Cyrtodiopsis dalmanni and the significance of this finding with regard to the biology of the species is discussed. In male Bibio, with its two distinct eye regions, the predominant short-wavelength receptor (max. 346 nm) seems to be restricted to the large dorsal eye, the green receptor is only found in the small ventral eyepart. In Mecoptera the green receptor system is dominant.

Recordings of the ERG allow to get an idea of the capability of insect eyes within a short time and by way of simple means. Particularly for comparative physiological view this method is favourably employed. In the following we wish to report on such studies in which two topics were of main interest: (1) a comparison of the light sensitivity of the eyes of day- and night-active Hymenoptera, (2) a comparison of the spectral response curves in Diptera and Mecoptera (Scorpion-flies). Details of the methods employed will be published later (BURKHARDT, 1972; BURKHARDT und DE LA MOTTE, 1972).

1. Light sensitivity of the eyes of Hymenoptera

Provespa nocturna proved to be the most interesting species. The genus Provespa, to which it belongs, comprises three species described and is only to be found in Southeast Asia. Provespa flies solely at night between 19.30 and 23.30 hours and carries on its activity even in utter darkness (new moon, rainfall). Fig.1 shows the amplitude of the ERG plotted versus the logarithm of light intensity, the maximum stimulation energy being at 10^{-4} Watt/cm^2 (approx. 1000 Lux). For comparison the second curve presents the values of a day-active wasp, Vespa tropica, recorded under the same conditions. The considerably higher sensitivity of the eye of the night-active species becomes especially obvious at low intensities. The curve of Provespa does not approach the x-axis but turns asymptotically towards 0.8 mV. Scattered light of 3×10^{-10} Watt/cm^2 (approx. 0.003 Lux, illumination in a starry night) is causing this "residual" ERG. Fig.2 indicates the time

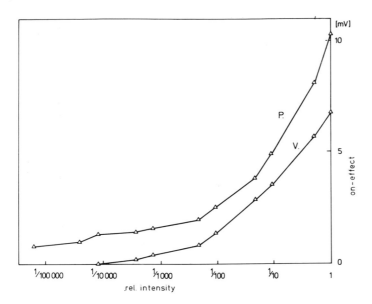

Fig.1. On effect of Provespa nocturna (P) and Vespa tropica (V) as depending on stimulus intensity. Abscissa: Light intersities in relative units, logarithmic scale, intensity 1 corresponds approximately to 10^{-4} Watt/cm^2. Ordinate: Amplitude of the on effect in mV

course of the dark adaptation. After every 1 minute of light adaptation at full stimulation energy the animal was subjected to a test stimulus of the same intensity and a duration of 100 msec at various intervals. We studied the period of the time between 1 sec and 20 min after break of the adapting light. The second curve is representing values for the honey bee and has been taken over from GOLDSMITH (1963). Compared with Apis the range of adaptation of Provespa is higher by one power of ten   adaptation proceeds more rapidly at the beginning and is lasting longer too. Yet, high sensitivity in the eye of Provespa is accompanied by much latency, fig.3: within the intensity range investigated, the latency of Provespa is three times as long as it is in the diurnal Vespa tropica.

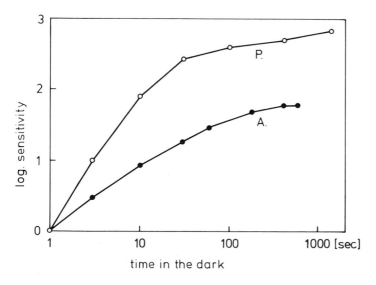

Fig.2. Time course of dark adaptation in Provespa nocturna (P) and Apis mellifica (A). Abscissa: Time after cessation of adapting light, log. scale. Ordinate: Sensitivity in log. units as calculated from response amplitudes and intensity cuves. The date for Apis have been replotted from GOLDSMITH (1963)

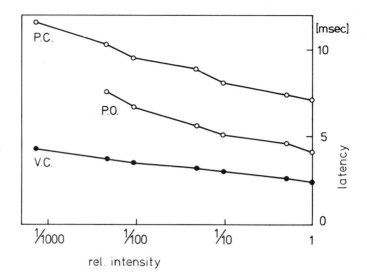

Fig.3. Latency in the com – pound eye of Provespa noctur – na (P.K.) and in the compound eye of Vespa tropica (V.K.) as depending on intensity and corresponding values for the median ocellus of Provespa (P.O.). Abscissa: Light inten – sity, log. scale, Ordinate: Latencies in msec

Like many other nocturnal Hymenoptera, Provespa possesses strikingly big ocelli so it was tempting to include them in the experiments. Fig.3 shows that the latency of the ocellus ERG, quite sur – prisingly, is about half as long as the one of the compound eye. The dependence of the ocellus ERG on the light intensity is given in fig.4. The curve of the on effect takes its usual sigmoid

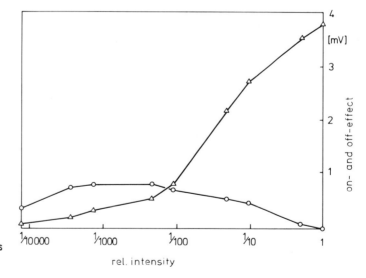

Fig.4. ERG of the median ocellus of Provespa nocturna as depending on stimulus in – tensity. Triangles: on effect, circles: off effect. Abscissa: light intensity, log. scale, Ordinate: response amplitudes in mV

course, its threshold, however, lies far above the corresponding one of the compound eye. Matters are altogether different with the off effect. The threshold appears at extremely low intensities, as may be extrapolated from the graph. The ocellar off effect reaches its peak at 1/1000 of the ma – ximum applied intensity and declines at further increasing stimulus energy. Corresponding results were obtained from experiments in which the duration of the stimulation instead of the light in – tensity has been altered, fig.5.

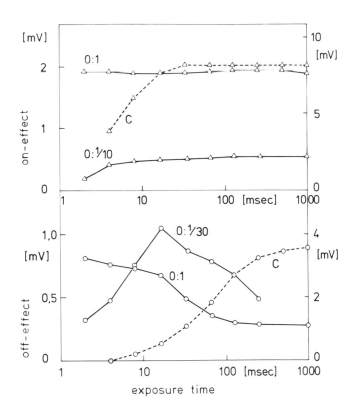

Fig.5. ERG of the median ocellus of Provespa nocturna (O, solid lines) and of the compound eye (K, dotted lines) as depending on stimulus duration. Upper graph: On effect, lower graph: Off effect. Abscissa: Stimulus duration in msec, log. scale. Ordinates on left side: Ocellar responses in mV, ordinates on right side: Compound eye responses in mV. Stimulus intensities as indicated at the curves

At high intensities the ocellar on effect is fully devel oped even at the shortest stimuli, to weaker stimuli it responds progressively, rising with duration, and reaches its saturation value with stimuli of about 10 to 30 msec duration. At high intensities the ocellar off effect is completely formed af ter short stimuli and falls off when the duration of stimulation is increased. With lower intensities it first rises with the duration of the stimulus, then reaches its maximum and drops with further in crease of stimulation time. RUCK (1961) reports of his experiments with the cockroach that the off effect presents the lowest threshold of all phases of the ocellar ERG. These and our observation most probably have some connection with the fact that impulse discharges in the ocellar nerves a inhibited through exposure to light and increased through darkening. The ocelli, therefore, are dark-detectors (for references see GOODMAN, 1970). In Provespa the ocelli very effectively signal to the CNS any break even of the most feeble and shortest stimuli.

2. Spectral response curves in the eyes of Diptera

Mainly Calliphora and Musca of the order of Diptera had been studied for their spectral response curves up to now (for references see ECKERT, 1970); both species show almost identical curves. The ERG holds four more or less clearly developed peaks at 360, 460, 490 and 620 nm. The peak at 620 nm is most probably due to transparency of the screening pigments to red light irradiation (BURKHARDT, 1962; GOLDSMITH, 1965). The maximum at 490 nm reflects the principal absorption peak of a green receptor system localized in the visual cells 1 - 6; the maximum at 460 nm represents the principal absorption peak of a blue receptor system localized in the visual cell 7 and probably 8 (LANGER, 1965). Minor maxima of both receptor systems in the ultraviolet contribute to the peak at 360 nm (BURKHARDT, 1962).

Lately we have been performing experiments with other species of Diptera. Cyrtodiopsis dalmanni, which we studied in Malaya and Frankfurt, shows apart from slight deviations the same response curves as Calliphora does, merely the red peak lies at 660 nm or above. This results in the possibility that quite remarkable ERG with stimuli of 690 resp. 790 nm can be released. This excessive red sensitivity - possibly to be made use of for phototactic reactions - might denote an adaptation to the biotope:under the jungle foliage the relative light intensity rises quickly above 650 nm owning to the decrease in absorption of chlorophyl there.

With Diptera in behavioural studies colour vision has been proved particularly in flower-visiting Syrphids (for references see KUGLER, 1970). Therefore we started to examine the spectral response curves of Eristalis. Three curves as results of individual experiments are shown in fig.6; Myiatropa florea, which was studied later on, does not differ distinctly.

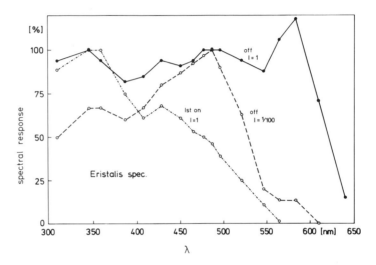

Fig.6.   Spectral response curves of Eristalis spec. Solid line: off effect, strong stimuli; broken line: off effect, same preparation as before, yet intensity lowered to one hundredth; half-dotted line: another preparation, strong stimuli, first negative phase of the on effect. Abscissa: Wavelength of equal quanta stimuli, ordinate: Responses in percent of the response at 350 or 490 nm resp

The off effect produces a response curve at high intensities similar to that of Calliphora, however, the blue maximum is shifted towards 430 nm, the pigment peak towards 585 nm. The pigment peak disappears as predictable when the intensity is lowered, but in addition the response to short wavelength falls off. From this results a response curve at low intensities, which resembles closely the green receptor curve of Calliphora. A second receptor system can be proved in Eristalis if instead of doing so with the off effect a small negative wave, preceding the on effect at high stimulation intensities, is being evaluated (AUTRUM, 1952; SCHATZ, 1971). This first wave has a spectral response curve with the principal maximum in the ultraviolet at 360 nm and a shoulder or minor maximum in the blue at 430 nm. The short wavelength receptor system of Eristalis obviously has a shorter latency and can be studied separately in the first wave of the ERG.

Even better than   is the case with Eristalis,the separation of a short-and a long-wavelength receptor system can be effected in Bibio marci (DE LA MOTTE, 1972). The Bibionidae belong to the Nematocera, yet they are closely linked to the Brachycera and equally possess non-fused rhabdomeres. The eyes of Bibio, the March fly, are divided, in particular the immensely big and turban-like eyes of the male are of striking size. An unobtrusive secondary eye is set ventrally below the main eye. The curves of fig.7 show the average of 43 individual measurings. The main eye of the male provedes a spectral response curve with the maximum in the ultraviolet at 350 nm, accompanied by a very flat shoulder at 430 nm in the blue; this curve bears much resemblance to the curve of the ultraviolet receptor of Ascalaphus (GOGALA, 1967). A response curve with a major peak at 520 nm and a minor one at 360 nm is found for the ventral or secondary eyes of the male and for wide areas of the female eye.

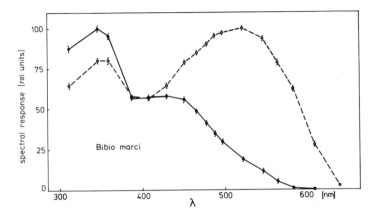

Fig.7. Spectral response curves of Bibio marci. Solid line: male, main eye. Broken line: male, ventral eye, and main part of the female eye. Abscissa: Wavelength of equal quanta stimuli. Ordinate: Responses in percent of the maximum response. Each point represents the average of 43 measurements, the standard errors of the mean are indicated by the vertical bars

As the ultraviolet receptor dominates in the main part of the very large turban eye of the male, and as Bibio appears in big swarms during blossoming time, the March fly may be an especially suitable object for spectrophotometrical and biochemical analyses of an ultraviolet visual pigment of insect eyes.

Among insects the Mecoptera are a key group, which is supposed to be near the common basis of Lepidoptera and Diptera. Therefore recently we started with experiments in the Scorpion-fly Panorpa cognata, which belongs to the Mecoptera. Fig.8 shows a spectral response curve derived from our first experiments. These measurements yielded only a type of response curves with a maximum in the green between 490 and 520 nm, the various curves showing a shoulder or a flat minor peak at 360 nm. If further experiments confirm that there is only a green receptor system present in Panorpa, this species would represent a favourable object for further studies of

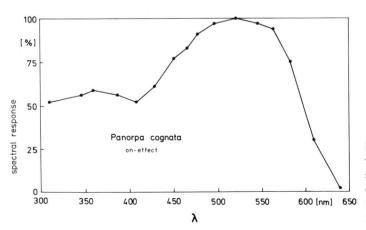

Fig.8. Spectral response curve of Panorpa cognata. Abscissa: Wavelength of equal quanta stimuli. Ordinate: Response in percent of the maximum response

this receptor type.

Conclusions: In summing up the above findings one may say that with all Diptera being observed this far there is evidence for the existence of a short- and a long-wavelength receptor system. The occurrence of a green receptor system seems to be typical, the main maximum lying between 490 and 520 nm accompanied by a minor maximum or shoulder in the ultraviolet. In addition a short-wavelength receptor is present with a principal maximum either at 360 nm or between 430 nm and 460 nm, the corresponding wavelength in blue or the ultraviolet resp. showing a shoulder or a minor peak. From this follows that in Diptera the prerequisites for at least dichromatic colour vision are existent, whereas it is not yet known — particularly as regards flower-visiting species — whether further receptor systems are involved.

The results as given in this brief report make us assume that comparative studies in this field will unearth rather interesting material and that consequently the studies in hand will prove useful for more extensive and more valuable analyses.

References

AUTRUM, H.: Ueber zeitliches Auflösungsvermögen und Primärvorgänge im Insektenauge. Naturwissenschaften 139, 290-297 (1952).
BURKHARDT, D.: Spectral sensitivity and other response characteristics of single visual cells in the arthropod eye. Symp. exp. Biol. 16, 86-109 (1962).
BURKHARDT, D.: Elektrophysiologische Untersuchungen am Diopsidenauge. Z. vergl. Physiol. (in Vorb., 1972).
BURKHARDT, D., DE LA MOTTE, I.: Untersuchungen über die spektrale Empfindlichkeit von Dipteren- und Mecopterenaugen. In Vorb. (1972).
ECKERT, H.: Diss. Freie Univ. Berlin (zit. n. Kirschfeld, Naturwissenschaften 58, 201-209) (1971).
GOGALA, M.: Die spektrale Empfindlichkeit der Doppelaugen von Ascalaphus macaronius Scop. (Neuroptera, Ascalaphidae). Z. vergl. Physiol. 57, 232-243 (1967).
GOLDSMITH, T.H.: The course of light and dark adaptation in the compound eye of the honey bee. Comp. Biochem. Physiol. 10, 227-237 (1963).
GOLDSMITH, T.H.: Do flies have a red receptor? J. gen. Physiol. 49, 265-287 (1965).
GOODMAN, L.J.: The structure and function of the insect dorsal ocellus. Advances Insect Physiol. 7, 97-195 (1970).
KUGLER, H.: Blütenökologie. 2. Aufl., G. Fischer Verl., Stuttgart (1970).
LANGER, H.: Spektrometrische Untersuchung der Absorptionseigenschaften einzelner Rhabdomere im Facettenauge. Verh. dt. zool. Ges., Jena 1965. Zool. Anz. Suppl. 29, 329-338 (1966).
MOTTE DE LA, I.: Ueber die spektrale Empfindlichkeit des Facettenauges der Märzfliege Bibio marci L. In Vorb. (1972).
RUCK, P.: Electrophysiology of the insect dorsal ocellus. I. Origin of the components of the Electroretinogram. J. gen. Physiol. 44, 605-627 (1961).
SCHATZ, B.: Ueber die spektrale Empfindlichkeit des Schwebfliegenauges: Elektrophysiologische Untersuchungen an Myiatropa florea. Diplomarbeit, Naturwiss. Fak. Univ. Frankfurt, 1971.

2. Wavelength-specific Behavioral Reactions in Drosophila melanogaster

R. Schuemperli
Department of Zoology, University of Zurich, Switzerland

Abstract.   In a Y-tube apparatus (fig. 1), one arm illuminated by a white or monochromatic light of constant intensity, the other by monochromatic lights of various intensities, the dependence of the spontaneous phototactic choice behaviour of females of Drosophila melanogaster on intensity and spectral composition of the stimulus lights has been tested.

The slope of the reaction–intensity curves (insert fig. 3) depends on the wavelength of the variable monochromatic light: the shorter the wavelength, the steeper are the curves (fig. 2). This dependence increases with increasing intensity. Therefore, the uv peak of the action spectrum (fig. 3) at about 350 nm gets relatively higher with increasing intensity compared to the peak in the visible at 490 nm. (Peak height ratio uv to visible in series 1 8.3 : 1, in series 2 with tenfold intensity 35.5 : 1).

The slope of the reaction–intensity curves depends also on the spectral composition of the constant light: the more uv the comparative light contains, the steeper is the slope of the curves (fig. 4). Wavelength-specific central processing is taken into account for interpretation of these results.

1. Introduction

In Hymenoptera (bees: FRISCH, 1915; DAUMER, 1956; and others; ants: KIEPENHEUER, 1968; WEHNER and TOGGWEILER, 1972) and in all probability in Lepidoptera (SWIHART, 1969, 1970), colour vision is proved, while   there is no evidence of the ability of Diptera to discriminate between different colours. This evidence would be of special importance, for the receptors of flies are well known from microspectrophotometry (LANGER and THORELL, 1966) and electrophysiology (AUTRUM and BURKHARDT, 1960, 1961; BURKHARDT, 1962) and we are informed of the organization of their visual pathway, at least in the peripheral ganglia (KIRSCHFELD, 1967; BOSCHEK, 1970, 1971; STRAUSFELD, 1970, 1971) better than in other groups of insects.

In preliminary investigations for training experiments, which should give an answer to the question of colour vision in Diptera, the spontaneous phototactic behaviour of Drosophila and its dependence on wavelength and intensity has been more closely analysed.

2. Apparatus (fig. 1) and Method

The light of a Xenon high-pressure lamp (XBO 450 W/1, Osram) is collected by two quartz lenses, passes heat absorbing filters (KG 1, Schott) and is reflected by aluminium surface mirrors (Alflex A, Balzers) to the front discs (uv-transmitting plexiglass 218, Röhm & Haas) of a Y-tube apparatus. Each light beam can be varied in spectral composition and intensity by interference filters and quartz glass neutral density filters (Balzers). The filter boxes containing these filters can be interchanged in position. A black screening keeps stray light away from the apparatus.

Light intensities were measured with a compensated thermopile (CA 1, Kipp & Zonen) connected to the micrograph BD 5 (Kipp & Zonen), at the point of decision of the Y-tube for each interference filter used. On the relative intensity scale, $10g \, I \, \lambda = 0$ corresponds to an intensity of $2.9 \cdot 10^3 \, \text{erg.sec}^{-1} \cdot \text{cm}^{-2}$. The spectral transmission of the interference and neutral density filters was measured with a spectrophotometer Spectronic (Bausch & Comb) and these values taken

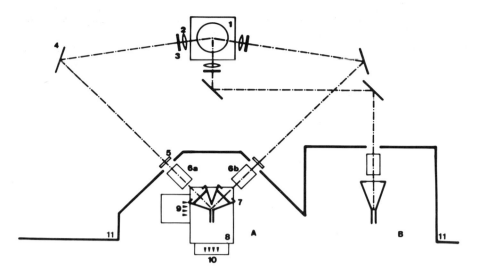

Fig. 1.A. Double-choice apparatus. 1. Xenon lamp; 2. quartz lenses; 3. heat-absorbing filters; 4. aluminium mirrors; 5. quartz–glass neutral density filters for adjusting the two light beams to equal intensity; 6. filter boxes; 7. Y-tube apparatus; 8. red light illumination; 9. 10 ventilation; 11 screening. B. Additional apparatus for adaptation and training

as a basis for the intensity calculations of filter combinations. Females of Drosophila melanogaster wild type reared under constant conditions, were used as experimental animals on the 14th day, i.e. 4 days after emergence, after one hour of dark adaptation at room temperature.

3. Results

Series 1. The flies had to choose between a constant intensity of white light in one arm and varying intensities of monochromatic lights in the other arm of the Y-tube. The constant intensity of the white light in the relevant region of the spectrum between 200nm and 700 nm was 1 erg sec^{-1} cm^{-2} (Spectral intensity distribution see LABHART, this volume). This intensity, compared to the less attractive monochromatic light of 617nm. just produced a fifty-fifty reaction.

For 18 wavelengths, situated between 311nm and 617nm, the reaction-intensity curves (r-i curve were determined (insert fig. 3). The slope of the r-i curves, as already shown in earlier experiments (WEHNER and SCHUEMPERLI, 1970), depends on the wavelength of the variable light. The regression line, calculated from the slopes of the r-i curves between the choice tendency lines for 40% and 60% differs significantly from zero (fig. 2). The slopes of the r-i curves above 539nm were not used for this linear regression, because of an indicated rerise of the slopes, and because these wavelengths had not been attractive enough to be tested in series two with tenfold intensity of the white light.

Series 2. In this second series, the white light intensity was increased by the factor 10 to 10 erg sec^{-1} cm^{-2}. The r-i curves of these experiments show the same, but significantly ($p < 0.05$) increased dependence on the wavelength of the variable light. The steeper r-i curves in the uv region of the spectrum indicate a quicker rise of the attractivity of uv light with increasing intensity than that of visible light. So, an action spectrum should have a relatively higher uv peak when obtained with higher intensity.

An action spectrum (fig. 3) is calculated from the r–i curves by intersecting these curves with the 50% choice tendency line and determining the appertaining intensity of the variable light. All these intensities of monochromatic lights are at the given experimental conditions as attractive as the constant white light, their reciprocal values are a criterion for this attractivity.

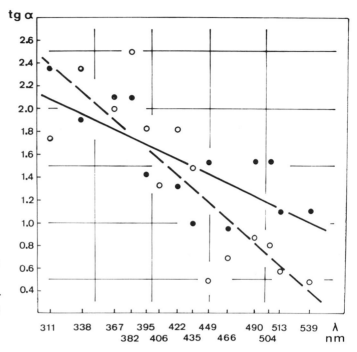

Fig. 2. Dependence of the slopes of the reaction–intensity–curves on the wavelength of the variable light. Full signatures: Series 1. Standard intensity 1 erg.sec^{-1} cm^{-2}. Open signatures: Series 2. Standard intensity 10 erg.sec^{-1} cm^{-2}. Ordinate: slope of the r–i curves between the choice-tendency lines for 40 and 60%; abscissa: wavelength of the variable light

The confidence limits for these values are constructed as shown in the insert of fig. 3. The corresponding peaks of both action spectra are situated at about 350nm in the uv and near 490nm in the visible. The proportion of the peak height uv to visible is, for series 1, 8.3 : 1, for series 2 35.5 : 1. In earlier experiments with a slightly different apparatus, a proportion of 72.5 : 1 has been found (WEHNER and SCHUEMPERLI, 1970). As predicted from the slopes of the r–i curves, the relative height of the peak increases with increasing intensity.

Series 3. If a monochromatic light is used as constant light instead of white light, it can be shown, that the slope of the r–i curves depends also on the spectral composition of the comparative light.

As shown in fig. 4, the slope of the r–i curves with uv as variable light does not depend on the constant light; they all are very steep. But with 514nm as variable light, the steepness increases significantly (p < 0,0025), if uv (338nm), and decreases (p < 0,0005), if green (514nm) is used as constant light instead of white light. Thus, the more uv the constant light contains, the steeper are the r–i curves.

Series 4. In all experiments described, each animal has chosen once. The question arises, whether the deviation of the reactions of one animal coincides with the deviation of the reactions of a group of animals. This was tested in series 4, where each animal had to choose 8 times between a uv light (338nm) and a green light (514nm) of equal attractivity (48% to green in series 3), which were interchanged in position according to the pattern right, left, r, r, l, r, l, l, so that side-constant behaviour could be separated from wavelength-constant behaviour.

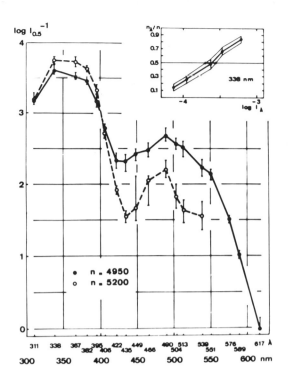

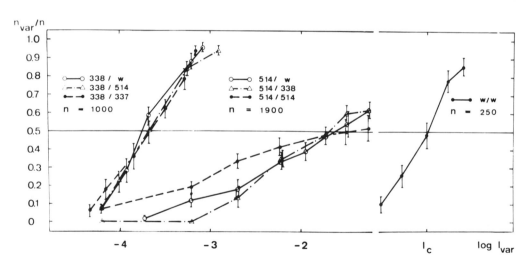

Fig. 3. Action spectra. Full signature: Series 1. 1erg.sec^{-1} cm^{-2} open signature: Series 2. 10 erg. sec^{-1} cm^{-2}. Ordinate: relative attractivity of the monochromatic lights; abscissa: wavelength; insert figure: reaction-intensity curve for 338nm with standard deviations. Construction of the confidence limits for the values of relative attractivity of the monochromatic lights

Fig. 4. Reaction-intensity curves with different constant lights of equal attractivity. Left: variable light: uv (338nm). Constant lights: white (w), uv (337nm), green (514nm); middle: variable light: green (514nm). Constant lights: white (w), uv (338nm), green (514nm); right: constant and variable light: white (w). I_c : Constant intensity of white light of series 2. Attractivity of the constant lights in these experiments is equal to the attractivity of this white light intensity

Fig. 5. Frequency distribution for animals choosing the right arm of the Y-tube 0 to 8 times out of 8 choices (fig. 5a), the green light 0 to 8 times out of 8 choices (fig. 5b), compared to the expected distribution (dotted lines) with the same mean values, independent choices supposed. Abscissa: Classes of 0 to 8 choices out of 8, ordinate: Frequency of these classes

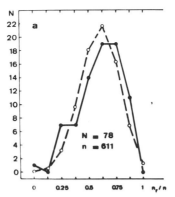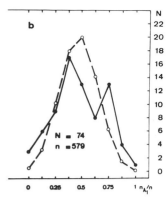

In fig. 5, the frequency distribution of the animals, which chose the right arm of the Y-tube 0 to 8 times out of 8 choices (fig. 5a) and the green light 0 to 8 times out of 8 choices, respectively (fig. 5b). The mean values are 60.2% to right (which is slightly significant, but has been constant in the previous experiments and so of no influence to the results), and 46.9% to the green light, which is not different from the 48% of series 3 with single courses.

A comparison of the frequency distribution found in these experiments with the expected distribution in the case of independent choices by X^2 test shows random distribution of left and right choices, but significantly ($p < 0.0005$) more wavelength-constant choosing animals than expected.

4. Discussion

Concerning the problem of colour vision, we must ask the question, which of the wavelength-specific reactions found can be explained by the properties of the receptors alone. Different slopes of r-i curves in different regions of the spectrum have been reported by several authors, and also correlated shifting of relative peak heights (HEINTZ, 1959; HASSELMANN, 1962). It would be exspected, that the characteristics of one receptor should have equal steepness in different regions of the spectrum, same stimulus efficiency presumed (Pigment migration should, if it influences the steepness, produce smoother characteristics for shorter wavelengths, where it is more effective; personal comment of FRANCESCHINI). Therefore, different slopes of the r-i curves would be expected to be produced by interaction of different receptor systems. But steeper characteristics in the uv indicate an increasing relative height of the uv peak with increasing intensity, as shown in the action spectra (fig. 3). In Diptereans, an increase in intensity should effect an increased influence of the retinular cells 7 and 8, for the central system is about 30 times less sensitive than the peripheral one (KIRSCHFELD and FRANCESCHINI, 1968; KIRSCHFELD and REICHARDT, 1970; ECKERT, 1971). An increased influence of the central system, which has no uv peak (LANGER and THORELL, 1966; ECKERT, 1971), should produce a decrease of the relative uv peak height. So, neither the increasing intensity, nor its direct cause, the different slopes of the r-i curves, can be explained by changing influences of the two receptor systems of the fly only. It is even more difficult to explain the dependence of the slopes of the r-i curves on the spectral composition of the constant light by the properties of the receptors only. The presence of uv seems to activate a mechanism which improves intensity discrimination. This presupposes a wavelength-specific nervous processing. There are different hypotheses to explain the wavelength-constant behaviour of the experimental animals in series 4: (1) The spectral sensitivity of the receptors might be different in different animals. This would produce different attractivities of the same lights for the different animals, but an average choice tendency of about 0.5. (2) Different adaptational states of the experimental animals would have the same effect. Though there is no directed change of the choice tendency from course 1 to course 8 and all animals had been handled in the same way, this possibility cannot be excluded. (3) Central preferences could explain the findings.

To sum up, the indicated possibilities of wavelength-specific central processing of the information from the receptors show a common tendency: an increased height of the uv peak of the action spectrum compared to the spectral sensitivity of the receptors known from intracellular and microspectrophotometric measurements and an improved intensity discrimination in the presence of uv. This might be of ecological importance insofar as in the natural environment of Drosophila, to discriminate between bushes, trees and so far independent of the actual illumination, the discrimination of colour shades is critical, which should be optimal in the green and blue-green region of the spectrum, where the spectral sensitivity curves of the peripheral and the central receptors intersect, meanwhile the intensity discrimination in the blue-violet and uv would be important for the orientation after the polarization pattern of the sky.

Acknowledgement

The work was financially supported by a grant awarded to RUDIGER WEHNER by the Fonds National Suisse de la Recherche Scientifique, No. 3.315.70.

References

AUTRUM, H.J., BURKHARDT, D.: Die spektrale Empfindlichkeit einzelner Sehzellen. Nat. wiss. 47, 527 (1960).

AUTRUM, H.J., BURKHARDT, D.: Spectral sensitivity of single visual cells. Nature 190, 639 (1961)

BOSCHEK, C.B.: On the structure and synaptic organization of the first optic ganglion in the Fly. Z. Naturforsch. 25b, 560 (1970).

BOSCHEK, C.B.: On the fine structure of the peripheral retina and lamina ganglionaris of the fly, Musca domestica. Z. Zellforsch. 118, 369-409 (1971).

BURKHARDT, D.: Spectral sensitivity and other response characteristics of single visual cells in the arthropode eye. Symp. Soc. Exp. Biol. 16, 86-109 (1962).

DAUMER, K.: Reizmetrische Untersuchungen des Farbensehens der Bienen. Z. vergl. Physiol. 38, 413-478 (1956).

ECKERT, H.: Die spektrale Empfindlichkeit des Komplexauges von Musca. Kybernetik 9, 145-156 (1971).

FRISCH, K.: Der Farbensinn und Formensinn der Biene. Zool. Jb. Abt. allg. Zool. Physiol. 35, 1 - 182 (1915).

HASSELMANN, E.-M.: Ueber die spektrale Empfindlichkeit von Insektenaugen in verschiedenen Helligkeitsbereichen. Zool. Jb. Physiol. 69, 537-576 (1962).

HEINTZ, E.: La question de la sensibilité des abeilles à l'ultraviolet. Ins. Soc. 6, 223-229 (1959).

KIEPENHEUER, J.: Farbunterscheidungsvermögen bei der roten Waldameise Formica polyctena. Z. vergl. Physiol. 57, 409-411 (1968).

KIRSCHFELD, K.: Die Projektion der optischen Umwelt auf das Raster der Rhabdomere im Komplexauge von Musca. Exp. Brain Res. 3, 248-270 (1967).

KIRSCHFELD, K., FRANCESCHINI, N.: Optische Eigenschaften der Ommatidien im Komplexauge von Musca. Kybernetik 5, 47-52 (1968).

KIRSCHFELD, K., REICHARDT, W.: Optomotorische Versuche an Musca mit polarisiertem Licht. Z. Naturforsch. 25b, 228 (1970).

LANGER, H., THORELL, B.: Microspectrophotometric assay of visual pigments in single rhabdomers of the insect eye. Wenner-Gren Center Int. Symp. Ser. 7, Symp. Publ. Div. 145-149 (1966).

STRAUSFELD, N.J.: Golgi studies on Insects: The optic lobes of Diptera. Phil. Trans. Roy Soc. London, B 258, 135 – 223 (1970).

STRAUSFELD, N.J.: The organization of the insect visual system. Z. Zellforsch. 121, 377 – 454 (1971).

SWIHART, S.L.: Colour vision and the physiology of the superposition eye of a butterfly. J. Insect. Physiol. 15, 1347 – 1365 (1969).

SWIHART, C.A., SWIHART, S.L.: Colour Selection and learned feeding preferences in the butterfly Heliconius charitonius. Anim. Behav. 18, 60 – 64 (1970).

SWIHART, S.L.: The neural basis of colour vision in the butterfly, Papilio troilus. J. Insect Physiol. 16, 1623 – 1636 (1970).

WEHNER, R., SCHUEMPERLI, R.: Das Aktionsspektrum der phototaktischen Spontantendenz bei Drosophila melanogaster. Rev. Suisse Zool. 76, 1087 – 1095 (1970).

WEHNER, R., TOGGWEILER, F.: Verhaltensphysiologischer Nachweis des Farbensehens bei Cataglyphis bicolor. (Formicidae, Hymenoptera) Z. Vergl. Physiol. 77, 239 – 255 (1972).

3. A Preliminary Report on the Analysis of the Optomotor System of the Bee – Behavioral Studies with Spectral Lights

W. Kaiser and E. Liske
Department of Zoology, Institute of Technology, Darmstadt, Germany

Abstract.   The results described here indicate that colour differences play no role within the "Funktionskreis" of the bee's optomotor system. It is thus also likely that flesh flies are not colour-blind. If it really is possible to demonstrate that only one receptor type is involved in the optomotor system of the bee, then the lack of colour-specific reactions is easily explained. The results obtained up till now strongly support this idea.

1. Introduction

The experiments we describe in this report were inspired by a study published by KAISER in 1968. He showed that differences in wavelengths are of no (or very little) importance to the optomotor reactions of Phormia. This finding makes two hypotheses plausible: Hypothesis 1: the animal is colour-blind. Hypothesis 2: colour does not matter in the "Funktionskreis" of the optomotor system. The well-founded assumption that Phormia possesses two receptor types makes colour vision probable. Thus Hypothesis 2 gains in plausibility. The question cannot be settled with finality, however, until positive proof is available that flesh flies have colour vision. For this reason it seemed worthwhile to study the question by using honey bees, whose colour vision has been extensively investigated. The central issue of this report is the importance of colour in the "Funktionskreis" of the bee's optomotor system. In parallel to the behavioral experiments, KAISER is approaching the problem by making single unit recordings from the bee's optic lobe (see report in this volume).

2. Material and methods

For our experiments we used a rotating cylinder-type arrangement consisting of two hollow truncated cones set into each other and coated with MgO on their inner surfaces (fig. 1). The light source was a 450 Watt Xenon lamp with exits for two light beams. The upper beam, against which the other was always compared, illuminated the inside of the outer cone's surface (that is the background); the lower beam, the test beam, illuminated the inside surface of the inner cone, which was slotted. The bee, suspended in the middle of this arrangement, could thus see, for example, UV stripes in front of a green background. When the inner slotted cone is rotated, the impression of a rotating two-colour striped pattern is created.

The torque developed by the bee in fixed flight was registered. KUNZE (1961) has shown that the torque exerted during fixed flight is an appropriate measure of reaction. A small piece of pasteboard was glued vertically onto the head and thorax of the bees so that they could be inserted into a mechanical measuring system. In this system the angular torque exerted by the bee had a maximum value of 0.77°, which is equivalent to a torque of 15,5 mg·cm.

3. Results and their significance

a) Colour-specific reactions.   In order to be able to judge whether colour-specific reactions are in fact present, the possibility of reactions made on basis of light-intensity differences must be ruled out. It is known that the optomotor reaction of the bee is dependent on contrast: strong contrasts induce a strong reaction, weaker contrasts induce smaller reactions.

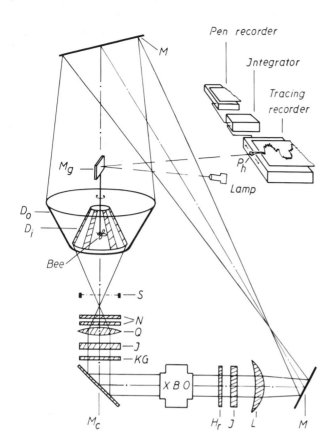

Pen recorder

Jntegrator

Tracing recorder

M

M_g

P_h'

Lamp

D_o
D_i

Bee

S

N
Q
J
KG

XBO

M_c

H_r J L M

Fig. 1. Schematic experimental arrangement. D_o outer wall of drum, D_i inner striped drum, H_r heat reflecting filter, I interference filter, KG heat absorbing filter, L lens, M mirror, M_c cold mirror, M_g galvanometer mirror, N neutral density filter, P_h photocell, Q quartz-glass lens, S shutter, XBO Xenon arc. From KAISER (1968); with alterations

The following experiment, (which can be performed in our experimental arrangement) is the logical consequence of these considerations: Two wavelengths between which bees can differentiate very well in training experiments are chosen. The shutter in the test beam is closed, producing a pattern of black stripes on an illuminated background. With this maximal contrast, the bee should display a maximal reaction. The shutter is then opened and the intensity of the test beam is slowly raised by small steps and the reaction of the bee for each step measured. Thus the contrast between the fore- and back-ground gradually decreases as their brightnesses approach one another. The contrast is zero when the two brightnesses are exactly equal; further increases in the intensity of the test beam produce increasing contrast but with opposite sign. If, in such an experiment, a minimum reaction occurs at a sharply defined intensity of the test beam, then one can assume that at this point, the two colours have matching subjective brightnesses for the bee. The intensity of the test beam at which this reaction occurs can be defined as the matching intensity. This minimum point on the response curve is of particular importance: reactions to different colour combinations of equal animal-subjective brightness would represent colour-specific reactions. If, however, the minimum reaction is zero, then it must be assumed that the optomotor system is independent of colour. Such a curve with a minimum at zero has been obtained using the colour combination 394 nm/ 509 nm, although bees can differentiate extremely well between these colours in training experiments.

Our results to date indicate that, in all likelihood, the bee shows no colour-specific optomotor reactions. This finding is contrary to those of SCHLEGTENDAL (1934) and MOLLER-RACKE (1952). The latter, using the optomotor walking reaction, was able to show that bees, in a two-colour striped rotating drum could differentiate between colour combinations having the same bee-subjective brightness. Our experiments have clearly demonstrated that bees have a highly contrast-sensitive optomotor system. On the basis of our results we believe that it is very likely that the colour-specific responses demonstrated by MOLLER-RACKE were in fact due to small differences in brightness, because she probably could not change the colour intensities by sufficiently fine steps. This supposition is also supported by the earlier literature: SCHLIEPER (1927) and MOLLER-RACKE herself found that in the colour-grey rotating drum the reaction was extinguished at a specific intensity of grey. This is a surprising result, since from our present knowledge about the colour vision of bees, we can assume that the grey papers used by these workers appeared as coloured papers to the bees: it is however possible to obtain fine intensity differences in grey papers.

b) Spectral sensitivity of the optomotor system.   An extremely exact spectral sensitivity curve can be obtained from the relative numbers of light quanta needed to achieve matching. The curve is as yet still incomplete, but it shows a remarkable similarity to the green-receptor sensitivity curve obtained by intracellular measurements by AUTRUM and von ZWEHL (1964). Thus it can be supposed that only the green receptor is involved in the optomotor system of the bee. More investigations are needed to show what contribution, if any, is made to the system by other receptor types.

Acknowledgements.   We are much obliged to the Deutsche Forschungsgemeinschaft for supporting this study within SFB 45 and under File No. 741,29. We thank Professor Dr. H. Autrum and Professor Dr. D. Burkhardt for the loan of some apparatus.

References

AUTRUM, H. and v. ZWEHL, V.: Die spektrale Empfindlichkeit einzelner Sehzellen des Bienen-auges. Z. vergl. Physiol. 48, 357-384 (1964).
KAISER, W.: Zur Frage des Unterscheidungsvermögens für Spektralfarben: Eine Untersuchung der Optomotorik der königlichen Glanzfliege Phormia regina Meig. Z. vergl. Physiol. 61, 71 - 102 (1968).
KAISER, W.: A preliminary report on the analysis of the optomotor system of the honeybee-single unit recordings during stimulation with spectral lights (see report in this volume).
KUNZE, P.: Untersuchung des Bewegungssehens fixiert fliegender Bienen. Z. vergl. Physiol. 44, 656-684 (1961).
MOLLER-RACKE, I.: Farbensinn und Farbenblindheit bei Insekten. Zool. Jb., Abt. allg. Zool. u. Physiol. 63, 237-274 (1952).
SCHLEGTENDAL, A.: Beitrag zum Farbensinn der Arthropoden. Z. vergl. Physiol. 20, 545-581 (1934).
SCHLIEPER, C.: Farbensinn der Tiere und optomotorische Reaktionen der Tiere. Z. vergl. Physiol. 6, 453-472 (1927).

4. A Preliminary Report on the Analysis of the Optomotor System of the Honey Bee – Single Unit Recordings During Stimulation with Spectral Lights

W. Kaiser
Department of Zoology, Institute of Technology, Darmstadt, Germany

Abstract. Extracellular single unit recordings were made from directionally specific movement-sensitive neurones in the optic lobe of the honey bee. It is highly likely that these neurones belong to the optomotor system. When these neurones were stimulated by a moving striped pattern consisting of two monochromatic spectral colours of equal animal-subjective brightness, no colour-specific effect could be detected. The conclusion thus follows that the optomotor system of the bee is independent of its "colour vision system", and that the former system does not transfer colour information. The spectral sensitivity of these movement-sensitive neurones is very similar to the intracellularly-recorded green receptor spectral sensitivity curve: the green receptor thus has a dominant influence on the optomotor system and is perhaps the only receptor type connected to it.

1. Introduction

For the fleshfly, Phormia, it is known from behavioral experiments that the optomotor system is not capable of distinguishing colour (KAISER, 1968), although it is likely that this insect has colour vision. The honey bee, on the other hand, has a well established ability to perceive colour, and is thus a very suitable animal with which to investigate the rôle of colour in the optomotor system. The other question of interest is the circuitry of the optomotor system.

It has previously been demonstrated that the bee has a system of directionally specific, movement-sensitive neurones within its optic lobe (KAISER and BISHOP, 1970). When the results from this work are considered in conjunction with earlier behavioral optomotor studies on bees, using rotating black-and-white striped patterns (KUNZE, 1961), the supposition that these movement-sensitive neurones belong in fact to the optomotor system becomes highly credible. It is therefore reasonable to suppose that the reactions of the neurones in response to stimulation with coloured striped patterns can be directly compared to the behavioral reactions of the bee in a coloured striped drum (see KAISER and LISKE, this volume).

2. Methods and experimental procedure

Extracellular recordings were made as described previously (KAISER and BISHOP, 1970). The optical arrangement used for the present experiments is in principle very similar to that used for the behavioral investigations: stripes of one monochromatic colour can be moved in front of a uniform background of a different colour. The light source is a 900 Watt Xenon arc; the spectral colours are produced by using a variety of interference filters ranging from 318 nm to 654 nm.

The 539 nm background beam (green) remains constant in intensity and wavelength throughout each experiment – it is the standard against which the test beam, whose intensity and wavelength can be varied, is compared. The intensity of the test beam at any given wavelength is increased from zero in small steps by the use of combinations of neutral density filters; thus a point is eventually reached at which both beams appear to the animal to be equally bright – this is defined as the matching point. The question then is whether the neurone reacts to this two-colour pattern.

3. Results

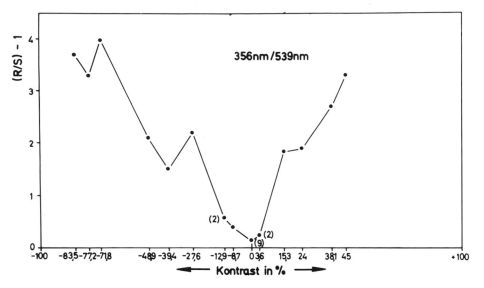

Fig. 1. Reaction curve of an optomotor neurone stimulated by a moving striped pattern of the colour combination 356 nm/539 nm. Those points measured more than once show the number of measurements in brackets. Negative contrasts correspond to foreground darker than background, vice versa for positive contrast. Ordinate represents reaction; details in text

a) Transfer of colour information. Fig. 1 shows the result for the spectral colour-combination 356 nm (UV) and 539 nm (green). It is very likely that the curve goes through 0, i.e. for a very precisely defined ratio between the intensities of the test and standard beams, the level of activity of the unit during stimulation (pattern movement) cannot be distinguished from the spontaneous level. In other words, the neurone did not react to the colour differences in the pattern. The reaction, represented on the ordinate, is calculated by dividing the number of spikes counted during the 15 sec. stimulus period (R) by the number of spikes counted during the immediately preceding 15 sec. period of spontaneous activity (S) and then subtracting 1. The resulting expression equals zero if the activity during pattern movement is the same as spontaneous activity. The abscissa represents the intensity contrast between background and foreground – zero contrast coincides with the minimum point on the curve.

b) Spectral sensitivity. A very precise spectral sensitivity curve can be obtained from the inverse of the relative quantum numbers required to achieve matching. Fig. 2 shows such a curve (closed circles) obtained from one unit. The majority of points on this curve lie exactly on the curve obtained earlier by combining the results from several units in different animals (not shown); the remainder showed only small variations. This indicates the exactness and reproducibility of the matching point method, and is a further support for my belief that I always record from the same neurones.

To my surprise, the spectral sensitivity curves I obtained are very similar to that for the green receptor, obtained from intracellular measurements by AUTRUM and von ZWEHL (1964). My surprise was due to the fact that BISHOP (1970), recording from the same fibres, but using different stimulation, obtained a curve completely unlike the green receptor curve. BISHOP measured latencies after stimulation with flashes from a stationary source. I hope to be able to

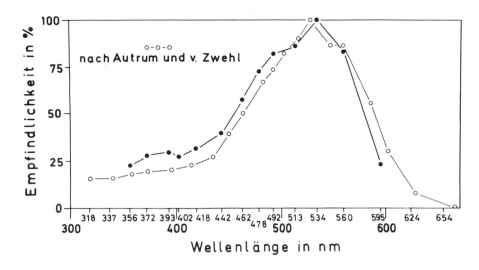

Fig. 2. Closed circles: spectral sensitivity curve obtained from a single neurone of one bee. The ends of the spectrum have not yet been investigated. Open circles: intracellularly measured sensitivity curve of the bee's green receptor (from AUTRUM and von ZWEHL, 1964). Ordinate: sensitivity in %; abscissa: wavelength in nm

explain this discrepancy by means of co-operative experiments with BISHOP. It should be emphasised however, that the spectral sensitivity values obtained up till now in the behavioral optomotor experiments are strikingly similar to mine.

Further experiments using my method are needed to definitely establish whether the curve is in fact a one-receptor curve. Such experiments include selective chromatic adaptation and experiments at threshold, as well as the use of higher light intensities. It can be stated with certainty, however, that the green receptor has a dominant influence in the bee's optomotor system.

4. Discussion

Should it be really possible to prove that the movement-sensitive neurones of the bee are indeed connected to only one receptor type - the green receptor - then an astounding similarity will have been established with the findings of MICHAEL (1968), who has made single unit recordings from the optic nerve of a mammal, the Mexican ground squirrel. The ground squirrel possesses an all-cone retina containing at least two types of receptors with different spectral sensitivities. According to MICHAEL, movement-sensitive neurones in the retina of this animal are connected to only one receptor type, namely the green receptor. The blue receptor exerts no influence on these cells. The apparent inability of these neurones to react to a moving striped pattern, the two colours of which have the same animal-subjective brightness, with a detectable electrophysiological response, demonstrates that in all probability, the optomotor system of the bee is independent of its "colour vision system"; i.e., the optomotor system does not play a rôle in the transfer of colour information. The results obtained so far in behavioral optomotor experiments support this conclusion.

Acknowledgements. I would like to thank the Deutsche Forschungsgemeinschaft for its generous support of this work within SFB 45 and under File No. 741,29. I would also like to thank my wife, Jana Steiner-Kaiser, for her constant co-operation; she, together with my colleagues Prof. H. Markl and Dr. R. Menzel, receive my thanks for many stimulating discussions dur' course of the work. Prof. G. Franke, E. Leitz Co., made valuable suggestions about th' system.

References

AUTRUM, H. and von ZWEHL, V.: Die spektrale Empfindlichkeit einzelner Sehzellen des Bienen-
auges. Z. vergl. Physiol. 48, 357 - 384 (1964).

KAISER, W.: Zur Frage des Unterscheidungsvermögens für Spektralfarben: Eine Untersuchung der
Optomotorik der königlichen Glanzfliege Phormia regina Meig. Z. vergl. Physiol. 61,
71 - 102 (1968).

KAISER, W. and BISHOP, L.G.: Directionally selective motion detecting units in the optic lobe
of the honeybee. Z. vergl. Physiol. 67, 403 - 413 (1970).

KAISER, W. and LISKE, E.: A preliminary report on the analysis of the optomotor system of the
bee - behavioral studies with spectral lights. In "Information processing in the visual system
of arthropods". Zurich (1972).

KUNZE, P.: Untersuchung des Bewegungssehens fixiert fliegender Bienen. Z. vergl. Physiol. 44,
656 - 684 (1961).

MICHAEL, C.R.: Receptive fields of single optic nerve fibers in a mammal with an all-cone
retina. II: Directionally, selective units. J. Neurophysiol. 31, 257 - 267 (1968).

5. Spectral Sensitivity and Wavelength Discrimination in Cataglyphis bicolor (Formicidae)

F. Toggweiler
Department of Zoology, University of Zurich, Switzerland

Abstract. By training experiments it could be proved that Cataglyphis bicolor does possess a colour sense. A spectral sensitivity curve of the spontaneous photopositive reaction was obtained by ranging the intensities of the spectral colours until they were equally attractive to the ants. By means of training procedures to 9 wavelengths in the range 341 nm $\leq \lambda \leq$ 626 nm, the accuracy of the wavelength discrimination could be examined (λ_+-functions). From these 9 λ_+-functions a $1/\Delta\lambda$-function was obtained by measuring the wavelength interval $\Delta\lambda = |\lambda_+ - \lambda_t|$, where λ_t is the wavelength chosen by a reaction frequency of 0.4 in comparison to λ_+. Two maxima of the $1/\Delta\lambda$-function are found: 380 and 550 nm. Even if the longwave filters possess UV subsidiary maxima, the transmission values of which are more than 10^{-4} of the main maxima, the λ_+-function is remarkably influenced by these subsidiary maxima in the ultraviolet range. The photoreisomerisation of an ultraviolet visual pigment may be due to the high ultraviolet sensitivity under these conditions.

1. Introduction

Principally, colour perception can be studied by two methods: by experimental training of the animals and by examination of the receptors (intracellular recordings for spectral sensitivity, selective spectral adaptation in ERG and electron microscopic studies), combined with electro-physiological leads from the central ganglia. With regard to the latter methods it must be observed that, although a positive finding on stimulating lights of different wavelengths does offer a clear proof of the presence of colour receptors, this does not give any indication how far the animal is able to make use of that information. If a conclusive proof is required that the insect is able to distinguish colours, training experiments are the only possible method. Since v. FRISCH's work on bees (1914, 1915), experiments have not only been attempted on bees (DAUMER, 1956; MENZEL, 1967; HELVERSEN, 1971), but also on other insects at various times and with varying degrees of success, to prove in this way that insects have colour sense, e.g. by ILSE (1928), SWIHART and SWIHART (1970) working on butterflies and TSUNEKI (1950) and KIEPENHEUER (1968) on ants. ILSE (1928) as well as SWIHART and SWIHART (1970) proceeded essentially along the same lines. In their experiments butterflies, trained to recognise a particular flower colour, have to choose between flowers of different colours. ILSE's experiments do not prove with complete certainty that butterflies can distinguish colours, because the colours used, reflecting in definite wavelengths, are a very uncertain factor as one regards the relative degree of brightness experienced subjectively by the insects. As already known, however, this factor has a great influence on the spontaneous behaviour of insects and on their ability to distinguish colours after training (MENZEL, 1967; HELVERSEN, 1970). – SWIHART and SWIHART arrive at a significant difference in the visiting frequency between the butterflies' spontaneous choice and the flower colour learned in training experiments. Leads from central nervous system also confirm the assumption that butterflies as well as bees have a colour sense. But detailed informations on the actual course of the experiments and the colours used in the test are not given.

Experiments with ants have so far brought only few results. The experiments of TSUNEKI (1953) failed because of method difficulties and because the animals used (Camponotus spec., Leptothorax spec.) are apparently not trainable. KIEPENHEUER (1968) on the other hand was able to prove that wood ants (Formica polyctena) are capable of distinguishing clearly between two ligh of different wavelengths (UV and green) independently of their subjective brightness, but as he used the mass homing method, no more detailed experiments can be put forward on that line.

2. Methods

As has already been indicated, it is essential to be certain first, with the help of spontaneous tests, about the animals' sensitivity to brightness in the given situation. These results may be obtained with the aid of a discrimination curve of white light intensities. By this it is also made sure that the animals make a positive phototactic response. Then it is possible to visualize the subjective brightness of the colours to be examined. In order that these results may be comparable one with another, not only must the different pairs of colours match each other in brightness, but all the colours must be mutually adapted in intensity to the spontaneous behaviour of the animals. In practice this may be done by showing the animals in a test situation two lights of different wavelengths. The wavelength which is spectrally more effective will then be more frequently visited and may therefore be progressively reduced in intensity until both lights are visited with equal reaction frequencies. This process should be repeated until all the colours used are equally attractive in any combination. If it is assumed that the animal cannot distinguish colours, then after this adjustment all the colours used would appear for the ant to a certain wavelength.

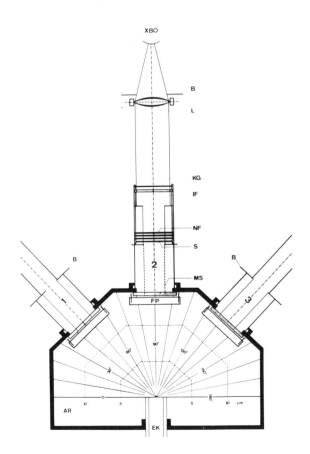

Fig. 1. Test apparatus and arena (AR).
XBO, Osram XBO 450 W/1;
B, diaphragm;
L, lens;
KG, IF, NF, filters;
MS, ground–glass screen;
EK, channel

Scale 1 : 4

During the following training period the animals learn not only a certain wavelength but also the brightness associated with it. This influence may be definitely excluded by successively re-ducing the intensity of the colour being learned during the course of the experiment; in this way the wavelength remains the only constant parameter.

As experimental animals hunters of North African desert ant, Cataglyphis bicolor, were used. This species has the advantage of being easy to train and also of making a positive phototactic reaction. Training to respond to a light of a certain wavelength was achieved with small pieces of cooked white of egg. All the experiments were carried out in an area 20 x 30 cm² (for further details see WEHNER and TOGGWEILER, 1972; apparatus see fig. 1).

3. Results

a) Spectral sensitivity and wavelength discrimination.   In the first experiment the spontaneous behaviour of the animals was tested on a white light of varying intensity as compared with one of constant intensity. Compared with the bee in a similar situation the ant's ability to distinguish intensity was poorer by the factor 2.6. This shows that the intensity discrimination in Cataglyphis is probably poorer than it was assumed by TSUNEKI (1953) for Camponotus and Leptothorax.

According to the process described above, all the colours used were brought to the same degree of subjective brightness. The reciprocal value of the intensities, measured absolutely, give a standard for the spectral sensitivity of the tested ants. The range $341 \text{ nm} \leqq \lambda \leqq 626 \text{ nm}$ was tested. The line of the curve corresponds roughly to the spectral sensitivity of the bee, but is somewhat displaced towards the longwave end of the spectrum.

The colour sense of Cataglyphis bicolor was tested by comparing spontaneous choices made after food training. In order to eliminate the influence of the brightness of the training colours their intensity was gradually reduced from I_o by 0.5, 1.0 and 1.5 logarithmic units. The difference of the two reactions is statistically highly significant ($p = 10^{-5}$ to $< 10^{-6}$, in a few cases $p < 0.03$; n = 50). With this, clear proof is given that another insect besides the bee possesses a colour sense (fig. 2).

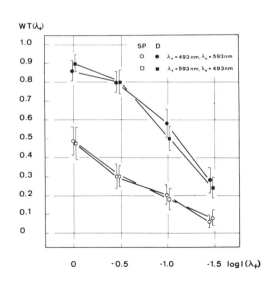

Fig. 2. Reaction-frequency (WT) before (O) and after (●) training in variable inten-sities of

$\square : \lambda_+ = 493, \lambda_t = 593$ nm
$\triangle : \lambda_+ = 593, \lambda_t = 493$ nm

This colour sense was subsequently determined quantitatively by measuring the reaction to the lights of varying wavelengths and the same subjective brightness, after the ants had been trained to one wavelength. In this way the accuracy of wavelength discrimination was tested between 341 and 626 nm. By means of $|\lambda_+$-functions for 9 different λ_+-values a $\Delta\lambda$-function was determined, representing the intervals ($\lambda_+ - \lambda t$) for a reaction frequency of 0.6, i.e. λ_+ was chosen to a ratio $n(\lambda_+)/_n(\lambda_+) +_n(\lambda_t) = 0.6$ when λ_+ and λ_t were simultaneously presented. The $1/\Delta\lambda$-function shows two maxima, in the ultraviolet (380 nm) and in the green region (550 nm) of the spectrum (WEHNER and TOGGWEILER, 1972). From these results it would be possible at first view, to postulate a colour perception mechanism, consisting of three receptors with sensitivity maxima at 350, 440 and 580 nm. This corresponds to the spectral sensitivity. In the spectral sensitivity curve of the phototactic behaviour the sensitivity peak at 580 nm could be explained by the transmission spectrum of the screening pigment. That effect, however, could not be responsible for the accuracy peak of wavelength discrimination at 550 nm. Therefore the question of a two - or three - receptor system in Cataglyphis bicolor cannot be answered at this point of the investigation.

b) The influence of subsidiary maxima in the transmission spectra. Since for the first time alternative colours from the whole range tested were compared with one training colour, the interference filters used had to be examined for possible subsidiary maxima. And in fact, especially in the shortwave range (300 to 400 nm) subsidiary maxima were found for the longwave filters. Since in this range the ants are up to 100 times more sensitive than in the longwave one, the

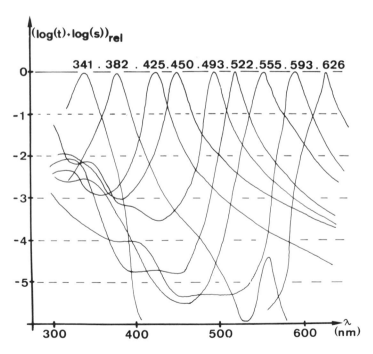

Fig. 3. "Subjective transmission" of the interference filters. $\log(t) + \log(s)$ = transmission logarithm, superposed upon the logarithm of spectral sensitivity. To make comparable, the curves are normalized

subsidiary maxima found are correspondingly more effective. To determine them, the logarithmic spectral sensitivity was superposed upon the transmission logarithm. The maxima of these curves were brought to the same height for reasons of comparison. This "subjective transmission" is shown in fig. 3. In spite of the fact that the subsidiary maxima have only a height of not more than 10^{-4} of the main maxima, the colour perception of Cataglyphis bicolor is highly influenced by these maxima. Theoretically one would expect that either the λ_+-function (wavelength-discrimination function) is flattened in the direction towards the subsidiary transmission maximum or that a second minimum of the λ_+-function occurs near the λ-value of the subsidiary maximum.

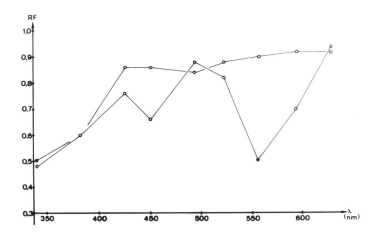

Fig. 4. The λ_+-function for λ = 555 and 341 nm.

RF = reaction frequency (further details see text)

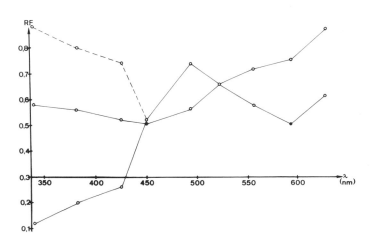

Fig. 5. The λ_+-function for λ = 593 and 450 nm.

RF = reaction frequency

As figs. 4 and 5 show, the latter is true. For λ_+ = 555 nm there is a strikingly clear descending branch in the shortwave range to RF < 0.5. Furthermore, the λ_+-curve has the same slope in the range of the shortwave alternatives λ_t = 341 nm (fig. 4). Thus the reaction of the animals is exactly the same as if they had been trained simultaneously to 555 nm and to 341 nm. The line of the second curve (λ_+ = 593 nm, fig. 5) is similar to the one described above, except that the reaction frequencies sink still lower in a shortwave alternative light. The surprisingly strong effect of the low transmission subsidiary maxima may be due to the fact that the sensitivity to UV is increased by the simultaneous presence of longwave light, due to the photoreisomerisation of the ultraviolet visual pigment by lights of the blue and green part of the spectrum (SCHWEMER, GOGALA and HAMDORF, 1971; HAMDORF, GOGALA and SCHWEMER, 1971).

Acknowledgement. The work was financially supported by a grant offered to RÜDIGER WEHNER by the Fonds National Suisse de la Recherche Scientifique, No. 3.315.70.

References

DAUMER, K.: Reizmetrische Untersuchungen des Farbensehens der Biene. Z. vergl. Physiol. 38, 413 - 478 (1956).

FRISCH, K. v.: Versuche zum Nachweis des Farbensinnes bei angeblich total farbenblinden Tieren. Verh. Dtsch. Zool. Ges. 24 (1914).

FRISCH, K. v.: Der Farbensinn und Formensinn der Biene. Zool. Jb. allg. Zool. Physiol. 35, 1 - 182 (1915).

HAMDORF, K., GOGALA, M. and SCHWEMER, J.: Beschleunigung der "Dunkeladaptation" eines UV-Rezeptors durch sichtbare Strahlung. Z. vergl. Physiol. 75, 189 - 199 (1971).

HELVERSEN, O. v.: Zur spektralen Unterschiedsempfindlichkeit der Honigbiene. Dissert. Albert-Ludw.-Universität, Freiburg i.B. (1971).

ILSE, D.: Ueber den Farbensinn der Tagfalter. Z. vergl. Physiol. 8, 658 - 692 (1928).

KIEPENHEUER, J.: Farbunterscheidungsvermögen bei der roten Waldameise. Z. vergl. Physiol. 57, 409 - 411 (1968).

MENZEL, R.: Untersuchungen zum Erlernen von Spektralfarben durch die Honigbiene. Z. vergl. Physiol. 56, 22 - 67 (1967).

MENZEL, R.: Das Gedächtnis der Honigbiene für Spektralfarben. I. Kurzzeitiges und lang-zeitiges Behalten. Z. vergl. Physiol. 60, 82 - 102 (1968).

SCHWEMER, J., GOGALA, M. and HAMDORF, K.: Der UV-Sehfarbstoff der Insekten: Photo-chemie in vitro und in vivo. Z. vergl. Physiol. 75, 174 - 188 (1971).

SWIHART, C.A. and SWIHART, S.L.: Colour Selection and Learned Feeding Preferences in the Butterfly. Anim. Behav. 18, 60 - 64 (1970).

TSUNEKI, K.: On Colour Vision in Two Species of Ants. Jap. J. Zool. 11, 187 - 221 (1953).

WEHNER, R. and TOGGWEILER, F.: Verhaltensphysiologischer Nachweis des Farbensehens bei Cataglyphis bicolor. Z. vergl. Physiol., in press (1972).

6. ERG of Formica polyctena and Selective Adaptation

H. Roth and R. Menzel
Department of Zoology, Institute of Technology, Darmstadt, Germany

Abstract. The electroretinogram (ERG) of Formica polyctena is a monophasic negative potential, which manifests peaks by "light-on" and rarely by "light-off" at higher intensities and some wave-lengths. The selectrive spectral adaptation curves suggest that there are only two different types of colour receptors: an uv receptor (λ_{max} = 361 nm) and a wide green receptor (495 nm – 531 nm). The results are discussed on the basis of behavioral studies concerning colour vision in wood ants.

1. Introduction

Today we know that many insects are able to recognize colours. The social high-developed Hymenoptera are most thoroughly investigated. By training experiments von FRISCH (1915) was able to demonstrate the existence of a colour sense in the orientation of the honey-bee. Since the experiments of AUTRUM (1964) we are especially well informed about colour-receptors of bees. On the other hand there are only a few studies about colour-vision in ants. TSUNEKI (1950) demonstrated that during an orientation to chromatic light there is a large intensity dependence. KIEPENHEUER (1968) tried to eliminate this factor in his colour training experiments. He found a colour discrimination between broad-banded uv and green.

The photo-receptors have not yet been analysed. This study is an attempt to characterize the colour receptors of Formica polyctena with mass potential in recordings of dark-, yellow-, blue- and uv-adapted animals.

2. Material and method

Ants from the wood near the institute were used in the first experiments, later on others from an artifical nest in the cellar, all being hunters of 8 mm length.

The ant was fixed on a small block with wax-collophonium so that all legs at the tarsi and at the coxi as well as the head stuck firmly on the block. The left eye was looking up, the gaster was movable. The recording was always made from the left compound eye. The recording electrode and the indifferent electrode were always newly chlorinated silver wires. A glass capillary micropipette was filled with RINGER solution and had a tip diameter of 6 - 8 ommatidia (85 μ - 100 μ). It was used as different electrode and pushed on through a slightly greater hole flat under the cornea, so that the tip was in the middle of the eye. The electrode was connected with the high ohm entrance of the DC amplifier GRASS P 16. The grounded indifferent electrode was inserted through a small hole in the lymphatic sack behind the brain. The signals were recorded at the beginning with the BRUSH MARK 220 and later on photographed. A timer controlled the course of the experiments. The stimulating light and the adapting light were projected together on a Y-tube (SCHOTT/MAINZ), which is permeable for uv and mixes both colours and illumined a frosted plexiglass plane (Ø 10 mm) uniformely.

After the preparation the animals remained for half an hour in the darkness. Then the 18 filters of the filter wheel (line and double line filters) were inserted in the beam of the Xenon lamp. Each filter was inserted for one second. Every fifteen seconds the filter was changed; the first sequence began with 337 nm and the following with 650 nm. The intensity was varied over three

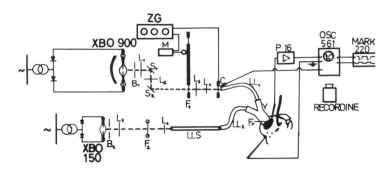

Fig. 1. Experimental con-
struction: testing beam: XBO
900 Xenon lamp 900 W, B_1
shutter, L_1-L_4 quartz lenses,
S_1-S_2 cold-light mirrors,
F_1 filter wheel with 18 line
and double line filters,
C magnetic shutter with
photo-diode, LL_1 light guide
permeable for uv, M motor,
ZG timer.

Adapting beam: XBO 150 Xenon lamp 150 W, B_2 shutter, L_5-L_6 quartz lenses, F_2 different adapting filters, LLS light-guide stick, LL_2 light guide, both permeable for uv; Y Y-tube; Fr frosted pane of plexiglass, P 16 kathode-follower, GRASS P 16, OSC 561 oscillograph, MARK 220 quick-writer

logarithmic unities in half logarithmic steps. All experiments were performed between 12.00 h and 19.00 h at room temperature.

The intensity of the adapting light was adjusted in such a way that the response was equal to that of 512 nm and 10% grey filter. The registered potentials ranged from 2 mV to 16 mV. Before and after every selective adaptation a dark adaptation curve was registered. The selective adap-tation was not interrupted during the exposure with the testing wave-lengths. To control the constancy of the potentials for the experimental time of three hours the measurements were re-gularily interrupted by the control wave-length 512 nm and 10% grey filter.

The spectral sensitivity was calculated from the intensity dependence for all eighteen testing wave-lengths. The beam of each wave-length was often measured and registered as logarithms in relative quanta on the abscissa of the intensity diagram. The standard reaction for the conver-sion into sensitivity varied from preparation to preparation, because it must cross the linear phase of the intensity curve. However, the standard reaction is for dark adaptation and corresponding spectral adaptation the same. The spectral sensitivity was expressed in percent of the maximal sensitivity for every animal. All sensitivity curves of the same experimental situation yield the mean values of spectral sensitivity.

3. Results

By exposure of large parts of the eye rectangular monophasic potentials can be recorded. At "light-on" the potential immediately becames negative and then builds a plateau. At "light-off" a slow come-back to the dark potential follows. Depending upon the intensity and the wave-length the plateau is surpassed by an on-peak. An off-peak is seldom. By using high intensities, double on-peaks can be obtained.

The selective adaptation was performed with an UG-11 filter of 336 nm ($\triangle \lambda$ = 43 nm) and a double line filter of 442 nm ($\triangle \lambda$ = 13.2 nm) and an edge filter of 570 nm, which only trans-mitted wave-lengths longer than 570 nm. After each adaptation the sensitivity curves were calculated from the intensity curves. The sensitivity curves obtained after selective adaptation and the corresponding sensitivity curves obtained after dark adaptation were each normalized in the percentage of their maximum and averaged.

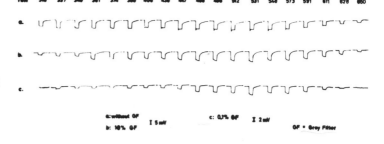

Fig. 2. Mass potentials of Formica polyctena at different wave-lengths and different intensities. The potentials were registered from one animal only

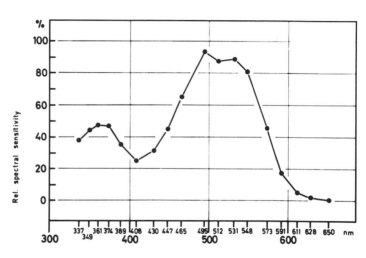

Fig. 3 shows the mean curve of the spectral sensitivity after dark adaptation, which is computed from sensitivity curves of 25 different preparations. A wide maximum in the blue-green field with a small peak at 495 nm can be observed. The maximum in uv (λ_{max} = 361 nm) is in comparison about half as high. In the individual preparations both maxima can variate between 495 nm and 531 nm resp. between 361 nm and 374 nm. The minimum is always at 408 nm. Above 628 nm the sensitivity drops more than three logarithmic decades

UV receptor: After adaptation with an UG-11 filter the UV second maximum disappeared and the green maximum shifted from 495 nm to 531 nm (fig. 4a).

Green receptor: Adaptation with an edge filter of 570 nm in the long wave region of the sensitivity leads to a selective reduction of the green maximum (fig. 4b). Whereas after dark adaptation it is **twice** as high as the UV maximum, after spectral adaptation it is reduced to half of the UV maximum. The UV maximum at 361 nm remains the same. The minimum is shifted from 408 nm to 430 nm.

The same spectral sensitivity curve can be obtained after an adaptation with 442 nm (fig. 4c).

Because the blue and yellow adaptation produces an identical sensitivity curve, one has to draw the conclusion that the wide green maximum originates from one receptor system only. This receptor is a green receptor. A blue receptor was not be found in the eye of Formica polyctena.

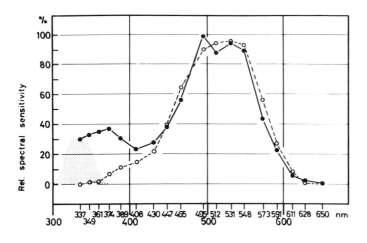

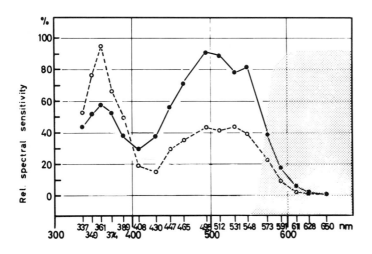

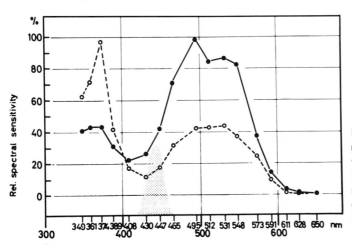

Fig. 4a – c. Solid lines: mean curves of the spectral sensitivity after dark adaptation. Dotted lines: mean curves of the spectral sensitivity after selective spectral adaptation. The shaded areas show the transmittance of the three selective adapting filters

4. Discussion

The spectral sensitivity curve after dark adaptation measured by the recording of mass potentials is similar to those found for other species of insects (MENZEL, 1971, literature). In most spectral sensitivity curves the uv maximum lies between 350 nm and 360 nm; that of the ant is at 361 nm. Only the study of HAMDORF, GOGALA and SCHWEMMER (1970) describes an uv maximum in Ascalaphus at 340 nm.

As also found in other insects, the green maximum in the ERG of the ant is relatively wide and ranges with the different animals from 495 nm to 531 nm. BENETT, TUNSTALL and HORRIDGE (1967) as well as AUTRUM and KOLB (1968) discussed different reasons for this phenomenon.

The selective chromatic adaptation experiments, described in this study, show that the uv maximum and green maximum of the spectral sensitivity curves each are based on a one-receptor system. Therefore with a high probability the ants have a dichromatic color sense. Our behavioral experiments to analyse the colour sense of the Formica polyctena and the studies of KIEPEN-HEUER (1968) proved that the ants can discriminate between the colours uv and green. In a dichromatic visual system the descrimination of colours is limited to a small region, and a grey point must be assumed. Our behaviour experiments, which should clarify these questions, are not yet complete.

References

AUTRUM, H.J. and KOLB, G.: Spektrale Empfindlichkeit einzelner Sehzellen der Aeschniden. Z. vergl. Physiol. 60, 450 - 477 (1968).

BENETT, TUNSTALL and HORRIDGE, G.: Spectral sensitivity of single retinula cells of the Locusta. Z. vergl. Physiol. 55, 195 - 206 (1967).

GOGALA, M., HAMDORF, K. and SCHWEMMER, J.: UV-Sehfarbstoff bei Insekten. Z. vergl. Physiol. 70, 410 - 413 (1970).

KIEPENHEUER, J.: Farbunterscheidungsvermögen bei der roten Waldameise Formica polyctena FOERSTER. Z. vergl. Physiol. 57, 409 - 411 (1968).

MENZEL, R.: Ueber den Farbensinn von Paravespula germanica F.: ERG und selektive Adaptation. Z. vergl. Physiol. 75, 86 - 104 (1971).

TSUNEKI, K.: Some experiments in colour vision in ants. Jour. Fac. Sci., Hokkaido Univ. Ser. 4 Zool. 10 (1950).

VI. Pattern Recognition

1. Pattern Modulation and Pattern Detection in the Visual System of Hymenoptera

R. Wehner
Department of Zoology, University of Zurich, Switzerland

Abstract.   In the bee, Apis mellifica, and the ant, Cataglyphis bicolor, pattern modulation and pattern detection are studied by optical methods (1) as well as by pattern discrimination tests using the training procedure (2). (1) Pattern modulation: The directional-intensity function measured by means of the light flux in a single rhabdom is characterized by a half-width of $\triangle \varrho = 2.6°$ (Apis) resp. $7.8°$ (Cataglyphis). In both insects, a better correlation between the directional-intensity function and the divergence angle can be found in the horizontal direction (z-axis) than in the x- and y-axes, respectively, meaning contrast transfer to be more complete in the horizontal direction. Some evaluations of contrast modulation of an array of ommatidia are given for light spots and stripe patterns by means of the optical data. (2) Pattern detection: (a) Using a simple antisymmetric intensity distribution, a vertically presented black-and-white disk with a specifically inclined contrast line, a topological central nervous representation of the visual field of the eye can be proved. (b) That system is shown not to be isotropic in function in so far as the middle lower part of the frontal visual field is most decisive for detecting contrast changes. Therefore, position-dependent weighting factors have to be introduced in any model of pattern recognition in bees. (c) The head or   eye position is nearly kept constant in the flying bee, as shown by high-speed cinematography during pattern discrimination tests. Therefore, the receptor coordinates are stabilized with respect to a space-constant system. (d) The mechanism of correlating two patterns by means of overlapping and non-overlapping areas in a space-constant system, where the patterns are not allowed to rotate, cannot be the only principle in data processing due to pattern recognition: for the direction of a stripe is also detected irrespective of the sign of contrast changes (as during training or completely reversed). Therefore reduction of information may be caused by visual generalization processes.

If we now deal with the possibilities of correlations between neurophysiological and behavioral studies, the problems of pattern recognition offer a well-suited example for the necessity of corre-lating input-output functions with the more fundamental work in neuroanatomy and neurophysiolo-gy. This correlation, however, must not consist in prematurely transferring concepts of one discipline to the other. To avoid such an unsuitable transfer of concepts, drawn from electrophysiological work done on special neuronal networks within the visual pathway and applied to the interpretation of behavioral studies, two examples will be presented first. (1) When the high-flicker fusion frequen-cies were found in insects, some earlier experiments performed with bees on pattern recognition (summary of literature in WEHNER, 1967) were   discussed by means of a mere detection of contrast frequencies. According to that theory, insects would only be able to discriminate between figures differing in their contents of contours per square unit, irrespective of other pattern featu-res, e.g. the direction of the contours, the amount and distribution of black and white areas within the visual field etc. As previous and more detail ed experiments on pattern discrimination in bees have shown (WEHNER, 1967, 1972; SCHNETTER, 1968; and the following contributions of SCHNETTER, CRUSE and ANDERSON in this volume), even elementary pattern discrimination tasks are not solved by the bees by only using the relative length of contours within a particular figure. As we now know, pattern recognition in bees does not merely consist in an overall integration of contrast frequencies of the visual units involved.

(2) Another example is put forward by the concepts of detector neurons (master neurons). Since at different levels of the visual pathway of vertebrates particular neuronal units have been found (first by LETTVIN, MATURANA, PITTS and Mc CULLOCH, 1959), responding to various specialized stimulus configurations, ethologists have immediately defined pattern detectors in any case, where an insect species showed some spontaneous preferences to special trigger features (JANDER, 1964; JANDER and VOLK-HEINRICHS, 1970; JANDER and SCHWEDER, 1971). Irrespective of the fact that there only exist a few electrophysiological recordings from higher order neurons in the visual pathway of insects, and irrespective of the fact that even in vertebrates the concepts of master neurons are not incontestable at all (GRUESSER and GRUESSER-CORNEHLS, 1970), the prediction of specialized central nervous visual detectors seems prematurely to be made. If central interneurons in the visual pathway are thought of as concerned with the analysis of features of patterns, the question for pattern discrimination tests has not to be to give a classification of all kinds of detectors possibly involved, but it has to be, how and by which neuronal mechanisms is the unit triggered by special pattern features.

The analysis of pattern recognition in insects, i.e. mainly in Diptera and Hymenoptera, is nowaday put forward on different levels of data processing: (1) by means of optical methods on the level of stimulus transfer within the dioptric systems, (2) by means of neurophysiological and neuroanatomical methods in the course of the visual pathway, in order to study some mechanisms involved in pattern recognition, i.e. lateral inhibition, directional selectivity or topology of the visual field, and (3) by behavioral approaches in analyzing input-output functions of the whole visual system. Here we want to deal with experiments performed by means of the first and the last methods.

1. Pattern modulation

If pattern detection mechanisms will be studied, it must first be clear how the patterns are modulated by the raster of the dioptric systems in the compound eye. As in the eyes of Hymenoptera the visual units consist in single ommatidia, the interommatidial inclinations between adjacent ommatidia (divergence angle $\Delta \varphi$) and the directional-intensity function of a single ommatidium (lens admittance function according to VARELA and WIITANEN, 1970; half-width $\Delta \varrho$) are the parameters of the optics of the apposition eye responsible for pattern modulation. Both parameters were calculated for bees and ants, Cataglyphis bicolor (WEHNER, EHEIM and HERRLING, 1971; EHEIM and WEHNER, 1972; EHEIM, this volume). Just recently LAUGHLIN and HORRIDGE (1971 have measured the angular sensitivity curve by intracellular recordings from single visual cells of the worker bee and have found a half-width of $\Delta \varrho = 2.6°$, exactly coinciding with our results.

If one knows these two parameters of the optics of the compound eye, pattern modulation as done by the lens systems can be fully described in terms of contrast transfer and resolving power of the array of the dioptric systems. If we consider, for example, a system of 3 ommatidia A, B and C containing point light sources in the optical axes of A and C, contrast modulation can be quantitatively determined in that array of lens systems. In the course of that procedure the light flux ϕ is calculated from the ordinate of the Gaussian directional-intensity function for each position of th two lights and for each of the 3 ommatidia. In our first calculation (fig.1, upper part), the light sources separated by the angle α are always positioned in the optical axes of A and C by this α and $\Delta \varphi$ being varied simultaneously (abscissa). The light flux ratio ($\phi_A - \phi_B$)/ ϕ_A , where ϕ_A ϕ_C , then describes the amount of contrast modulation. Where the ratio becomes zero, ($\Delta \varphi$)$_0$, no contrast is to be seen by the array of the 3 ommatidia. Negative values represent contrast reversal, whereas the maximal value 1 is due to total contrast transfer. As in Apis contrast reversal occurs within the range of $0° \leq \Delta \varphi \leq 1.2°$, the interommatidial inclinations in the x- and y-axi of the central honeybee eye (measured by PORTILLO, 1936) scarcely exceed the zero point ($\Delta \varphi$)$_0$. Therefore, internal contrast is reduced to nearly 0.1 of the external contrast. The z-axi however, provides a better adaptation of the divergence angles to the Gaussian characteristics of the visual units. As in Cataglyphis $\Delta\varphi$ (z) also exceeds $\Delta\varphi$ (x,y), there is a better correlation between acceptance angles and divergence angles in the horizontal plane for both species of Hymen optera considered.

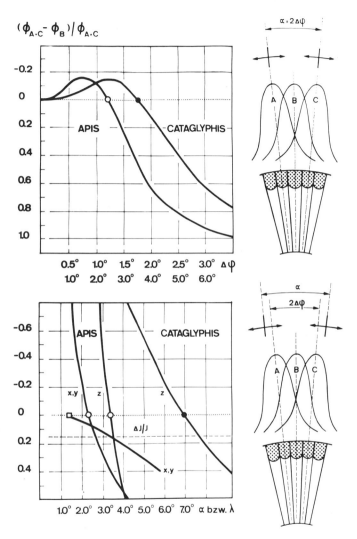

Fig.1. Contrast modulation in an array of 3 ommatidia in the compound eyes of bees, Apis , and ants, Cataglyphis.

A,B,C, light flux in the ommatidia A,B, and C, respectively, measured by means of the ordinate of the directional-intensity curve. Upper part: variation of the angular distance α of two light sources as well as the divergence angle $\Delta\varphi$. The upper abscissa refers to Apis, the lower one to Cataglyphis. Lower part: variation of the angular distance α resp. the period λ of an equidistant stripe pattern. x, y, z , axes of the ommatidial array (see BRAITENBERG, 1967; WEHNER, EHEIM and HERRLING, 1971). The value of $\Delta I/I$ is taken from LABHART (this volume)

Secondly, contrast modulation can be approximated by variation of the angle α between the light sources, when the interommatidial inclinations $\Delta\varphi$ are kept constant (fig.1, lower part). According to the procedure mentioned above, the light flux ratios were determined for varying angles α and for varying pattern wavelengths λ of an equidistant stripe pattern. (In the latter case the directional-intensity function had to be rotated about the axis of the Gaussian characteristic of the visual unit.) The values of α and λ , where the external contrast is completely lost, are marked by the open (Apis) and filled **symbols** (Cataglyphis). The effective threshold for contrast modulation, however, depends on the sensitivity in discriminating different light intensities. For the highest intensities used in our experiments on the photopositive reaction of worker bees, a $\Delta I/I$ of 0.13 was found (LABHART and WEHNER, in press; LABHART, this volume). By means of that value $\Delta I/I$ and the function describing contrast modulation for varying angles α , or λ , predictions can be made on the detection of light points and stripe patterns by the visual system of the bee resp. ant. The steeper the contrast modulation curve extends beyond the α - resp. λ - scale, the less critical are variations in the intensity discrimination threshold.

With regard to the resolving power of a striped pattern characterized by the pattern period λ, the first zero-point of the interference function, $(\Delta \varphi)_o = \lambda/2$, is to be defined as the lower limit of the resolving power. Beyond this point no unambiguous information on the direction and wavelength of the moving pattern can be drawn from the external intensity function. Using this procedure, GOETZ (1964) for the first time has determined the interommatidial inclination by measuring optokinetic reactions in Drosophila flies.

The acceptance angles and the divergence angles enable us completely to describe pattern modu-lation as done by the array of lens systems. The image, formed by the dioptric apparatus on the receptor layer, can therefore be deduced unambiguously from the external contrast distribution. Within the receptor layer two further parameters are responsible for contrast modulation in single visual cells: intensity discrimination in terms of the $\Delta I/I$-threshold and the time-resolving pro-perties, characterized by the flicker fusion frequency. The lower the $\Delta I/I$-threshold is, and the higher the flicker fusion frequencies are, the more looks the internal contrast like the external one

2. Pattern detection

There is only little direct evidence for data processing at various levels of the visual pathway in Hymenoptera. For the first time BISHOP (1970) and KAISER and BISHOP (1970) were able to re-cord from higher-order interneurons of the visual system of the honeybee. The main properties of these units are to respond to changes of light intensity and selectively to movement of objects in a preferred direction and within the entire acceptance angle of the eye. By the interaction of four classes of those two-motion detecting units, movement in any given direction is simultaneous-ly encoded by increased and decreased intensities in these four types of units.

On account of the technical difficulties up to now involved in recording from the Hymenopterean visual pathway, the study of input-output relations by pattern discrimination tests seems to be the most successful method in the present course of analysis. As done by SCHNETTER (1968; this volu-me), CRUSE (this volume) and WEHNER (1969, 1972). models can be derived from the experimen-tal results in order to analyse the pattern parameters involved in pattern detection. The number of contrast changes per time unit, generated at the compound eye of the flying insect, was considered to be the decisive parameter as well as the ratio of overlapping and non-overlapping areas of two figures.

Here we will deal with three aspects of pattern detecting mechanisms: (a) the topology of the vi-sual field, (b) the question, whether the visual system, representing the visual surround, is isotro-pic in function, i.e. whether there exists a position-dependent ranking of stimuli (whereby the term position refers to the absolute position of the stimulus within the visual field as well as to the relative position within the pattern context), and (c) with the correspondence of the coordinates of the visual system with the coordinates of a space-constant system. Finally (d) we may consider the use of problem-orientated parameters by the visual system of the bee (MAZOCHIN-PORSHNY KOV, 1969; ANDERSON, this volume).

a) From Diptera it is known by histological work that the visual environment is mapped in a one to-one relationship at least up to the laminar and medullar level (BRAITENBERG, 1970; this volum As also shown by electrophysiological recordings from the fly's lobular region, the spike frequency of a single unit depends on the field position of the light stimulus (BISHOP, KEEHN and Mc CANI 1968; Mc CANN and DILL, 1969; MIMURA, 1971). In Hymenoptera, however, until now there is no experimental evidence for a central nervous representation of the visual field, neither by neuro anatomical and electrophysiological nor by behavioral work.

As deduced from pattern recognition tests in bees (WEHNER, 1972), a quantitative model can be found which describes the discrimination of identical figures only differing in their position in space. If two shapes - as CRUSE points out in this volume - are discriminated by means of a two-dimensional cross-correlation, the correlation coefficient of which corresponds to the largest common area between the two shapes, then the stimulatory values of the two shapes had to be identical, meaning that the shapes cannot be discriminated. In the cross-correlation concept the assumption is made that shapes offered in the horizontal plane are computed by the visual system of the bee in such a way that the common areas reach maximal values, presupposing rotation invariance. On the vertical plane, however, serving as pattern screen in all experiments dealt with here, that presupposition cannot be made. Otherwise differently inclined black-and-white disks should not be discriminated by the bees - but they are. Therefore we must conclude that the patterns are projected to a central nervous grid, which preserves the intensity distributions spatially according to a fixed coordinate system.

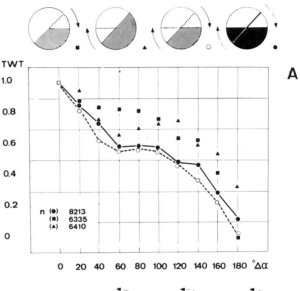

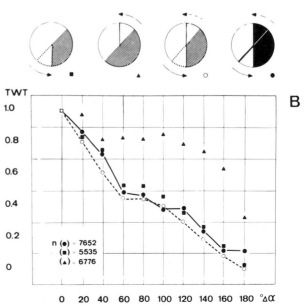

Fig.2. The discrimination of differently inclined black-and-white disks on a vertical pattern screen. Abscissa: deviations $\Delta\alpha$ of the contrast line in the test pattern from the training direction $\alpha_+ = 45°$; ordinate: stimulatory values (TWT) of the test patterns as calculated by means of the reaction frequencies to the test patterns, a probit transformation and a standardizing procedure. TWT (training pattern) = 1. The effects of varying angular positions of the contrast line (●) are plotted as well as the effects of positive (■) and negative (▲) contrast changes (see inset figures); o stimulatory values of the rotated disks as deduced from the ■ and ▲ values (see text). A clockwise, B counterclockwise rotation

When the disk is rotated around an axis centrally penetrating the pattern disk and perpendicularly orientated to the vertical screen, even slight angle differences $\Delta\alpha$ are measured by the bee's visual system according to fig.2 ($\bullet$-curves). The stimulatory values of the rotated black-and-white pattern disks can be approximated by the stimulatory values of the corresponding positive and negative contrast changes, tested separately ($\blacksquare$- and $\blacktriangle$-curves in fig.2) (WEHNER, 1972).

b) That approximation only holds when, besides the size, the position of the contrasting sectors inserted into the training pattern is also taken into account. When the training inclination of the contrast line is $45°$ (but not, when it is $0°$ or $90°$), the effect of the rotation of the disk as well as the effect of the insertions of black and white sectors on the particular stimulatory values is not invariant against the direction of rotation (clockwise or counterclockwise). Therefore, position-dependent weighting factors have to be introduced in any model of pattern recognition in bees.

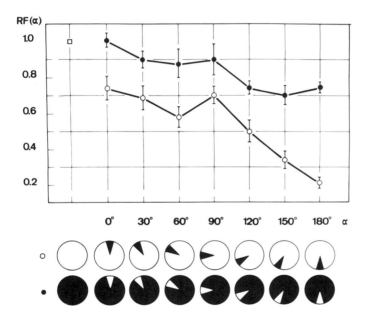

Fig.3. Position-dependent effects of positive ($\bullet$) and negative (o) contrast changes. The reaction frequencies RF ($\pm$ errors of mean values) are measured in competition to the training patterns (black resp. white screen). RF (training pattern) = 1 resp. 50 %. N = 464

These weighting factors can be roughly taken from the data in fig.3, which represent the effect of inserting a $30°$-wide black or . white sector into a white or black training disk in varying part of the visual field. According to that procedure, the middle lower part of the frontal visual field is found to be most decisive for detecting contrast changes. That negative contrast changes cause a stronger effect than positive ones, may depend on the totally white pattern screen, in front of which the disks were offered, and may therefore be not invariant against the test procedure. Additionally, for constant values of $\Delta\alpha$ one obtains the strongest effects of changing contrast in the areas near the contrast line. That statement can be drawn from experiments with varying training directions of the contrast line. It can be also demonstrated by a calculation which combines figs. and 3 and hence eliminates the effect of visual field position: a given area increment ΔA (insertion of black areas into previously white areas) decreases the stimulatory value of the training pattern less the farther it is positioned from the contrast line.

c) As the visual system of the bee is not isotropic in function with respect to pattern recognition mechanisms, position-dependent data processing may be related to receptor-constant or space-constant coordinates. In the latter case, information about the positions of the eye or head during pattern recognition must additionally be computed by mechanoreceptive control and processed with the visual information (compare the "space-constant fibres" in Crustaceans; WIERSMA and YAMA-

GUCHI, 1966).

If in the bee one half of the eye is covered with paint (simultaneously on both eyes, figs.4 and 5), contrast changes are only detected in that region of the visual field corresponding with the unpainted part of the eye. With the lower part of the eye covered, for example, bees are only able to compute contrast changes in the upper part of the visual field. At the beginning of each experimental series all bees were individually trained (60 reinforcements) to a black-and-white disk, the contrast line of which extended in the zero direction. By this the plateau level of the learning curve was reached in every case. After the following procedure of painting over special regions of the eye, for example its lower half, when the bees were kept in 4° C anesthesia, the same discrimination tests as before were offered to the bees. They succeeded in pattern discrimination to the

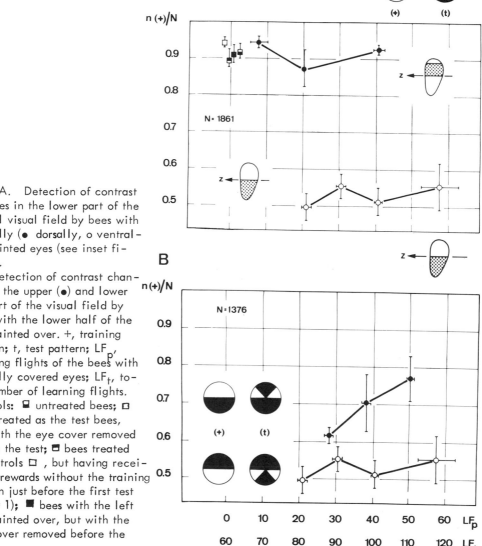

Fig.4.A. Detection of contrast changes in the lower part of the frontal visual field by bees with partially (● dorsally, o ventrally) painted eyes (see inset figures).

B. Detection of contrast changes in the upper (●) and lower (o) part of the visual field by bees with the lower half of the eye painted over. +, training pattern; t, test pattern; LF_p, learning flights of the bees with partially covered eyes; LF_t, total number of learning flights. Controls: ▤ untreated bees; ▢ bees treated as the test bees, but with the eye cover removed before the test; ▨ bees treated as controls ▢ , but having received 2 rewards without the training pattern just before the first test (LF_p = 1); ■ bees with the left eye painted over, but with the eye cover removed before the test

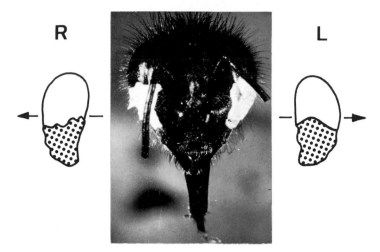

R   L

Fig.5.   Eye covers in a bee
used in the experiments of
fig.4B

Fig.6.   Scanning electron m
crograph of a bee's eye wit
the lower half painted over
Magnification 145x

same extent, compared with the tests before treatment, in all these cases, in which the test patterns had been derived from the training pattern by contrast changes in that region of the visual
field which projects to the uncovered part of the eye (compare o-values in figs. 4A and 4B).

In painting over the upper half of the eye (fig.4A, upper curve), a dorsal region had to be left free
for astromenotactic orientation. Otherwise, the bees would not appear at the training apparatus.
That dorsal region, however, cannot be decisive for the detection of stimuli within the lower visual field, as one may deduce from the experiments with the lower half of the eye painted over
(fig.4A and B, lower curves).

The question has now to be: how precise are the coordinates of the receptor system adjusted to
space-constant coordinates? Then we are able to decide between the two possibilities for a position-dependent detection of the visual patterns: (1) a constant adjustment of the receptor system
in space or (2) an evaluation of permanently varying deviations between receptor-constant and
space-constant coordinates.

By high-speed cinematography of single bees flying before the test apparatus, the inclination of
the eye or   head (frons, angle c in fig.7) as well as the inclination of the thorax (terga, angle
th in fig.7) were measured to the horizontal with the bees seen in lateral view. As one may deduce from the example presented in fig.8 the head and by this the eyes of the flying bee are kept in
nearly constant inclinations relative to the horizontal. If the deviations of the head (frontal line
in lateral view) from the zero direction are plotted in time intervals of 25 msec., directions c
and lengths r of the mean vectors are $\bar{c}(1) = 72,1°$, $r(1) = 1.00$; $\bar{c}(2) = 65,8°$, $r(2) = 1.00$; $\bar{c}(3) =$
$67,6°$, $r(3) = 1.00$; $\bar{c}(4) = 67,5°$, $r(4) = 0.99$; $\bar{c}(5) = 58,4°$, $r(5) = 0.99$ for 5 bees (n = 50 for each
individual). Therefore alternative (1) is proved to be true: the receptor coordinates are stabilized
with respect to a space-constant system.

F1 LD Z 59-62

Δt ≈ 20 msec

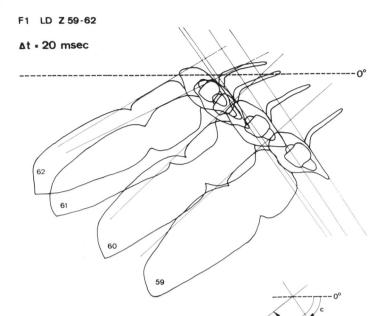

Fig.7. Measurements of the angular position of head (c; frons) and thorax (th; terga) by use of high-speed cinematography. Δt, time interval between succeeding positions (59–62) of the flying bee

d) Up to this point, one fundamental aspect of pattern recognition was taken into account: the capacity of computing the spatial position of the high-intensity and the low-intensity areas of a black and-white pattern disk. If the bees, however, are trained to a black stripe specifically inclined on a white screen, they are able to detect that direction in competition to the 90° counter-direction even under reversed contrast conditions, although the white stripe on the black screen, deviating by 90° from the training direction, leads to larger cross-correlation coefficients. One may assume at first view that the direction of the stripe is computed by the direction of its contours irrespective of the sign of contrast changes along an axis perpendicular to the stripe. That, of course, cann

F3   L₀ S5   Z 1-33

Δt ≈ 25 msec

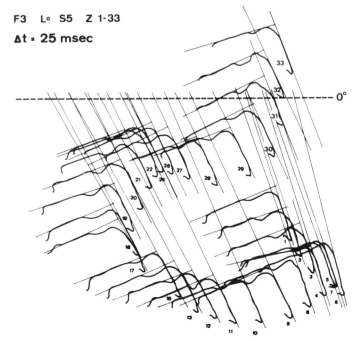

Fig.8. A 0.8 sec sequence of head and thorax positions of a flying bee in a pattern discrimination test. For explanations of angles see fig.7

be the only parameter used by the bees, for the reaction frequency to the trained direction as well as the reaction intensity (binary choices per time unit) are markedly lower under reversed contrast conditions (fig. 9). Therefore the hypothesis may be advanced that the black stripe is first processed – and stored – by its spatial distribution of black and white areas (see section 2a), whereas the contrast–invariant information about the direction of the stripe corresponds to a further step in data processing. In this context, the question arises, how information is reduced in the visual system of insects by generalization processes.

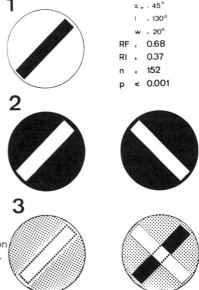

$$\alpha_+ = 45°$$
$$l = 130°$$
$$w = 20°$$
$$RF = 0.68$$
$$RI = 0.37$$
$$n = 152$$
$$p < 0.001$$

Fig.9.   Contrast–invariant detection of stripe inclination. Parameters of stripe geometry: α_+, training inclination; l, length; w, width; RF, reaction frequency to α_+; RI, reaction intensity under reversed contrast conditions; 1,training pattern; 2, test patterns; 3, positive and negative contrast changes in the test patterns (compare fig.2)

Acknowledgements.   The experimental work was financially supported by grant No. 3.315.70 of the Fonds National Suisse de la Recherche Scientifique.

References

BISHOP, L.G.: The spectral sensitivity of motion–detector units recorded in the optic lobes of the honeybee (Apis mellifera). J. comp. Physiol. 70, 374-381 (1970).
BISHOP, L.G., KEEHN, D.G., Mc CANN, D.G.: Motion detection by interneurons of optic lobes and brain of the flies Calliphora phaenicia and Musca domestica. J. Neurophysiol. 31, 509-525 (1968).
BRAITENBERG, V.: Ordnung und Orientierung der Elemente im Sehsystem der Fliege. Kybernetik 7, 235-242 (1970).
EHEIM, W.P., WEHNER, R.: Die Sehfelder der zentralen Ommatidien in den Appositionsaugen von Apis mellifica und Cataglyphis bicolor (Apidae, Formicidae; Hymenoptera). Kybernetik, in press.
GOETZ, K.G.: Optomotorische Untersuchung des visuellen Systems einiger Augenmutanten der Fruchtfliege Drosophila. Kybernetik 2, 77-92 (1964).
GRUESSER, O.J., GRUESSER–CORNEHLS, U.: Die Neurophysiologie visuell gesteuerter Verhaltensweisen bei Anuren. Verh. dtsch. zool. Ges. 64, 201-218 (1970).
JANDER, R.: Die Detektortheorie optischer Auslösemechanismen von Insekten. Z. Tierpsychol. 21, 302-307 (1964).
JANDER, R., SCHWEDER, M.: Ueber das Formunterscheidungsvermögen der Schmeissfliege, Calliphora erythrocephala. Z. vergl. Physiol. 72, 186-196 (1971).

JANDER, R., VOLK-HEINRICHS, I.: Das strauch-spezifische visuelle Perceptor-System der Stab-heuschrecke, Caraussius morosus. Z. vergl. Physiol. 70, 425-447 (1970).

KAISER, W., BISHOP, L.G.: Directionally selective motion detecting units in the optic lobe of the honeybee. Z. vergl. Physiol. 67, 403-413 (1970).

LABHART, T., WEHNER, R.: Die Unterschiedsempfindlichkeit für Lichtintensitäten bei Apis melli-fera. Rev. Suisse Zool., in press.

LAUGHLIN, S.B., HORRIDGE, G.A.: Angular sensitivity of retinula cells of dark-adapted worker bees. J. comp. Physiol. 74, 329-335 (1971).

LETTVIN, J.Y., MATURANA, H.R., PITTS, W.H., Mc CULLOCH, W.S.: What the frog's eye tells the frog's brain. Proc. Inst. Radio Engr. 47, 1940-1951 (1959).

MAZOCHIN-PORSHNYAKOV, G.A.: Die Fähigkeit der Bienen, visuelle Reize zu generalisieren. Z. vergl. Physiol. 65, 15-28 (1969).

Mc CANN, G.D., DILL, J.C.: Fundamental properties of intensity, form and motion perception in the visual nervous systems of Calliphora phaenicia and Musca domestica. J. gen. Physiol. 53, 385-413 (1969).

MIMURA, K.: Movement discrimination by the visual system of flies. J. comp. Physiol. 73, 105-138 (1971).

PORTILLO, J. DEL: Beziehung zwischen den Oeffnungswinkeln der Ommatidien Krümmung und Gestalt der Insektenaugen und ihrer funktionellen Aufgaben. Z. vergl. Physiol. 23. 100-145 (1936).

SCHNETTER, B.: Visuelle Formunterscheidung der Honigbiene im Bereich von Vier- und Sechs-strahlsternen. Z. vergl. Physiol. 59, 90-109 (1968).

VARELA, F.G., WIITANEN, W.: The optics of the compound eye of the honeybee, Apis mellifera. J. gen. Physiol. 55, 336-358 (1970).

WEHNER, R.: Pattern recognition in bees. Nature 215, 1244-1248 (1967).

WEHNER, R.: Der Mechanismus der optischen Winkelmessung bei der Biene, Apis mellifera. Verh. dtsch. Zool. Ges., Zool. Anz. Suppl. 33, 586-592 (1969).

WEHNER, R.: Dorsoventral asymmetry in the visual field of the bee, Apis mellifera. J. comp. Physiol. 77, 256-277 (1972).

WEHNER, R., EHEIM, W.P., HERRLING, P.L.: Die Rastereigenschaften des Komplexauges von Ca-taglyphis bicolor (Formicidae, Hymenoptera). Rev. Suisse Zoll. 78, 722-737 (1971).

WIERSMA, C.A.G., YAMAGUCHI, T.: The neuronal components of the optic nerve of the cray-fish as studied by single unit analysis. J. comp. Neurol. 128, 333-358 (1966).

2. Experiments on Pattern Discrimination in Honey Bees

B. Schnetter
Department of Zoology, University of Würzburg, Germany

Abstract.   Honey bees can discriminate between different n-painted stars (fig. 1) offered on a
horizontal plane. If 2 of the 4 parameters are constant within one series of shapes, the results
can be described by at least one of these different parameters uniquely (figs. 3, 4, 5). If only
one of these values remains constant, one simple parameter does not describe the results (fig. 6a)
as expected after earlier experiments.

During the past 50 years pattern recognition in bees has been the subject of a fairly large amount
of research. For many years the abundance of contour length in a pattern was regarded as the
decisive form parameter. But even as early as 1933 experiments carried out by HERTZ showed
that black and white patterns are distinguished not only by the number of their contours but
also by their "characteristic contour shapes". However, it was not then possible to describe this
quantitatively by means of a geometric parameter. In recent years WEHNER (1966) has been able
to show that bees can distinguish differently inclined stripes offered on a vertical screen, and
he succeeded in isolating the position of the black areas as the relevant parameter for direction
information (WEHNER, 1969).

The search for the relevant parameters for pattern recognition in bees is also the principal
question for the following series of experiments. For these experiments I chose a system of radially symmetrical shapes (SCHNETTER, 1968) , most of them being n-pointed stars (fig. 1). When
these shapes are used their position in regard to the flight approach direction of the bees does
not have to be considered. In fig. 1, starting from a 4-pointed star, 4 examples of possible series
of shapes are drawn, and in each of these series 2 of the geometric parameters (F, $\propto$, n, d) are
constant; 3 of these parameters will suffice to describe each shape clearly.

The experiments presented here are based on the following methodic plan: step by step variation
of the different parameters in all reasonable dimensions leads to quantitative stimulus-reaction
relationships which may be expressed as mathematical functions. The question of the biological
mechanism involved, in particular the problem of the initial optical stimulus, will not be discussed here.

A single marked bee flies through the open window into the laboratory and finds on the slowly
revolving table 6 - 8 identical pairs of shapes (exposed photographic paper AGFA GEVAERT P 90).
As the bee's reward on the "positive" training shapes is merely 2 - 3 μl sugar-water, she must in
the course of one visit - both during the learning and the testing phases - fly on to these shapes
20 - 30 times before she can return to the hive with her crop filled. The bees are free to choose
both the direction and height of their flying approach. The number of approaches and the length
of sucking time are recorded by a polygraph.

In order to standardize the course of the experiments as much as possible and to keep all the
parameters constant, except the one being tested, it is necessary to control the height of the
flight approach, even if only indirectly. The discrimination M is, amongst other things, also a
function of the approach height of the bees. The bigger the shape, the higher the bee usually
flies (with optimal training results) between 2 approaches. If, however, she stays below a minimum
height depending on the size of the shapes - for most of them this was about 50 mm - she will not
learn to distinguish optimally between two shapes even after several hour's training. Therefore

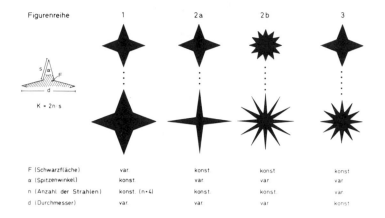

Fig. 1. 4 series of shapes in the experiment described, in each of which 2 of the 4 parameters (F, $\propto$, n, d) remain constant while the others are varied

the only bees to be further trained were those which distinguished particular pairs of test shapes with the certainty expected from previous series of experiments.

Fig. 2 shows such a pair of shapes. In a reciprocal training experiment the learning curves of the bees reach the plateau after about 30 visits, a constant level of M = 0,92, independent of whether the bees are rewarded on the 4-pointed or on the 24-pointed star. Significant differences appear in the rate of rise of the learning curves. The ordinate value M gives the reaction of the bees in such a way that value 1 is equal to "100% positive distinction" and value 0 "no distinction" (SCHNETTER, 1968). Moreover, all the training experiments were able to be carried out reciprocally, and with very few exceptions the same discrimination resulted.

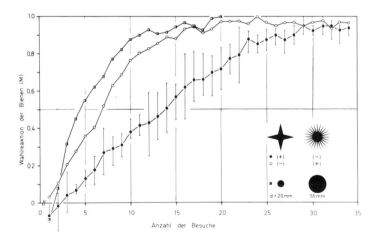

Fig. 2. Learning curves: ● mean values and standard deviation of the learning curves 4-pointed star (rewarded training shape +) and 24-pointed star; o reciprocal training

Experiment group 1: geometrically similar shapes (F, n = constant). Fig. 3 shows how well bees can distinguish between shapes of different sizes. As abscissa appears the relative difference of the parameter diameter (d) of the shapes. The ordinate is the choice frequency of the bees (M). The greater this relative difference, the better the shapes are distinguished, independent of whether the bees were trained on circular discs or on 4- or 8-pointed stars. Does that mean that the question of the relevant parameter is solved, at least for this kind of shape? Certainly not, for in the graphical presentation of the results it is immaterial what is chosen as parameter, whether diameter (d), plane (F), contour length (K), or even a plane function F^x, as described by WEHNER (1969) and CRUSE (1972). (cf. fig. 2 B). The clear arrangement of the parameter value remains in each case. (Insert shape in fig. 3).

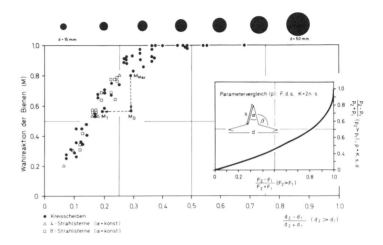

Fig. 3. Training experiments with geometrically similar shapes. △ 4-pointed stars (fig. 1, row 1); ●circular discs (7 of the 10 shapes are shown above); ▢ 8-pointed stars. M_{max} = choice frequency of the bees in a flight approach height of more than 15 cm, M_C less than 15 cm. M_T corresponds to M_D, after transformation of the parameter; shape measurements in the sight-angle linear coordinate system (WEHNER, 1969)

If, as has already been said, the certainty of distinction should be a function of the flight approach height, it should be possible to prove this clearly in geometrically similar shapes. The discrimination is M_{max} = 0.85, if in experiments using the absolutely largest pair of shapes (fig. 3, d > 75 mm) the values are taken at first only of the flight approaches from a height of more than 15 cm above the experimental table. But if the approaches under 15 cm are included, this value drops to M_D = 0.55 and is thus significantly less than the rest of the comparable values. But M_D moves towards M_T if the following is taken into account: as the bee flies over these large shapes at a relatively low level, it is not the actual shape measurements which are considered for the parameter calculation, but only that pattern which appears on the ommatidia screen at the point selected by the bee. Both shapes are therefore so transformed that the sight angle under which they appear to the bee's eye may be read linearly on them.

The results of the experiment groups 2a and 2b are seen in fig. 4, in which the choice frequency of the bees is shown against the relative difference of the angle $\propto$. The results of a series of 8-pointed stars are also drawn here. It will be seen that the discrimination M can be described as function of the parameter $\propto$, independent of n for all n-pointed stars (n = 4...12). If on the other hand the parameter contour length, or the plane function F^x mentioned previously, is chosen as abscissa, the result is a collection of measurement curves in which the different slopes may be represented as function of n. In this way these results confirm earlier tests on star shapes (SCHNETTER, 1968).

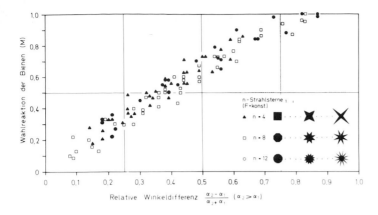

Fig. 4. Training experiments to the shape series 2a: ▲ 4-pointed stars and 2b: ☐ 8-pointed stars; ○ 12-pointed stars

How does the variation of n within a series of shapes affect the reaction of the bees? An answer to this is given in experiment group 3. As can be seen, here too the discrimination may be described as function of the parameter α - with equal success, however, also as function of the contour length (not shown in fig. 5). Both measurement curves in figs. 4 and 5 are significantly different in their initial slope. When the angle difference is relatively equal, pairs of shapes with varying numbers of points n but with constant diameter d are better distinguished than pairs of shapes with constant n but variable d.

Without considering to which parameter this difference may actually be attributed, it must be expected that this series of n-pointed stars will not be adequate for answering the question posed above. Further series of experiments should be included, in which α and n may be varied independently of each other: an experiment group with 4-pointed against 12-pointed stars with pairs of shapes in which one of the pair belongs to the series 2a (fig. 1) and the other to 2b.

In fig. 6a again the choice frequency of the bees is shown against the relative difference of the acute angle α, but only those pairs of shapes are considered in which the angles $\alpha \lessgtr 90°$ and thus may appear in both series of shapes: e.g. the shape pairs 4-pointed star: dodecagon ($\alpha = 150°$) are missing.

As can be seen, discrimination is at a minimum in the area of similar angles ($\alpha_4 = \alpha_{12}$) and increases from here symmetrically to the Y-axis on both sides. It can further be seen that the parameter of relative angle difference is not sufficient to describe the behaviour of the bees accurately. Thus the certainty of distinction for α_4, $\alpha_{12} = 20°$ is M = 0.70 and is therefore significantly higher than M = 0.45 for α_4, $\alpha_{12} = 90°$. These values may be confirmed in further series of experiments (n variable, α constant). Other series of experiments show that when the relative angle difference is equal the bees are less able to differentiate between 4- and 6-pointed stars than between 4- and 8-pointed ones, and these again less than between the 4- and 12-pointed ones discussed in fig. 6.

To sum up, it may be said that if in a series of experiments 2 of the 4 parameters (F, α, d, n) are constant, the results may be clearly described by at least one of the variable parameters. But if only one geometric size is constant, then a single parameter is no longer sufficient, although this was expected on the basis of earlier experiments. The next step in these investigations will be to find a formula which will include a description of all experiments known up to date. But this evaluation will be discussed elsewhere.

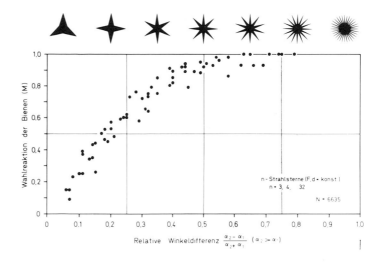

Fig. 5. Training experiments of shape series 3 (fig. 1). 7 of the 12 shapes are shown

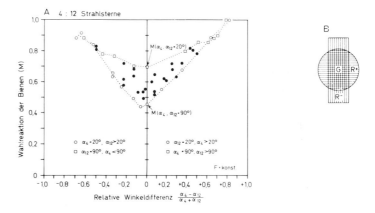

Fig. 6. A. Training experiments 4-pointed and 12-pointed stars; α_4 and α_{12} angles of the 4- and 12-pointed stars. B: calculation of F^x. The shapes are so placed one upon the other that the common area G is at its maximum. From G and the remainder areas R^+ and R^- F^x may be calculated (cf. CRUSE, 1972, and WEHNER, 1969)

References

CRUSE, H.: Zum Formensehen der Honigbiene. Dissertation, Stuttgart (1972).

HERTZ, M.: Ueber figurale Intensitäten und Qualitäten in der optischen Wahrnehmung der Biene. Biol. Zbl. 53, 10 - 40 (1933).

JANDER, R.: Ueber die Ethometrie von Schlüsselreizen, die Theorie der telotaktischen Wahl-handlung und das Potenzprinzip der terminalen Cumulation bei Arthropoden. Z. vergl. Physiol. 59, 319 - 356 (1968).

SCHNETTER, B.: Visuelle Formunterscheidung der Honigbiene im Bereich von Vier- und Sechs-strahlsternen. Z. vergl. Physiol. 59, 90 - 109 (1968).

WEHNER, R.: Der Mechanismus der optischen Winkelmessung bei der Biene (Apis mellifica). Zool. Anz., Suppl.-Bd. 33, 586 - 592 (1969).

WEHNER, R. and LINDAUER, M.: Zur Physiologie des Formensehens bei der Honigbiene. I. Winkelunterscheidung an vertikal orientierten Streifenmustern. Z. vergl. Physiol. 52, 290 - 324 (1966).

3. A Qualitative Model for Pattern Discrimination in the Honey Bee

H. Cruse
Department of Biology, University of Trier-Kaiserslautern, Germany

Abstract. Experimental results on the pattern discrimination of the honey bee can be described by a difference function, defined for every pair of shapes. This function consists of two terms. The first term is an application of a two-dimensional cross-correlation, the second one is a comparison of the contour length of the shapes and is a qualitative measure for the number of alternating stimuli on the ommatidial elements produced by movement of the shapes.

When an image of a figure is formed by the optic apparatus of an eye on the receptor layer, this two-dimensional image will be translated by the receptors into a three-dimensional intensity-locus function. Of these three coordinates, one plane is given by the arrangement of the receptors, the third coordinate by the amplitude of the excitation of the different receptors, evoked by the intensity of the different points of the image. This intensity-locus function ("I-L function") of a shape is either immediately processed, e.g. stored, or the information is reduced in great measure by generalisation of a few specific parameters of this I-L function. Accordingly, in pattern recognition, either the perceived I-L function is compared with a stored I-L function or the measured parameters with the values of stored parameters.

One hypothesis often found in literature on the nature of this processing assumes that two shapes are compared by computing a two-dimensional cross-correlation, with the coefficient of this correlation giving the difference between the two shapes. The value of this coefficient, irrespective of the norming factors, is equal to the value of the largest common volume of both I-L functions. In the discrimination of shapes of the same contrast, with I-L functions of the same amplitude, the correlation coefficient corresponds to the value of the largest common area G (fig.1). The exclusive application of a cross-correlation coefficient would mean, however, that parts of shapes which do not overlap this common area G have no influence on the capacity of discrimination. Thus, for example, all shapes which totally cover a comparative shape would not be distinguished from this shape, because the common area being considered always corresponds to the one which the comparative shape would show when covered by itself. That this is not true, however, concerning the pattern discrimination of the honey bee is shown by the results of many experiments (CRUSE, 1972, SCHNETTER, preceding contribution, WEHNER, 1969). The hypothesis should therefore be extended by considering, besides the correlation coefficient, other qualities of the correlation function of the two I-L functions. As a first approximation, we shall attempt to include in the description of the difference between two shapes, the non-overlapping areas R^+ and R^- of the positive and the negative shapes (fig.1). One could use $U = f(R^+, R^-, G)$ as a function describing this difference. In addition to this, WOLF (1935) and other authors could show that, at least in spontaneous-choice experiments, the number of stimuli which a shape generates on the compound eye when the bee flies over it is the decisive measure. Although it is not known how these stimuli are measured, the length of contour of a shape seems to be a qualitative measure. Therefore a second term of the form $f(K^+, K^-)$ should be added to the difference function (K^+ and K^- are the lengths of contours of the positive and negative shape). Because the spontaneous tendency (WOLF, 1935) as well as the I-L function depend on the value of the contrast A the latter should appear in both terms: $U = C_1 f_1 (R^+, R^-, G, A^+, A^-) + C_2 f_2 (K^+, K^-, A^+, A^-)$, where C_1 and C_2 are weighting factors.

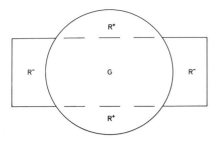

Fig.1. The rewarded (positive) shape (circle) and the unrewarded (negative) shape (rectangle) are laid on each other in such a way, that the common area G reaches a maximum. Then the non-overlapping areas of the positive shape R^+, and the negative shape, R^-, can be determined

In order to be able to formulate this hypothesis more exactly, and at the same time to prove it, several forms of difference functions have been examined using data taken from the literature and our own experiments, initially with no respect to the contrast. A good fit was found between the difference function

$$U = \left| C_1 \frac{R^+ + R^-}{G} (G + R^+) + C_2 (\log K^+ - \log K^-) \right|$$ and the experimental data. This is shown first

with results of SCHNETTER (1968) for discrimination of rectangles of different inclination and area. WEHNER (1968, 1969) (fig.3.) can be described by this difference function. Since the shapes tested in the experiments of WEHNER were shown to the bees on a perpendicular wall, the bees do not discriminate them rotation-invariantly, as they do shapes shown on a horizontal plane. This has to be taken into account in determination of the areas R^+, R^- and G. In the same way results of own experiments with different shapes, e.g. stars,checkerboards and concentric annular rings (CRUSE 1972), can be described by this difference function (fig.4.). I shall refer to some deviations as well as the meaning of the weighting factors later. Because this function is asymmetrical corresponding with the experimental results, one cannot define a metric in the pattern discrimination as HELVERSON (this volume) does with colour discrimination.

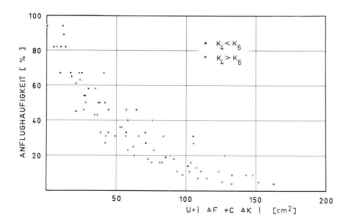

Fig.2. Results of SCHNETTER (1968). The frequency of choices (Anflughäufigkeit) on the negative shape is plotted against the value of the difference function U computed for every pair of shapes

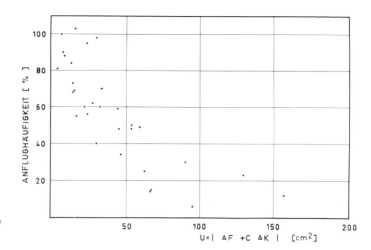

Fig.3. Results of WEHNER
(1968, 1969). Coordinates see
fig. 2

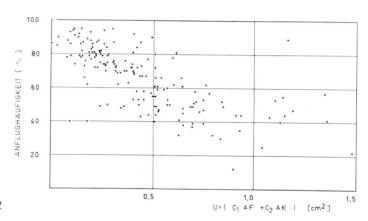

Fig. 4. Results of CRUSE
(1972). Coordinates see fig. 2

Fig. 5. Schematic cross-section through the superimposed
intensity-locus functions of
the positiveshape (full lines)
and the negative shape (dashed lines)

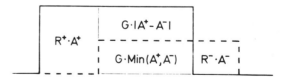

If one is to apply the difference function on the discrimination of shapes of different contrast, one has to take into account, as stated above, the three-dimensional nature of the I-L functions. As shown by a schematic cross-section through two superimposed I-L functions of different contrast A^+ and A^- (fig. 5), the value of the common space is $G \cdot Min(A^+, A^-)$, where A^+ and A^- are values of the contrast of the positive and the negative shape. After standardisation, this value corresponds with the correlation coefficient $G \cdot A^+ \cdot A^-$. So the difference function could be extended for shapes of different contrast in the following way:

$$U = C_1 \frac{R^+A^+ + R^-A^- + G A^+ - A^-}{G \; Min \, (A^+, A^-)} \quad A^+ (G + R^+) + C_2(A^+ log \, K^+ - A^- log \, K^-)$$

In fact experimental results could be described with this difference function, both when the shapes had the same contrast, which was changed in different presentations (fig.6), and when the shapes had different contrast themselves (fig. 7) (CRUSE, 1968).

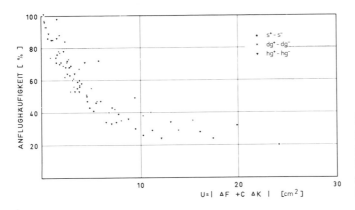

Fig. 6. Results of CRUSE (1968). Positive and negative shapes have the same contrast. Coordinate see fig. 2. Circles: black shapes; crosses : dark grey shapes; triangles: light grey shapes

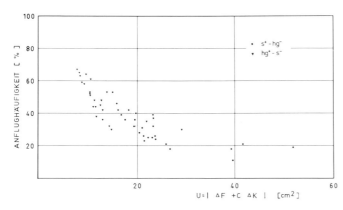

Fig. 7. Results of CRUSE (1968). Positive and negative shapes have different contrast. Coordinates see fig. 2. Circles: positive shape light grey, negative shapes black; triangles: positive shape black, negative shapes light grey

A comparison of figures 2 and 7 with figure 4 shows that the standard deviation in figure 4 is much higher. Since the method in all experiments was practically the same - several different shapes presented at the same time on a horizontal plane - the only difference between the two series of experiments was that in figs. 6 and 7 only sixpointed stars, and in fig. 4 very different kinds of shapes were used. So one could assume, that the kind of shape influences the choice of form parameters. In fact, assuming that the bees are actually applying the discussed difference function, it can be shown that the weighting factors C_1 and C_2 are chosen differently in different experiments (CRUSE, 1972). Probably, however, with respect to the results of MAZOCHIN-PORSHNYAKOV (1969), besides the difference function the bees apply other parameters according to the problem to be solved. How difficult the proof of the use of such problem-orientated parameters is, can be shown by the results of WEHNER (1971). He measured the discrimination of pairs of stripes of different inclination and different (within pairs, however, equal) length. A pair of very short stripes of different inclination can hardly be discriminated much better. WEHNER concluded, therefore that during the experiment something has changed in the bees' interpreting apparatus. These results, however, can also be qualitatively described by the difference function. Table 1 shows the comparison between the computed and the measured discriminations (for simplicity, I put C_2 = 0). On the one hand, the quantitative deviations with the short stripes could be explained by a lower intensity of choices. On the other hand, results of SCHNETTER (preceding contribution) can only be described after changing the difference function:

$$f_1\ (R^+, R^-, G)\ \text{to}\ f_1'\ (R^+, R^-, G) = \frac{R^+ + R^-}{G}$$

With this somewhat different function the results of WEHNER (1971) can be described quantitatively (table 1).

Table 1. The discrimination of stripes of equal width (50°), different length (shown in angular grades) and different inclination ($+ 45^\circ$ and $- 45^\circ$, shown by the sign) from a positive stripe of the length 130° and the inclination $+ 45^\circ$ (WEHNER, 1971). A comparison of measured and computed discriminations.

neg. shape	discrimination (%)		
	WEHNER (1971)	$\frac{R^+ + R^-}{G}$	F^+ $\frac{R^+ + R^-}{G}$
110 +	100	100	100
110 -	4	6	5
90+	100	100	100
90 -	8	10	9
70+	100	100	100
70 -	51,5	34	47

From these results one can formulate the hypothesis that one part of the pattern discrimination of the honey bee could be reduced to a two-dimensional cross-correlation of the two shapes to be compared. Hereby not only the coefficient, but also other qualities of the correlation function will be used by the bees. Another part of the pattern discrimination could be a comparison of the number of alternating stimuli in the ommatidial elements produced by the movement of the shapes relative to the flying bee.

References

CRUSE, H.: Zum Einfluss der Abhebung auf das Formensehen der Honigbiene. Staatsexamensarbeit, Freiburg i. Brsg. (1968).
CRUSE, H.: Versuch einer quantitativen Beschreibung des Formensehens der Honigbiene. Kybernetik (in Vorbereitung) (1972).
MAZOCHIN-PORSHNYAKOV, G.A.: Die Fähigkeit der Bienen, visuelle Reize zu generalisieren. Z. vergl. Physiol. 65, 15 - 28 (1969).
SCHNETTER, B.: Visuelle Formenunterscheidung der Honigbiene im Bereich von Vier- und Sechsstrahlsternen. Z. vergl. Physiol. 59, 90 - 109 (1968).
WEHNER, R.: Die Bedeutung der Streifenbreite für die optische Winkelmessung der Biene (Apis mellifica). Z. vergl. Physiol. 58, 322 - 343 (1968).
WEHNER, R.: Der Mechanismus der optischen Winkelmessung bei der Biene (Apis mellifica). Zool. Anz., Suppl. Bd. 33, 586 - 592 (1969).

WEHNER, R.: The Generalisation of Directional Visual Stimuli in the Honey Bee, Apis mellifera. J. Insect Physiol., 17, 1579 - 1591 (1971).

WOLF, E.: Der Einfluss von intermittierender Reizung auf die optische Reaktion von Insekten. Naturwiss. 23, 369 - 371 (1935).

4. The Ability of Honey Bees to Generalise Visual Stimuli

A. Anderson
Department of Zoology, University of Edinburgh, Great Britain

Abstract.   No evidence is found that the bee can form a concept of 'triangularity' as suggested by MAZOCHIN-PORSHNYAKOV. The amount of contour per unit area of a shape appears to be of overwhelming importance in the ability of the bee to make discriminations, as well as in its spontaneous preference for certain shapes. Bees can be made to utilize other parameters by trai-ning along a decreasing continuum of contour per unit area. Thus there may be attentional me-chanisms in bees similar to those found in vertebrates.

1. Introduction

In recent years several studies have suggested that insects may possess learning capabilities not inferior to those seen in vertebrates. Perhaps the most striking of these investigations is the work of MAZOCHIN-PORSHNYAKOV (1969a, b) on concept formation in honey bees. He shows that bees can learn to discriminate a triangle from a square and then generalise this discrimination to other triangles of different size, orientation, colour, background colour and form of outline. This level of performance is not reached by many mammals; rats do not generalise from an upright triangle to one rotated through ninety degrees (HEBB, 1958).

Early investigations into the form vision of bees were unsuccessful in attempts to train bees to discriminate a triangle from a square (HERTZ, 1929). However, MAZOCHIN-PORSHNYAKOV attributes this difference to the training shapes used (MAZOCHIN-PORSHNYAKOV, 1969c). He uses multiple figures containing many small sub-units; a triangle is built up from a set of small triangles and a square from a set of small squares (see fig. 5).

Unfortunately, MAZOCHIN-PORSHNYAKOV misses out one very important control from his ex-periments. He fails to show that the bee can discriminate the different shapes which it appears to generalise between. In this context, generalisation implies that the animals see the set of shapes as being different but at the same time in some respect similar or belonging to the same class of · objects. It may be that the honey bee has no concept of 'triangularity' but is responding to some common factor in the triangles, and cannot tell them apart.

The classical papers on form perception in bees showed that the figural intensity of shapes was the most important parameter in enabling them to make discriminations (HERTZ, 1929, 1930, 1931; ZERRAHN, 1933; WOLF, 1934, 1937). Figural intensity is a measure of the amount of contour per unit area of a figure. Highly divided shapes have high figural intensities, whereas plain figures have low figural intensities. It was shown that bees could discriminate divided figures on the basis of this parameter and that they had a strong spontaneous preference for shapes containing high amounts of contour per unit area. As the shapes used by MAZOCHIN-PORSHNYAKOV were multiple ones, they were of high figural intensity. It is possible that the bee performed the discriminations on the basis of this parameter alone.

The research reported in this paper seeks to investigate the ability of the honey bee to discrimi-nate triangles and squares and to generalise to other triangles and squares. Both simple and mul-tiple figures are used and particular attention is paid to ensuring correct controls.

2. Materials and methods

a) Apparatus.   The bees (Apis mellifica) were trained to fly in through window of a laboratory to receive reward of 2 molar sucrose. The training figures are shown in fig. 5. They were on rectangular cards 10.5 cm wide and 12.5 cm high. Each figure was cut out of black cardboard and backed with yellow fluorescent paper so that it appeared very bright against a black background. The shapes were presented either in the horizontal plane laid flat on a table or in the vertical plane on a special apparatus. This consisted of a board with a row of holes drilled in it. Each hole leads into a separate perspex chamber at the back of the board. The stimulus cards could be attached to the board by means of a "Velcro" strip. Bees were trained to fly up to the stimuli, alight on them and then crawl through holes cut in their centres, through the board and into the chambers behind. Here they found either sugar solution (reward) or water. In experiments with stimuli in the horizontal plane the sugar reward was placed in a small transparent dish above the stimulus card.

b) Training and testing procedures.   During discrimination training the two figures were displayed side by side and only the triangle rewarded. The positions of the two cards are frequently interchanged. On a test neither of the figures is rewarded and the number of approaches made to each is measured for a five minute period. Approach is defined as the bee alighting on the shape or flying up to it so close that its legs or antennae touch it or come within approximately one millimetre of doing so.

3. Results

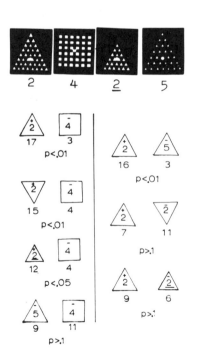

Fig.1. A series of discriminations. The shape number is shown inside the figures. The numbers beneath the figures show the number of bees that approached each shape and the p value on a chi–squared test

a) A group of 20 marked bees were trained to discriminate triangle 2 from square 4 and then tested on a series of generalisations (fig.1, column 1). In this series of generalisation tests only the initial choice, not the total number of approaches, of each of the 20 bees was measured.

The bees were given 30 minutes training between each generalisation test. They were found to significantly prefer the triangle, even when its orientation and size were changed, but not when a triangle of different sub-unit size was put in its place (shape 5). It was then shown that the bee could not discriminate any of the figures it had apparently generalised to, although it did discriminate the two triangles of different sub-unit size (fig. 1, column 2). The bees were trained for two days before testing in an attempt to make them learn to discriminate these shapes. Thus, although the results of MAZOCHIN-PORSHNYAKOV are verified, an additional control shows the bee has not formed a concept of 'triangularity'.

b) The ability of the bee to discriminate triangles and squares was investigated further. Many pairs of figures of differing stimulus dimensions were used.

Firstly, for each pair of figures the spontaneous preference of the bee was recorded. Twenty bees visited each pair of shapes, both figures in each pair being rewarded. After one hour the number of approaches made to each shape was measured for a five-minute test period. The set of shapes could be divided into two groups on this basis; in one group the bees had definite preferences for one or other of the shapes in each pair while in the other group bees were found to choose randomly between the two shapes of each pair.

Secondly, the triangle of each pair was reinforced and twenty bees allowed to visit each pair of figures. After one hour a five-minute approach test was made with neither figure rewarded. If the period of differential reinforcement had had no effect upon the choice of the bee, then training was continued for up to 8 days.

Fig. 2. In this set of shapes the bees could discriminate any of the triangles from any of the squares. They could not, however, generalise this discrimination from any triangle to any other in the set

All the triangles and squares seen in fig. 2 could be discriminated from one another. This was also the group in which there was always a pre-training preference for one or other of the figures in each pair. Once a discrimination had been established, the effect of replacing the triangle with another from the set was tried. There was never any generalisation to any of the other figures. In the discriminations shown in fig. 3 the bees had no spontaneous preferences and could not be trained to tell the figures apart. Thus, the effect of differential reinforcement appears to be only to modify pre-existing preferences, either it enhance or diminish them.

c) As mentioned in the introduction, the spontaneous preference of bees is held to be dependent upon their figural intensities. This was tested by allowing 60 bees to visit the shapes seen in fig. 5. All the figures were rewarded and after one hour the number of approaches made to each shape was recorded for 90 minutes. The number of visits made to each figure was found to be closely related to figural intensity.

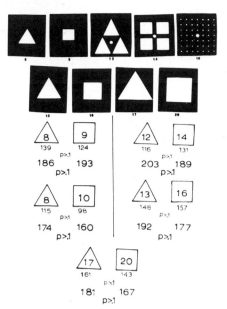

123-post training

Fig. 3. Bees were trained on each pair of shapes. The small figures show the number of approaches to each shape in a spontaneous preference test before training. The large figures show the bees' choices on an unrewarded test after 8 days training. P values from the chi-square test

d) As the parameter of figural intensity seems to determine the bees' spontaneous preference and its ability to discriminate shapes, it was decided to train along a continuum of decreasing figural intensity to see if the bee could be trained to ignore this parameter and use others. A group of 20 bees were trained over the series of discriminations seen in fig. 6. As training continues they begin to be able to discriminate a pair of test figures. Eight days direct training on this same pair had previously proved unsuccessful.

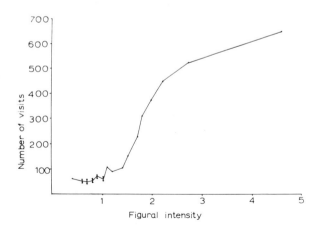

Fig. 4. The spontaneous preference of the bees is measured by allowing them to visit the set of shapes shown in fig. 5. All the stimuli are rewarded and the frequency of visits to each shape is recorded. Here the number of visits to each figure is plotted against their figural intensities

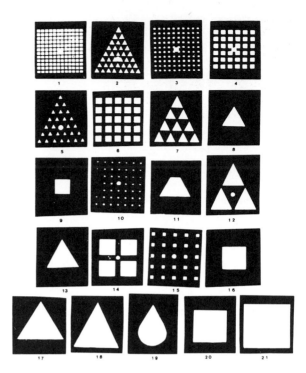

Fig. 5. The complete set of training figures

e) An attempt was now made to find out what parameters the bees were using to make this discrimination. Three groups of twenty bees were trained along the continuum and then given a
series of generalisation tests as shown in fig. 6. The first eight tests are from all three groups.
The last seven tests were performed with only one group of twenty bees. The bees do not generalise to triangles of different orientation but do to triangles of different size, even though these
figures were undiscriminated in experiment two. If the base of the triangle is altered, the discrimination persists, but if the top angle is cut off, then the bee no longer generalises to it.

4. Conclusions

The discrimination of shape appears to be largely dependent upon the parameter of figural intensity. No evidence was found which suggested that the bee could form a concept of 'triangularity'. By training along a continuum it is possible to shift the attention of the bee to other
parameters of shape than figural intensity. Attentional mechanisms of a similar nature have recently been demonstrated in the discrimination of stripes by honey bees (WEHNER, 1971) although
previously mechanisms of this sort had been thought to be restricted to vertebrates.

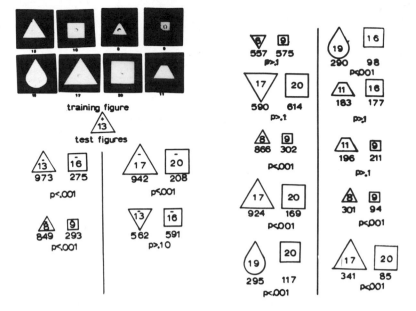

Fig. 6. A series of unrewarded discrimination tests. The bees receive 30 minutes training on the triangle 13 between each test. The figures show the number of approaches to each shape and the p value from the chi-square test

References

HEBB, D.O.: Alice in Wonderland or psychology among the biological sciences. In Biological and Biochemical Bases of Behaviour, ed. HARLOW, H.F. and WOOLSEY, C.N. pp 451 - 467. University of Wisconsin Press (1958).

HERTZ, M.: Die Organisation des optischen Feldes bei der Biene. Part 1: Z. vergl. Physiol. 8, 693 - 748 (1929). Part 2: Z. vergl. Physiol. 11, 107 (1930). Part 3: Z. vergl. Physiol. 14, 629 (1931).

MAZOCHIN-PORSHNYAKOV, G.A.: Die Fähigkeit der Bienen, visuelle Reize zu generalisieren. Z. vergl. Physiol. 65, 15 - 28 (1969b).

MAZOCHIN-PORSHNYAKOV, G.A.: Generalisation of visual stimuli as an example of solution of abstract problems by bees. Zool. Jb. 48, 1125 - 1136 (1969a).

MAZOCHIN-PORSHNYAKOV, G.A.: Insect vision, pub. Plenum Press, New York (1969c).

WEHNER, R.: The generalisation of directional visual stimuli in the honey bee, Apis mellifera. J. insect. Physiol. 17, 1579 - 1591 (1971).

WOLF, E.: Das Verhalten der Bienen gegenüber flimmernden Feldern und bewegten Objekten. Z. vergl. Physiol. 20, 151 (1934).

WOLF, E.: Flicker and the reaction of bees to flowers. J. Gen. Physiol. 20, 511 (1937).

ZERRAHN, G.: Formdressur und Formunterscheidung bei der Honigbiene. Z. vergl. Physiol. 20, 117 - 150 (1934).

5. First Steps in a Behavioral Analysis of Pattern Discrimination in Diptera

W. Reichardt
Max-Planck-Institute of Biological Cybernetics, Tübingen, Germany

Abstract.   Visual fixation of objects is studied in Musca domestica using the closed-loop method. The fixation process is shown to be due to relative movements of patterns, i.e. vertically arranged black stripes, with regard to the fly's compound eyes. The cause of the stripe transport in the closed-loop apparatus is the symmetric torque fluctuation and the asymmetric induced torque response to progressing and regressing motion. Only the lower parts of the compound eyes contribute to the fixation process.

During the past years investigations have been undertaken in our laboratory with the aim to determine and to analyze spontaneous preferences of flying Diptera for elementary patterns in their optical environment, REICHARDT and WENKING (1969), REICHARDT (1970), REICHARDT (1971).

1. Experimental conditions

During the experiments a test fly is suspended by a torque-compensating device and mounted on the axis of a cylindrical panorama. The output voltage of the compensator, which is proportional to the instant flight torque signal generated by the test fly during fixed flight, is coupled to a fast servomotor whose shaft carries a ringpotentiometer and a panorama. The ringpotentiometer serves as an indicator of the angular position ψ of the cylindrical panorama during the tests. The panorama is homogeneously illuminated and carries on its inner surface the individual test patterns. Under these operational conditions a torque signal D, generated by a test fly, results in an angular velocity $d\psi/dt$ of the panorama.

Consequently we are simulating only those retinal displacements of pattern images which result from a rotation of the animal in free flight. Moreover, this limiting case is the better fulfilled in free flight the more distant the optical environment is positioned from the fly, since the amount of every retinal displacement due to translatory relative motions is   inversely proportional to that distance.

2. Results

When a single black stripe is mounted vertically on the panorama and the fly's torque signal is coupled to the servomotor drive system, the panorama is rotated until the stripe has reached the position $\psi = 0$, which is defined by the fly's direction of flight. The fly fixates the object or, as we may say, the object reaches an equilibrium state at the position $\psi = 0$. This specific behaviour is observed also with individual grey stripes in a large contrast range if the average brightness of the panorama is kept at a high level. Conversely, this also holds for the case of contrast inversion, namely when the stripe is brighter than its surrounding.

If the pattern consists of two identical, vertically oriented stripes separated by an angle $\Delta\psi$, the test flies fixate an imaginary center line between the two stripes if $\Delta\psi$ amounts to less than 40 degrees. Also under these conditions there is one equilibrium position for the two stripe configuration. The test flies behave as if the two stripes form a single pattern. The behaviour of the flies is quite different for separation angles $\triangle \psi > 40$ degrees. They are able to fixate two imaginary lines which are positioned in the neighborhood of one or the other stripe. These are fixated with equal probability. If $\triangle \psi_+$ describes the angular separation between the two fixation positions then $\triangle \psi_+$ is always smaller than $\triangle \psi$, except for the case that $\triangle \psi$ amounts to 180 degrees where $\triangle \psi_>$ equals $\triangle \psi_>$. Under the conditions $\triangle \psi > 40$ degrees there are two equilibrium states for the configuration; that is to say, the test flies behave as if the two-stripe pattern has broken into two parts.

An interesting finding is that only the lower parts of the compound eyes – below the equatorial line – contribute to the fixation of objects or patterns whereas the upper parts – above the equatorial line – do not.

Except for the latter finding, the results described so far are based on, and can be described in terms of, the following elementary observations: (1) The test flies generate torque signals irrespective of their optical environment if the environment does not move with respect to the fly's compound eyes. The histogram of this torque signal is nearly gaussian. (2) If the object of pattern exposed to a test fly is stationary but illuminated with flickering light, no significant changes, either in the center of gravity or in the shape of the torque histogram is observed. From (1) and (2) it follows that the fixation of objects or patterns must be due to relative movements of these configurations with regard to the fly's compound eyes. (3) This can be shown when a single stripe panorama is moved clock- and counterclockwise around a position ψ with a small swing amplitude of $\pm \varepsilon$ degrees. Under these experimental conditions the test flies try to turn towards the direction of the position ψ, indicating that the progressing stripe motion (from front to back with regard to one of the compound eyes) elicits a stronger induced torque signal than the regressing stripe motion (from back to front).

From to the findings summarized in (1) and (3) it follows that the cause of the stripe transport is the symmetric torque fluctuation and the asymmetric induced torque response to progressing and regressing motion. The stripe reaches an equilibrium state at the fixation position $\psi = O$ as this position is defined by the symmetry line between the two compound eyes. This is the point where, when the stripe passes across it, regressive movement turns into a progressive one.

The resultant torque signal responsible for the stripe transport into the position of fixation depends strongly on ψ. The amount increases approximately linearly with increasing ψ, reaching extrema for $\psi = \pm 20$ degrees and falling off to zero at $\psi = \pm 180$ degrees.

The ψ dependence of the fly's resultant torque signal is in accordance with the findings reported in connection with the two-stripe experiments if one takes into consideration that the two eyes of the fly with their associated parts of the visual system operate independently with respect to the motion perception. For stripe separations $\triangle \psi < 40$ degrees, the ψ dependence of the resultant torque leads towards a stable fixation of the center line between the two stripes since the two stripes are positioned between the two extrema at $\psi = \psi \pm 20$ degrees, whereas for separation angles $\triangle \psi > 40$ degrees this position is a labile one as the two stripes are positioned beyond the extrema. Under the latter condition two stable positions exist when one of the two stripes is positioned between $\psi = \pm 20$ degrees and the other beyond the extrema of the resultant torque characteristics. The breaking up of a pattern is therefore a consequence of a stability criterion caused by the ψ dependence of the fly's resulting torque signal.

The experimental results of our investigations suggest that there exists a many-to-one corresponden between pattern and the ψ-dependent torque characteristics which determine the fixation behavio of the flies and their abilities to discriminate patterns.

The account presented here is a summary version of an article to be published in "Die Natur-wissenschaften" in summer 1972.

References

REICHARDT, W., WENKING, H.: Optical detection and fixation of objects by fixed flying flies. Die Naturwiss. 56, 424 – 425 (1969).
REICHARDT, W.: The insect eye as a model for analysis of uptake, transduction, and processing of optical data in the nervous system. The Neurosciences, Second Study Program, F.O. Schmitt Editor. The Rockefeller University Press, New York (1970).
REICHARDT, W.: Visual detection and fixation of objects by fixed flying flies. Pattern recognition in biological and technical systems. Springer-Verlag Berlin, Heidelberg, New York (1971).

6. Intraaxonal Visual Responses from Visual Cells and Second-order Neurons of an Insect Retina

F. Zettler and M. Jaervilehto
Department of Zoology, University of Munich, Germany

Abstract.   In the Calliphora eye axonal potentials are recorded from R1–6 and R8 cells as well as from the monopolar neurons L1–2. Additionally, the cells studied electrophysiologically and the recording sites are identified by intracellular injection of Procion–Yellow and subsequent histological treatment. Potentials from both types of cells (from retina and lamina) propagate along the axon with no measurable decrement. The temporal transducing properties of the synapse R1–6 / L1–2 are studied by subtracting the presynaptic R1–6 potentials from the postsynaptic L1–2 potentials, both evoked by sinusoidally modulated lights of various frequencies and different modulation degrees. As the angular sensitivity curve of L1–2 extends beyond a narrower angular range than the one of R1–6, evidence for lateral inhibitory interactions is presented in the lamina.

Due to morphological criteria there can be distinguished two types of visual cells in the insect retina. The cells of the one type, the long visual cells, project with a long axonal process to the 2nd optic ganglion, the medulla. The cells of the other type have only a short axon, which terminates in the 1st optic ganglion, the lamina ganglionaris (CAJAL and SANCHEZ, 1915).

The visual information of the short–type cell is transmitted to the medulla via interneurons. Interneurons, which connect the lamina with the medulla in the dipteran eye, are the monopolar neurons L1 to L5 and several types of centrifugal fibres (TRUJILLO–CENOZ, 1965; BRAITEN-BERG, 1967; STRAUSFELD, 1971).

The following results, as far as concerning interneurons, only refer to a certain type of mono-polar neurons, the L1–2 neurons. Every cartridge in the lamina has one L1 and one L2 neuron, and both of these neurons are synaptically contacted by all of the 6 short–type visual cells R1 to R6 of a cartridge (TRUJILLO–CENOZ, 1965; BOSCHEK, 1971).

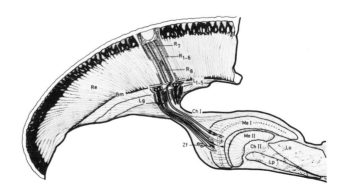

Fig. 1. The retina of Calliphora erythrocephala (longitudinal section). The lateral extent of the detailed ommatidium and neuroommatidium (cartridge) is exaggerated. Bm, base-

ment membrane; Ch, Chiasma; Lg, lamina ganglionaris; L1-5,monopolar neurons of a cartridge; Lo, lobula; Lp, lobula plate; Me, medulla; Re, retina externa; R1-6, short visual cells; R7 and R8, long visual cells; Zf, centrifugal cells

To record intraaxonal potentials from visual cells and monopolar neurons glass capillaries of no more than 0.1μm tip diameter and filled with a 5% Procion-Yellow solution were used. The stimulus consisted of a punctiform light source which could be moved equidistantly around the fly's head. With the exception of measuring angular sensitivities, all potential recordings were carried out at the maximal effective position of the light source.

1. Axonal potentials of the two types of visual cells

Within the axons of both visual cell types not nerve impulses but depolarizing light-induced potentials were found. Form and amplitude of these potentials were affected by the stimulus intensity (fig. 2). At low intensities the R8 potentials showed a considerable noise which is characteristic for intracellular potentials, recorded from visual cell somas, irrespective of their type (SCHOLES, 1969; ZETTLER and JAERVILEHTO, 1970). The axonal R1-6 potentials, on the contrary, were smoothed. From these findings it must be concluded that the membrane properties which accomplish the conduction of potentials along the axons are considerably different for the two different cell types.

The light sensitivity of both visual cell types showed no remarkable differences; especially the threshold sensitivity of R8 was as high as that of R1-6.

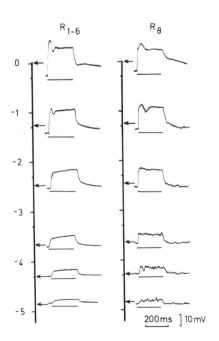

Fig. 2. Light-evoked axonal potentials from visual cells. The recording site of R1-6 potentials was 50 μm and that of R8 potentials 150 μm distant from the basement membrane. Various stimulus intensities are indicated by the vertical logarithmic scale (JAERVILEHTO, 1971)

2. Identification of cell type and recording site

Those cells studied electrophysiologically were identified by intracellular injection of Procion-Yellow in connection with a subsequent histological treatment. The identification of the recording site was accomplished by extracellular dye accumulation, by tissue deformation caused by the electrode, or by a direct histological identification of the electrode tip.

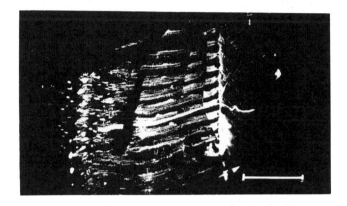

Fig. 3. Visual cell of type R1–6 (JAERVILEHTO and ZETTLER, 1971). Scale 100 μ m

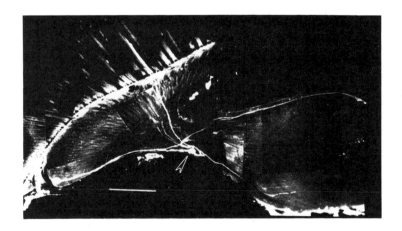

Fig. 4. Visual cell of type R8 and monopolar cells of type L2 and L3, stained by successive injection of Procion-Yellow. Scale 100 μ m

3. Axonal potentials of the monopolar neurons L1-2

Like the axons of the visual cells, the axons of monopolar neurons L1 or L2 did not exhibit nerve impulses, but rather hyperpolarizing potentials whose form and amplitude could be varied by the light stimulus intensity (AUTRUM, ZETTLER and HAERVILEHTO, 1970; JAERVILEHTO and ZETTLER 1971).

These potentials (fig. 5) propagate along the axon with no measurable decrement. In the case of passive spread, one should expect a considerable damping, especially of the very fast on-effect (up to 30 mV/ms at highest intensities). Therefore, we think that the axonal propagation of the postsynaptic L1-2 potentials is not established by passive spread, but rather by an active mechanism of the axonal membrane (ZETTLER and JAERVILEHTO, 1971).

It is near at hand that this somewhat unusual coding and axonal propagation of nervous information provides the L1-2 axons with a high channel capacity (fig. 6a).

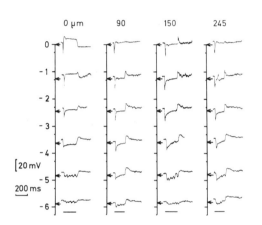

Fig. 5. Light-evoked axonal potentials from L1-2 neurons, recorded at different distances from the synapse. Various stimulus intensities are indicated by the vertical logarithmic scale. The absolute intensity values are the same as the corresponding ones in fig. 2

4. Temporal synaptic information processing

By stimulating with sinusoidally modulated light of various frequencies, the transduction of light intensity to presynaptic R1-6 potentials and the transduction of light intensity to postsynaptic L1-2 potentials were measured. By subtracting the first curve from the second we got the temporal transducing properties of the synapse R1-6 / L1-2 at a certain adaptation level (fig. 6b). This level was at about -4 of the relative intensity scale of figs. 2 and 5.

The temporal transducing property of this synapse is found to be qualitatively the same as that for the synapse visual cell - excentric cell in the Limulus eye (KNIGHT, TOYODA and DODGE, 1970). It results in a frequency-dependent gain of 3 times at low frequencies and 8 times at high frequencies (dashed curve of fig. 6b).

5. Spatial synaptic information processing

The angular sensitivity of R1-6 potentials and of L1-2 potentials was measured under the same stimulus conditions. By translating the potential values of a certain cell via the characteristic curve of this cell into intensity values, one gets the angular dependence of the effective light intensity (WASHIZU, BURKHARDT and STRECK, 1964) for both cell types (fig. 7). These curves should be identical for the two types, if we assume synaptical inputs of L1-2 neurons coming only from visual cells R1-6 of a cartridge (ZETTLER and JAERVILEHTO, 1972).

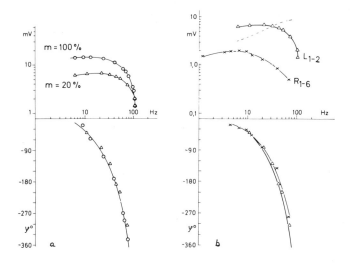

Fig. 6. a. Amplitude and phase-lag of L1-2 potentials versus frequency of a sinusoidally modulated light of different modulation degrees m. b. Amplitude and phase-lag of pre- and postsynaptic components versus frequency of a sinusoidally modulated light. Modulation degree and ground stimulus were the same for both cells. The dashed line is the gain-frequency curve of the synapse

Obviously the light intensity is effective to L1-2 neurons in a considerably narrower angular range than to visual cells R1-6 (fig. 7). This means that L1-2 neurons must have additional synaptic inputs from some lateral components which are inhibitory in function.

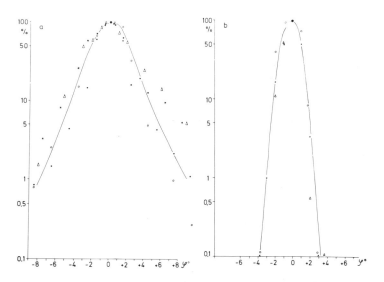

Fig. 7. Visual fields of the presynaptic R1-6 component (a) and the postsynaptic L1-2 component (b). The effective light intensity (ordinate) was defined to be 100% at the maximum for every level of the individual cell

References

AUTRUM, H., ZETTLER, F., JAERVILEHTO, M.: Postsynaptic potentials from a single monopolar neuron of the ganglion opticum I of the blowfly Calliphora. Z. verg. Physiol. 70, 414 – 424 (1970).

BOSCHEK, C.B.: On the fine structure of the peripheral retina and lamina ganglionaris of the fly Musca domestica. Z. Zellforsch. 118, 369 – 409 (1971).

BRAITENBERG, V.: Patterns of projection in the visual system of the fly. I. Retina–lamina–projections. Exp. Brain Res. 3, 271 – 298 (1967).

CAJAL, S.R., SANCHEZ, D.: Contribution al conocimiento de los centros nerviosos de los insectos. Trab. Lab. Invest. Biol. (Madrid) 13, 1 – 168 (1915).

JAERVILEHTO, M.: Lokalisierte intrazelluläre Ableitungen aus den Axonen der 8. Sehzelle der Fliege Calliphora erythrocephala. Diss. Universität München, 1971.

JAERVILEHTO, M., ZETTLER, F.: Localized intracellular potentials from pre- and postsynaptic components in the external plexiform layer of an insect retina. Z. vergl. Physiol. 75, 422 – 440 (1971).

KNIGHT, B.W., TOYODA, J., DODGE, F.A.: A quantitative description of the dynamics of excitation and inhibition in the eye of Limulus. J. gen. Physiol. 56, 421 – 437 (1970).

SCHOLES, J.: The electrical responses of the retinal receptors and the lamina in the visual system of the fly Musca. Kybernetik 6, 149 – 162 (1969).

STRAUSFELD, N. J.: Organization of the insect visual system (light microscopy). I. Projections and arrangements of neurons in the lamina ganglionaris of Diptera. Z. Zellforsch. 121, 377 – 441 (1971).

TRUJILLO-CENOZ, O.: Some aspects of the structural organization of the intermediate retina of Dipterans. J. Ultrastruct. Res. 13, 1 – 33 (1965).

WASHIZU, Y., BURKHARDT, D., STRECK, P.: Visual field of single retinula cells and interommatidial inclination in the compound eye of the blowfly Calliphora erythrocephala. Z. vergl. Physiol. 48, 413 – 428 (1964).

ZETTLER, F., JAERVILEHTO, M.: Histologische Lokalisation der Ableitelektrode. Belichtungspotentiale aus Retina und Lamina bei Calliphora. Z. vergl. Physiol. 68, 202 – 210 (1970).

ZETTLER, F., JAERVILEHTO, M.: Decrement–free conduction of graded potentials along the axon of a monopolar neuron. Z. vergl. Physiol. 75, 402 – 421 (1971).

ZETTLER, F., JAERVILEHTO, M.: Lateral inhibition in an insect eye. Z. vergl. Physiol. 76, 233 – 244 (1972).

7. Ethometrical Investigations into the Spatial Interaction within the Visual System of Velia caprai (Hemiptera, Heteroptera)

H. W. Meyer
Department of Zoology, University of Frankfurt/M., Germany

Abstract.   In spontaneous binary choices between visual prey dummies conclusions are made from the relative frequencies of choice (= stimulus values) to the spatial interaction within the visual system of the water bug Velia caprai. As the zero point of the frequency distribution is 5.0 probit (= 50% reaction), deviations to higher or lower indicate excitation or inhibitation resulting from stimulus variation. When two discs are stimulating simultaneously with a constant horizontal angular distance of 10.3^0, lateral inhibition arises in relation to their size. With increasing horizontal interstimulus distance, stimulus value changes periodically. Four or five inhibitory maxima alternating with ineffective ranges could be recorded. Power and spread of the lateral inhibition depend on stimulus intensity, interstimulus distance and spatial sensitivity distribution in the eye.

1. Introduction and testing method

One of the main functions of the visual system is the detection of significant objects in the environment. On the way from simultaneous scanning at receptor level to central pattern classification, visual data reduction and extraction of invariants are important processes in the afferent system. Since GRANIT (1962) lateral inhibition is expected to be general property of neuronal networks. In the visual system it is a well-known principle to diminish redundant information and thus at the same time to increase the contrast, as shown in many investigations into Limulus and vertebrates. In insects, too, lateral inhibition has been proved at various stages of processing using the electrophysiological method (PALKA, 1967; ZETTLER and JAERVILEHTO, 1972) and has frequently been demonstrated in behavioural reaction to visual patterns (JANDER, 1964; MEYER, 1969, 1971a).

The response of a particular element within a sensory network is proportional to the sum of all excitatory and inhibitory influences of the periphery acting upon it. By using small dots as testing stimulus , characteristic of single perceptor units, the lateral influences may be expected to be of little account. This provides a method for investigating the transfer properties of single visual channels by measuring their input–output relationship. The elementary case of interaction between parallel exciting currents can be studied by simultaneous stimulation of two centers of perception.

In the water bug Velia caprai we found a very suitable insect for proving information transmission within an intact biological system. Velia possesses a release mechanism for the capture of its prey which reacts with high sensitivity and selectivity to spatial and temporal qualities of visual stimulus pattern if accompanied simultaneously by attractive vibration from the same direction. An experimental set-up was developed that allows the testing of all relevant figure parameters while vibration remains constant (fig. 1). It consists of two coupled vibrating needles above which small discs, as prey dummies, can be fixed at one or at both sides. In spontaneous binary choices, as the needles produce similar wave patterns and furthermore exert no visual stimulus, the relative frequency of telotactic turns to the varied figure (=stimulus value) is measured in comparison with a standard figure or with the pure vibration. The obtained values in percentage choice frequencies are then transformed into probit. This procedure is proved to lineari e the ordinate scale over a large range (JANDER, 1968; MEYER, 1971a).

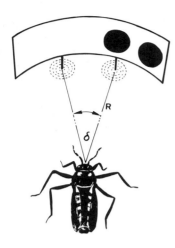

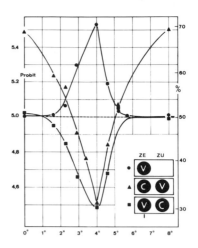

Fig. 1. Experimental set-up for binary
choices between prey dummies. Stimulus
distance R, optimal angle of decision
and vertical angular distance between
the disc centers and the horizon = const

Fig. 2. Relationship between the relative
stimulus values of black discs (ordinate) and
their diameter in angular degrees (abszissa)
when compared with the pure vibration. a:
size of the central figure varied; b: size of
the added figure varied while the central figure remains constant (= 4.0° $\emptyset$); c: reversed
program of b

By using this method we found interacting structures with a low degree of complexity in the visual
system of Velia called Central Flickering Detectors, CFD. They consist of a central unit that
is inhibited in the case of simultaneous stimulation by at least 4 surrounding units. CFD's can arise
locally in the frontal part of the eye when a visual unit becomes activated by vibration in coincidence with its main directional sensitivity. The perceptor raster has the same spatial properties
as the ommatidial raster.

In this approach the threshold distance of measurable horizontal inhibitory transfer is studied as a
function of the local excitation level in the eye.

2. Experimental data

Firstly the size of a black disc before a white background is increased stepwise (Fig. 2a). When
compared with the pure vibration the zero point of the frequency of choice distribution is 5.0
probit. Deviations to higher or lower values indicate excitatory or inhibitory effects caused only
by the figure. The stimulus value becomes highest when the optical mark is presented close to the
horizon above the needle. With regard to former experiments and considerations, a disc in this most
effective position (= central figure) projects into the maximum of the Gaussian directional characteristic of one activated visual unit. The curve obtained is an optimum distribution. Three valu-
ating dispositions can be distinguished; with increasing disc diameter an exciting range above the
measuring threshold up to an optimum is followed by an inhibitory range, that ends at an ineffective range. From this angle onwards stimulus value becomes independent from size.

Simultaneous stimulation by means of a second disc of constant size (= additional figure) falling into a laterally neighbouring perception center reverses the sign of evaluation without changing the absolute values (= hierarchy of effectiveness) (fig. 2b). The same result is obtained when the additional figure is varied in size while the other is kept constant (fig. 2c). Significant differences in choice frequency for comparable pairs of discs, however, indicate that the power of lateral inhibition is not only a function of size, but also of space.

Systematic testing of all combinations of two sizes between 1.7° and 8.0° Ø shows a maximum of interaction with black discs of 4.0° Ø (fig. 3). This is better illustrated by a model built up with the measured sensitivity curves (fig. 4). It makes clear that under test conditions reciprocal inhibition is limited between 2.5° and 4.6° Ø.

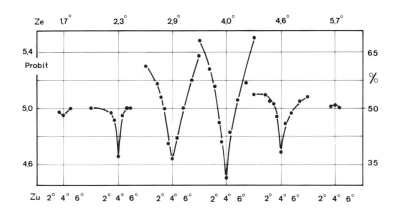

Fig. 3. The relative stimulus values (ordinate) as a size-varied second figure is added to the constant central figure (both black discs) in relation to its size (lower abscissa: diameter in degrees). Upper abscissa: the diameter of the central figure. The disc centers are seen with an angular distance of 10.3°. Standard is the vibration

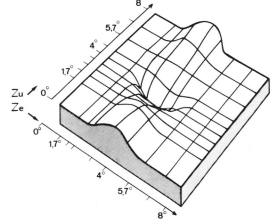

Fig. 4. Diagram showing the inhibitory interaction between two neighbouring perception centers as a function of the stimulus size. Ze, Zu = angular diameter of the central and the added figure. In this diagram the level at size of 1.7° ≤ Ze ≤ 5.7° corresponds with the zero level of the frequency of choice distribution (= 5.0 probit)

Provided that in a homogeneous raster of an infinite number of receptors each unit is inhibitorily connected with all the others, the coefficient of mutual inhibition depends exponentially on the distance between two units (REICHARDT and MAC GINITIE, 1962). All visual units participating in Velia's CFD inhibit each other (MEYER, 1971a), so that one of the assumptions is realized. On the other hand it is shown in Limulus (see RATLIFF, 1968) and vertebrates (GRUESSER et al., 1968; CREUTZFELDT et al., 1970), that the shape of a function describing inhibition with increasing interstimulus distance can be nicely fitted by a Gaussian distribution. Using the elementary simultaneous stimulation of two discs changed stepwise in position this question can be proved for Velia.

In the following experiment the horizontal angular distance between the centers of an optimal central figure (= white disc of $3,4°$ Ø) and an equal sized additional figure increases. The effect caused by the central figure is compensated by the standard figure, so that the zero level is 5.0 probit (fig

A short increase in attractiveness can be observed (for explanation see MEYER, 1971a), followed by several curve minima, indicating lateral inhibition. These inhibitory centers are arranged in surprisingly regular distances of exactly $10.3°$ and are separated from each other by zero ranges within the added figure loses its influence completely. With discs smaller or larger than the optimum, the intensity and spread of inhibition decrease, while the periodical changes in spatial sensitivity remain constant, even when the contrast is reversed (fig. 6). Optimum-sized white discs inhibit over a larger distance ($50 - 55°$) than black ones ($40 - 45°$). The curves in figs. 5, 6 represent sections through several approximately concentric inhibitory fields on the level of their centers (MEYER, unpubl.). By connecting the inhibition maxima in one of the curves we obtain the envelope of discrete, nearly Gaussian-shaped sensitivity distributions (fig. 7, upper graph). In the single-dimensional and stationary special case the following equation relates the lateral inhibition to the interstimulus distance.

$$R(i) = a \cdot e^{-bx} - C$$

$R(i)$ = Intensity of inhibition; a = coefficient influenced by contrast ($1.1 - 1.4$ probit); x = interstimulus distance (= angular distance between two perception centers); b = invariant with contrast and size, $= 0.35$; C = pattern constant.

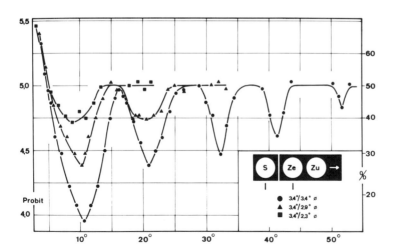

Fig. 5. Correlation between the stimulus values (ordinate) and the horizontal angular distance between the centers of two white discs (abscissa). Central figure (Ze) = $3.4°$Ø; additional figure (Zu) = $3.4°$; $2.9°$; $2.3°$Ø. The dotted line marks the zero level of the distribution

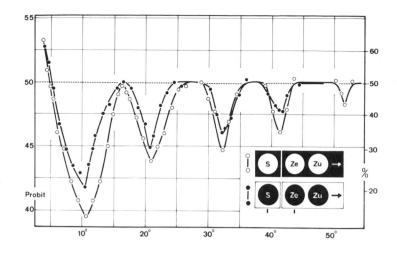

Fig. 6. Comparison of curves measured with optimal sized black (4.0°∅) or white (3.4°∅) central, additional and standard figures as in fig. 5

When additional figures smaller or larger than the optimal size are presented, the decrease in inhibition is uniform throughout the entire range, as shown by the constant shape of the curves. This also provides proof of the fact that ipsi- and contralateral inputs are compared subtractively within the CFD of Velia (MEYER, 1971a).

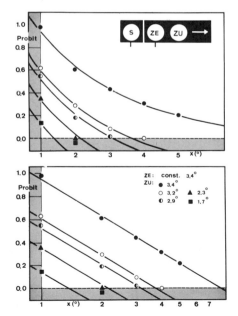

Fig. 7. Upper graph: The best approximation to a curve connecting the lowest stimulus values in fig. 5 is an exponential curve. Strength of inhibition (ordinate, in absolute probit values) in relation to the normalized angular interstimulus distance (abszissa). Lower graph: logarithmic scaling of the interstimulus distance linearize the curves. The grey stripes indicate their non-linear ranges

3. Discussion

In a previous paper it was shown that the most effective prey stimuli come from optical marks between 1.2° and a maximum of 4.6° in diameter. The fact that discs larger than the optimal size diminish the stimulus value, which finally becomes independent of size (fig. 2a), was traced to the lateral inhibition from the simultaneously stimulated periphery. The importance of these inhibitory surroundings in the CFD as regards signal transmission lies obviously in limiting the angular size on which the perceptor mechanism for prey stimuli answers with reaction. When the interstimulus distance increases stepwise, it can be concluded from the periodical changes in spatial sensitivity that the lateral inhibition extends over at least 4-5 unstimulated or weakly stimulated visual units. No investigations into the fine structure of the retina and the retina-lamina projection have as yet been made. The results described above give reason to suppose the existence of extensive lateral nervous interconnections.

Strength and spread of the lateral inhibition are dependent on the local flux rate (fig. 3, 5), a well-known fact from electrophysiological studies on the retina of vertebrates (e.g. BUETTNER et al., 1968; MAFFEI, 1968). Both become strongest when the frequency content of the scanned stimulus pattern is better matched to the system's transfer characteristics than that of other proved patterns. This seems to be the case with the optimal size (fig. 3, 4). The result that units of the examined equator-parallel row are most strongly activated by the same figure size, can be explained best by the assumption that receiving channels within a perceptorial chain are not specialized in frequency transfer, but possess the same filter characteristics.

Lateral inhibition is well known to diminish the spike frequency of neurons. If the inhibitory synapses converge before the band-pass filtering process (MEYER, 1971b), the resulting slowed impulse currents within the band width of the proved low-pass filter with an inhibitory output would block the prey-catching reaction more or less severely. As, however, the band width of exciting "high" stimulus change frequencies is very narrow as compared with the band width of inhibiting "low" frequencies, it is not understood why the size evaluation at simultaneous stimulation (fig. 2b, c) is practically inverse to that at single stimulation (fig. 2a). This finding and the other results from tests with flickering light (MEYER, unpubl.) suggest an evaluation reversal at higher level in the visual pathway.

The importance of a mechanism which is inhibited by synchronous stimulus changes in equator-parallel receptor chains of the eye is that it enables elongated and horizontally orientated contours, e.g. of an obstacle or potential enemy, to be detected and avoided.

Acknowledgement. The research was supported by a grant of the Deutsche Forschungsgemeinschaft.

References

BUETTNER, C., BUETTNER, U., GRUESSER, O.J., RACKENSPERGER, W.: Die Summation von zwei unabhängig voneinander ausgelösten Erregungen im rezeptiven Feldzentrum retinaler Neurone der Katze. Biokybernetik 2, 178 - 182 (1968).

CREUTZFELDT, O.D., SAKMANN, B., SCHAICH, H. and KORN, A.: Sensitivity distribution and spatial summation within receptive-field center of retinal on-center-ganglion cells and transfer function of the retina. J. Neurophysiol. 33, 654 - 674 (1970).

GRANIT, R.: Receptors and sensory perception. New Haven: Yale University Press (1962).

GRUESSER, O.J., VIERKANT, J. and WUTTKE, W.: Die räumliche Verteilung und die Ausbreitungsgeschwindigkeit der lateralen Hemmung in den receptiven Feldern der Katzenretina. Biokybernetik 2, 175 - 178 (1968).

JANDER, R.: Die Detektortheorie optischer Auslösemechanismen. Z. Tierpsychol. 21, 302 - 307 (1964).

JANDER, R.: Ueber die Ethometrie von Schlüsselreizen, die Theorie der telotaktischen Wahlhandlu und das Prinzip der terminalen Cumulation bei Arthropoden. Z. vergl. Physiol. 59, 319 – 356 (1968).

JANDER, R., HEINRICHS, I.: Das strauch-spezifische visuelle Perceptor-System der Stabheuschrecke (Carausis morosus). Z. vergl. Physiol. 70, 425 – 447 (1971).

MAFFEI, L.: Inhibitory and facilitory spatial interactions in retinal receptive fields. Vision Res. 8, 1187 – 1194 (1968).

MEYER, H. W.: Visuelle Auflösung des Beutefangs beim Bachwasserläufer Velia caprai. Zool. Anz. Suppl.-Bd. 33, 596 – 601 (1969).

–   Visuelle Schlüsselreize für die Auflösung der Beutefanghandlung beim Bachwasserläufer Velia caprai (Hemiptera, Heteroptera). 1. Untersuchung der räumlichen und zeitlichen Reizparameter mit formverschiedenen Attrappen. Z. vergl. Physiol. 72, 260 – 297 (1971a).

–   dto. 2. Untersuchung der Wirkung zeitlicher Reizparameter mit Flimmerlicht. Z. vergl. Physiol. 72, 298 – 342 (1971b).

PALKA, J.: An inhibitory process influencing visual responses in a fibre of the ventral nerve cord of locust. J. Insect Physiol. 13, 235 – 248 (1967).

REICHARDT, W., MAC GINITIE, G.: Zur Theorie der lateralen Inhibition. Kybernetik 1. 155 – 165 (1962).

ZETTLER, F., JAERVILEHTO, M.: Intrazelluläre Potentiale aus den Axonen von Sehzellen und sekundären Neuronen einer Insektenretina. Laterale Inhibition im ersten optischen Ganglion der Fliege. Im Druck.

8. Mechanisms of Orientation and Pattern Recognition by Jumping Spiders (Salticidae)

M. F. Land

School of Biological Sciences, University of Sussex, Brighton, Great Britain

Abstract. Jumping spiders capture flies by stalking them visually, and jumping when sufficiently close. They also distinguish prey from other jumping spiders. The first step in prey capture is for the spider to turn, using its four side eyes, to face the prey with its large movable principal eyes. It is shown that this turn can be made accurately in the absence of visual feedback by preventing the spider from turning its body, but allowing it to turn a cardboard ring instead. The angle through which the spider turns the ring is closely similar to the angle made by the stimulus and the body axis. Once a "side-eye" turn has been made the principal eyes normally observe the stimulus. These eyes have six muscles which can move the retinae up and down, side to side, and rotationally. Using anthophalmoscope techniquethe eye movements were observed while the spider watched stimuli. The eyes show spontaneous activity, tracking movements, saccades towards the stimulus, and scanning. During scanning the retinae fixate the stimulus with their central region, and move laterally back and forth across it at a frequency of 0.5 - 1 Hz. At the same time both retinae move conjugately through about $50°$ at a slower rate, 0.1 - 0.2 Hz. Previous authors had shown jumping spiders distinguish other spiders from prey by the positions and angles of the legs. It is suggested that scanning is a mechanism by which line detectors in the retinae are used to determine the inclination of contours present in a stationary stimulus.

1. Introduction

A jumping spider has two main tasks: to catch small insects and to find a mate. Unlike most other spiders, which build webs, jumping spiders rely almost entirely on vision in the performance of these tasks. Their strategy during prey capture is remarkably similar to that of a cat catching a mouse, in that they track down their prey - usually flies - in a sufficiently slow and stealthy manner to avoid being seen themselves until close enough to jump, after which the prey has no time to escape. This performance requires a visual system that is not only more acute than that of their prey, but is also competent to perform the several kinds of orientational manoeuvres needed while turning towards and then following moving targets. In addition, because jumping spiders are not themselves very different in appearance from their prey, the animals must have sufficiently well developed pattern recognising abilities to make this difficult distinction. In this paper I shall discuss first the anatomy and optics of the jumping spider's unique visual system, and the deal with some of the control systems involved in orientation, and with the mechanisms concerned with pattern recognition - especially the role of scanning movements of the retinae of the principal (AM) eyes.

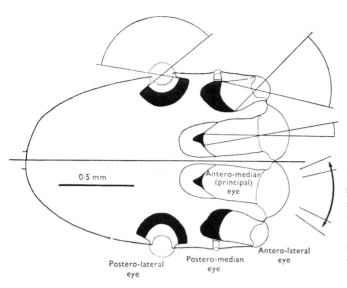

Fig.1. Diagrammatic longitudinal section of the prosoma of a jumping spider (Metaphidippus harfordi) showing the fields of view of the eyes in the horizontal plane. For the right antero-median eye the field of view is shown as it is extended by lateral retinal movements

Prey-capture and courtship in jumping spiders has been described many times (see HOMANN, 1928) for an account of the roles of the different eyes, CRANE (1949) for a discussion of species va riations, DREES (1952) for an analysis of the visual releaser mechanisms involved, or BRISTOWE (1958) for an easily-read introduction to Salticid behaviour. The events which lead to both kinds of behaviour can be summarised by the following diagram:

$$\text{Detect movement} \longrightarrow \text{Turn to stimulus} \longrightarrow \text{Run} \longrightarrow \text{Wait} \begin{cases} \nearrow \text{Dance } (\male) \\ \rightarrow \text{Watch } (\female) \longrightarrow \text{Mate or evade} \\ \searrow \text{Creep slowly} \longrightarrow \text{Jump on prey} \end{cases}$$

This sequence is wholly under visual control: contact chemoreception may play a part in the final stages of mating, but the remainder can be elicited by suitable artificial stimuli (DREES, 1952). HOMANN (1928) first demonstrated the roles of the various eyes in behaviour. He showed that blinding the side eyes (antero-lateral or AL and postero-lateral or PL - see fig.1) prevented the initial turn to a moving stimulus to the side or behind the animal, but did not prevent any of the subsequent behaviour provided that the stimulus lay within the field of vision of the principal (AM) eyes. When eye movements are taken into account, this extends to about $35°$ on either side of the body axis. Conversely, blinding the principal eyes totally prevented prey capture and courtship be haviour, but not turns towards stimuli behind and to the side. The side eyes are thus responsible for detecting moving objects and orienting the spider towards them, whereas the principal eyes are con cerned with the recognition of stimuli (which clearly occurs during the 'wait' period before the divergence of the two lines in the diagram), and with guiding the approach of the spider towards its prey or potential mate. The only evidence of interaction between the two systems is that the re tinae of the principal eyes may be directed towards stimuli seen by the AL eyes (LAND, 1969b); otherwise the two systems appear to act quite independently.

2. Anatomy and optics of the eyes

By analogy with human vision one can say that the four side eyes of the spider correspond to the peripheral retina - principally responsible for low-resolution movement detection - and the princi pal eyes to the fovea, concerned with following moving targets and using high acuity to resolve an identify stimuli. The advantage to the spider of separating 'peripheral' from 'foveal' vision is princi pally one of compactness: since high resolution is not required for the former, small eyes with short focal lengths are adequate, and the principal eyes in which high resolution is required can be made tubular instead of spherical. In fact, a spherical eye on the vertebrate pattern with the same focal length as the principal eyes would occupy most of the spider's head.

a) Anatomy.   This account is necessarily a summary: the reader is referred to LAND (1969a) for details of structure at the light microscope level and for references to the older literature, and to EAKIN and BRANDENBURGER (1971) for a detailed electron microscope study of the eyes.

The eight eyes are alike in possessing a cuticular cornea, a lens, a cellular 'vitreous' space and a retina. The main differences between them, apart from size, lie in the organization of the retina. In the lateral (AL and PL) eyes, and the small and probably vestigial PM eyes, the receptor cell bodies lie outside the eye and send their axons into the eye and across the surface of the retina. Each axon penetrates the retina, and from its sides arise a pair of highly-ordered microvillous rhabdomeres, and these remain present until the axon leaves the retina terminating in the synaptic complex of the first optic glomerulus. The receptive part of the axon, which may be as long as 200 um, is surrounded by pigment cells on all sides: the receptors can hardly fail to operate as optically isolated light-guides.

By contrast, in the principal eyes the receptor bodies lie behind the retina and the much shorter rhabdomeres are the terminations of dendrites (fig.2). A second difference is that the part of the retina containing the receptor endings is pigment-free, as in the human eye. The pigment-containing cells form a black background behind the receptive region, although these cells do send transparent processes into the region of receptor terminals to form a more or less continuous supportive feltwork.

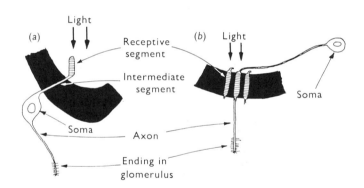

Fig.2a.   Structure of a receptor in a principal eye (AM), 2b.   in a side eye (AL or PL)

The most striking feature, however, of the principal eye retinae is that the receptors are arranged in four layers (figs. 3 and 4), one behind the other when viewed from the direction of the incoming light. The deepest two layers (1 and 2) occupy the whole of the boomerang-shaped retinal area, whereas the more superficial layers (3 and 4) occupy only the central region. Layer 4 differs from the rest in that its receptor terminals are short ovoids, oriented at right angles to the entering light: this is in contrast to the rod-shaped terminal segments of the other layers, oriented parallel to the visual axis. The distance separating layers 1 and 4 amounts to nearly 10% of the eye's focal length and this is too long for each retinal layer to receive a well-focussed image of the same plane in object space, unless the chromatic aberration of the eye is considered (see LAND, 1969a). From ophthalmoscopic measurements it has been shown that while layer 4 is too close to the lens to receive a resolved image at any wavelength, the distance separating layers 1 and 3 is approximately equal to that separating the image planes for violet and red light. In view of the spider's ability to discriminate colour (KAESTNER, 1950; CRANE, 1949) the most likely explanation for the layered structure of the retina is that the receptors of each layer contain different photopigments, and that each is situated in the focal plane for the appropriate wavelength of maximal spectral sensitivity. This situation if rather similar to the layering of fish cones (EBERLE, 1967) where there appears to be a correspondence between the different layers and the focal plances for different wavelengths

The other possible explanation for the observed layering is that the different layers receive images from different distances in object space, but while it is optically true that this does occur, it is unlikely that this is the explanation since maximum resolution is likely to be required only at relatively long distances (5 cm or more where identification usually occurs) and for an eye with a focal length as short as this (0.5 mm) all points more than about 3 cm from the eye are brought to a focus within a single receptor layer. The colour hypothesis thus seems the most likely explanation, and recently Dr. Robert DEVOE (personal communication) has strengthened this conclusion with the finding that single receptors in the retina can be shown electrophysiologically to have different spectral sensitivities: so far only two types have been found, one peaking in the violet and the other in the green region of the spectrum. This argument, however, does not hold for the rather different receptors of layer 4, which are conjugate with points behind the eye for all visible and near-visible wavelengths. It is possible that these cells analyse polarised light; this is reasonable since in each cell the microvilli are parallel and in a plane at right angles to the incident light – the condition for acting as an analyser (KIRSCHFELD, 1970). Since the cells of the other layers each have several rhabdomeres whose microvilli lie in different directions, the cells of these layers will not have this capability. Although there is no evidence for the use of polarised light by jumping spiders – none has been sought yet – it has been shown that in wolf spiders (Lycosidae) the principal eyes are responsible for polarotactic navigation (MAGNI et al., 1964, 1965).

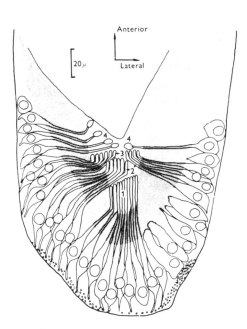

Fig.3. Horizontal section through the retina of the right antero-median eye of Metaphidippus aeneolus. The section is taken close to the centre of the retina and shows the four layers of receptor endings (1–4). Light would reach the retina from the top of the page

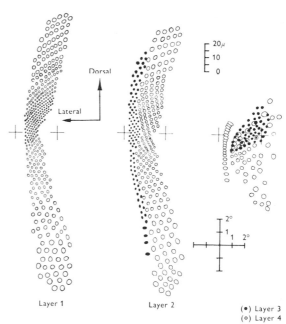

Fig.4. Complete reconstruction of the right AM retina of Metaphidippus aeneolus, showing the arrangement of receptor endings in each of the four layers. For layer 2, filled circles indicate thosereceptors whose cell bodies lie on the lateral side of the retina, open circles on the median side. The angular scale assumes a focal length of 512 um, that measured for a closely similar specimen. The numbers of receptors in each layer are: 1) 376, 2) 302, 3) 48 and 4) 68, total 794. The smallest receptor separation in this species is 11 minutes (layer 1), and 16 minutes (2). The small crosses show how the different layers superimpose

Layer 1

Layer 2

(•) Layer 3
(o) Layer 4

b) Fields of view and acuity.   Fig.1 shows the fields of view of the various eyes in the horizontal plane: the vertical extent of these fields is similar, except for the AM eyes whose fields are shown in fig.4. The fields of view were first described accurately by HOMANN (1928) from anatomical measurements, and have been confirmed by LAND (1971) using a technique in which orienting responses were observed with each eye blinded in turn. More recently direct ophthalmoscopy (see LAND, 1969b, for details of the method) has shown that the fields of the PL eyes in fact extend behind the animal, with a 1 – 2° binocular overlap: otherwise the earlier observations as shown in fig.1 were confirmed. The principal features are that the fields of view of the AL and PL eyes are virtually contiguous at approximately 42° from the body axis. The AL eyes have a field extending 12 – 15° across the midline, giving a binocular field in front of 25 – 30°.

It is interesting that the region of binocular overlap of the AL eyes is also the region of greatest acuity of the lateral eye system. The receptor separation on the body axis is 30 – 35 minutes, falling to nearly 1° in Metaphidippus, and 1.5° in Salticus (Epiblemum). In the PL eyes the receptor separation is virtually constant across the retina and is very similar to the values just given (1 – 1.5°). There are between 8000 and 16000 receptors in each PL eye (depending on the species) and 3000 to 6000 receptors in each AL eye. This contrasts with less than 1000 receptors in all four layers of each AM eye, and invites comparison with the human eye where the peripheral retina contains about 1.6×10^8 rods, and the rod-free fovea only about 3.4×10^4 cones. In both systems movement detection seems to be expensive compared to pattern recognition!

In the principal eyes the angular acuity varies from layer to layer and place to place, as shown in fig.4. It is greatest in the centre of layer 1 where the angular receptor separation is as small as 9 minutes in Phidippus johnsoni, and 11 minutes in Metaphidippus aeneolus. It is worth pointing out that such acuity, which is certainly required and used in interspecies recognition, is at least three times better than that of any insect, and better than any other invertebrate except for the cephalopod molluscs. In the body, the retinae of the principal eyes appear as thin boomerang-shaped strips of tissue in transverse sections of the head, with the convexities of the strips directed outwards: the visual fields in object space are reversed and inverted, giving a combined field for the two retinae which is cross-shaped (fig.13). For layer 1 the retina is little more than 1° wide at its narrowest point, i.e. about 6 receptors, by about 21° in vertical extent. During eye movements the two retinae may diverge by at least 10° - ophthalmoscopically the two halves of the cross (fig.13) are seen to separate - but the fields of view of the retinae have never been seen to overlap. When the eyes are at rest the inner edges of the fields of view are separated by 1-2°.

c) Eye muscles.   Movements of the principal eyes have been observed by DZIMIRSKI (1959) and LAND (1969b) and will be discussed later in the paper. Each eye can move horizontally and vertically, or a vector combination, and also rotate through up to 50° around the visual axis. Jumping spiders thus have the same 3 degrees of freedom as do human eye movements, although the muscular arrangement is quite different. The lens does not move, instead the six eye muscles move the eye tube, and hence the retina, which thereby 'scans' a stationary image. The six muscles were first described by SCHEURING (1913-14) and his description was confirmed and extended by LAND (1969b). The muscles are arranged as three antagonistic pairs (fig.5): two muscles ventral to the eye joining the eye tube about half-way between the lens and retina (1 and 2) - these pull the retina down, and laterally or medially: and a third pair (5 and 6) of muscles which partially encircle the eye, and are presumably responsible for the observed rotation movements.

Each eye muscle is a single motor unit, supplied by a single axon. The oculomotor nerve, one on each side, emerges from the rear of the brain and runs round the brain forwards to join the eye muscle from beneath the eye. The nerve contains only 6 axons, each of which enters one of the muscles just described (LAND, 1969b). This finding, which is astonishing in view of the complexity of the eye movements observed as well as the fact that the human visual system uses several thousand axons to manage the same number of degrees of freedom, has been confirmed by electron microscopy by EAKIN and BRANDENBURGER (1971).

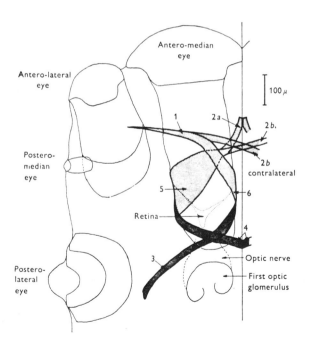

Fig.5.   Scale reconstruction of the muscles of the left antero-median eye of Metaphidippus aeneolus, as seen from above. The muscles lying dorsal to the eye are shown in darker stipple. The more ventral, anterior muscles (1 and 2) insert into the carapace of the clypeus just above and in front of the falces; the upper muscles (3 and 4) insert on the dorsal cephalothorax, laterally and medially

d) Physical limits of resolution.   It is useful when trying to determine the quality of the image on the retina of an eye to determine the physical limit to resolution imposed by the diffraction of pattern due to the size of the aperture. For a test-grating, the contrast of the image becomes zero when one cycle of the grating subtends an angle of λ/d, where λ is the wavelength of light and d the pupil diameter. Since there is no spatial information in the image for angles less than this, one would not expect the minimum receptor separation to be less than half this angle (the situation approximated in the human eye with a 2-3 mm pupil). If the receptor separation is substantially greater than half this limit, this represents either unused resolution, or poor optics, i.e. a substantial amount of geometrical aberration.

A comparison of the physical limit with the minimum receptor separation (M.R.S.) in each eye of Metaphidippus aeneolus (5 mm) is given below:

Table 1.   Limits of resolution in different eyes (Metaphidippus aeneolus).

Eye	Focal Length $-f_1$ (µm)	Diameter of Entrance Pupil (µm)	λ/d (min) ($\lambda = 0.5$ µm)	Min. Receptor Separation (min)	λ/d M.R.S.
A.M.	512	267	6.4	11 (layer 1)	0.58
A.L.	256	174	9.9	27	0.36
P.L.	174	174	9.9	59	0.17
Human	17'000	3'000 (typical)	0.57	0.45 (fovea)	1.27

The table shows that in no eye does the receptor separation come close to matching the potential optical resolution, when one would expect the figure in the last column to approach 2.0, although the A.M. eyes come closest to it. In the other eyes it is easily shown that this discrepancy is not caused by poor optics, because by ophalmoscopy through the spider's own lens it is possible to resolve details not only of the retinal mosaic, but even of the pigment granules in the cells surrounding the receptors. These stand out as bright (reflecting) spots in fig.6. Since the angular resolution of the optical system is the same when used to view the retina as it is when the retina is viewing the outside world, one is forced to conclude that the optical resolution is much better than the physiological resolution. The probable answer to this paradox is that in juvenile (second instar) spiders the situation is different in that the apertures of the lenses are 1/4 to 1/2 as large as in the adult, but that the receptors, which grow at the same rate as the rest of the eye, subtend the same angles as in the adult. Thus in juvelines the minimum receptor separation comes to within a factor of two of the physical limit, for all eyes. The adults' optics are a legacy of juvenile perfection.

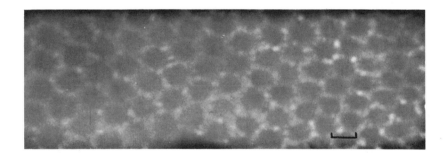

Fig.6.   Ophthalmoscope photograph of the retina of the A.M. eye of Salticus. The region seen is in the horizontal plane and centred about 17° lateral to the body axis. The scale is 1°, or about 4.5 µm on the retina. Note that whereas the receptors in this region are about 1° apart, it is possible to resolve individual pigment granules that are about 1/4° apart, which is close to the physical limit of resolution

Finally, because resolution of this order is attainable by eyes that are scarcely larger than the ocelli of many insects, it is very difficult to understand why insects went to the evolutionary trouble to develop their much larger and optically less perfect compound eyes.

3. Orientation

Jumping spiders orient towards objects in at least three situations: they will turn (1) towards moving targets to the side and behind (side eyes), (2) towards stationary targets that lie within 35° of the body axis (principal eyes) and (3) towards targets moving away from the midline, in the range $5-15^{\circ}$ from the body axis (tracking using the principal eyes). The second and third kinds of orientation have yet to be investigated in detail, and we shall concentrate here on the first kind – turns mediated by the side eyes (LAND, 1971, 1972).

The first question to be answered is whether the mechanism of orientation is based on an 'open' or 'closed' type of control system (see, for a discussion, MITTELSTAEDT, 1962). In the latter case the spider, having detected the presence of a target to one side or the other, could turn towards it until the object was seen to lie directly ahead; this would be a 'closed' control system since the animal relies on the consequences of its own motion - visual feedback - for the accurate execution of the turn. However, in the former case, 'open' control, the size of the turn made would be appropriate and unaffected by the relative movement of the target across the retinae once the turn has been initiated; the kind of instruction the retina would issue to the legs would be "turn left through x° and then stop", x being set by the initial retinal location of the stimulus.

In the case of turns mediated by the lateral eyes the system involved is an 'open' one: visual feedback is not required. This makes these turns like those of male fireflies turning towards females whose flash is complete before orientation takes place (MAST, 1912), or like human saccadic eye-movements where a saccade, once initiated, continues to completion, whatever subsequently happen to the stimulus that caused it. The evidence for this comes from the result of an experiment shown in figs. 7 and 8, where the spider was prevented from receiving visual feedback while turning.

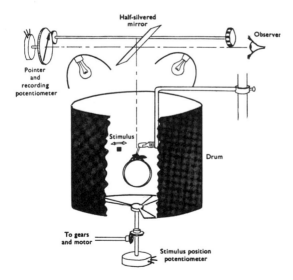

Fig.7. Apparatus for eliciting and recording turns made by jumping spiders in the absence of visual feedback. The spider is fixed by wax to a card support, and holds a light card ring which is free to turn. Turns are elicited by moving a small (5°) black stimulus, on the surface of a drum, through 5° at a speed of 25°/sec. These conditions were found to be optimal. Stimulus position, and the angle through which the spider turns its ring, are both measured using 360° potentiometers

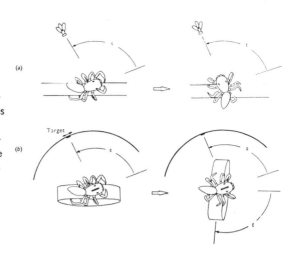

Fig.8. Diagram showing the general result of the experiment outlined in fig.7. In the 'real world' (a) the spider turns to a fly initially situated at an angle s from the spider's body axis, and the spider responds with a turn of angular magnitude t°. In the experimental situation (b), where the spider is free to turn its substrate but not move relative to its surroundings, it still makes a turn of angle t°, this time by turning the ring. In both cases t is approximally equal to s (see fig.9)

When, in the apparatus shown in fig.7, the spider turns its ring, it does not change its position relative to the stimulus. Fig.9 shows that in spite of this absence of visual feedback the turns that the animal makes are such as would have been appropriate and accurate in the 'real-world closed loop' situation of fig.8a. Perfect accuracy would be indicated on such a plot by a straight line through the origin with a slope of unity; most points lie very close to this line. Some, however, do not: a consistent feature of several hundred experiments performed using this technique was that a proportion of the turns made were small, and although still in the appropriate direction were insufficiently large to have brought the spider (in a closed-loop situation) round to face the stimulus. Unrestrained spiders show such turns also, and it is concluded that they represent an alternative mode of response: the spider either makes a turn appropriate to the stimulus angle (a complete turn), or one of 10 – 20° independent of the stimulus angle (partial turn).

At first sight, partial turns would seem to be pointless, since unlike complete turns they will not help the spider to see its prey with the principal eyes – the prerequisite for further behaviour. However, this is probably not so because of the way the spider habituates to repeatedly presented stimuli. If a stimulus is moved back and forth around the same point in space, and therefore around the same retinal cells, the probability of a response occurring diminishes very rapidly, so that after a few movements one can guarantee that no turn will be produced. However, if the stimulus is moved 5° or more to a new region of the retina, the probability of a response occurring is undiminished (LAND, 1971). The function of these small movements by the spider may therefore be simply to move the stimulus to a new patch of retina, unstimulated and therefore not habituated.

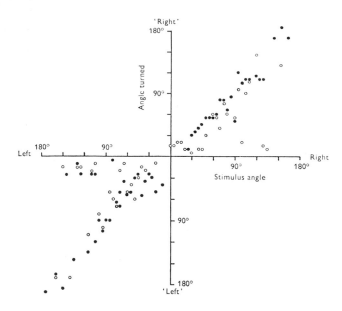

Fig.9. Plot, from a single fairly typical experiment, of the angle the spider turns its ring against the angle of the stimulus with respect to the body axis. The stimulus angle is taken as the position of the leading edge of the 5° black square at the midpoint of each 5° movement. Closed circles (●) are for stimuli moving right to left around the animal, open circles (o) left to right

Not all stimuli are equally effective in causing turns. The minimum requirements are that a stimulus (a black or white spot) must subtend at least 0.75° at the retina and move through an angle of at least 1° at a speed between 1°/sec. and 100°/sec. These figures can be translated into retinal dimensions when one recalls that receptors are situated about 1° apart in the PL eyes and in all but the axial region of the AL eyes (compare fig.6). To be effective, a stimulus must move from one receptor to the next in a period of time not less than 10 msec. nor more than 1 sec. To this can be added the result of an unpublished experiment using long (90°) black and white stripes in which the question was asked: are turns directed to the leading or trailing edge of the stripe? With black stripes the leading edge acted as the stimulus, and with light stripes the trailing edge, and there were no exceptions. One can thus add to the above conditions that the stimulus must always move in such a way as to cause sequential darkening of adjacent receptors. For stimuli much larger than 1°, or moving through a greater distance than 1°, there was little, if any increase in the probability of a turn occurring. Summation at the level of the sub-retinal movement detectors thus seems to be minimal, and the 'unit of movement detection' is probably two adjacent receptors.

Stimuli presented at different locations in the visual surroundings are not equally effective in eliciting turns, even though the optimal stimulus conditions appear to be the same throughout the lateral eye fields. Fig.10 shows that the probability of a turn occurring is least immediately in front of and behind the animal, and maximal, sometimes approaching unity, in the region 30 – 90 from the axis, i.e. just outside the field of view of the principal eyes. It is interesting that the same animal will, on different occasions, give probability curves of the same shape, but different overall magnitude (open and closed squares). The probability of a seen stimulus evoking a response thus depends on a number of factors: the retinal position of the stimulus, long-term factors of unknown nature which affect the response propability everywhere, and habituation, previously mentioned, which is position specific.

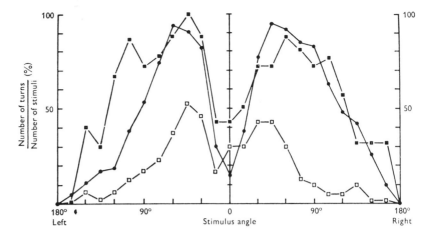

Fig.10.   Effect of stimulus position on the probability of a turn behind made. Equal numbers of stimuli were presented to each 15° region of the horizontal field of view. Filled
circles are for one female Metaphidippus harfordi and squares for one female M.aeneolus
examined on separate occasions – open squares 2 months later than the closed squares

Perhaps the most interesting problem which arises from the discovery of an open-loop orienting
system is the way in which it is calibrated: specifically, what is the nature of the instruction that
each part of the retina sends to the nervous system controlling the legs, and how is this obeyed?
A number of possible mechanisms are evaluated in a recent paper (LAND, 1972), and only the
principal conclusion will be given here. The positions of the legs are different at the beginning
and end of every turn, so there is no possibility that an instruction from a given retinal location
is obeyed by the legs the same way each time; the legmovements required for a particular turn must
vary with starting position. Furthermore, turns of the same size can be executed at speeds which
differ by a factor of at least ten, so that there is no way the retina could specify turn size in terms
of turn duration. The only feature of the turn which does seem to remain constant, over a wide
range of turning speeds and different conditions of load, is the angle turned by the spider in the
course of a single step. When all eight legs have stepped once (backwards on the side turned towards and forwards on the other) the spider turns through an average of 75 - 80°. Or to put this
another way, since the spider has eight legs, one of the legs will step every 9 - 10° turned. As a
working hypothesis, the most likely way the retina might specify turn size is to specify the number
of individual steps that are to be taken during the turn, and to rely on the control system of each
leg to insure that step length remains constant – as it does indeed appear to do.

Finally it is worth mentioning that the orientational movements mediated by the principal eyes
(ii and iii in the introduction to this section) appear to be based on 'closed' rather than 'open'
control systems. This is shown qualitatively by the performance of spiders in the same apparatus
(fig.7) when stimuli lie within the fields of view of the unblinded principal eyes. The animal then
often makes many (up to 10) small (about 10°) turns in quick succession, producing in total a turn
many times larger than would have been appropriate. The conclusion from this is that the principal   eyes do require visual feedback (reafferent cancellation) for the proper control of turning,
in contrast to the lateral eyes, which do not.

4. Pattern recognition

a) The problem: DRESS (1952) showed that jumping spiders recognise other jumping spiders by the
presence of legs making appropriate angles with each other (fig.11). To be taken for another spider
and evoke courtship behaviour, a model must have a central 'body' and a series of 'legs' on each

side making angles of roughly 25 - 30° with the vertical. The more legs, the more effective the model. In contrast, objects classified by the spider as prey, and jumped on, could take a wide variety of forms: the only common feature being that they must have moved, and must lie within a given size range. The task of the spider can thus be described as follows: "if it moves, find out whether it has legs in the right places; if it does, mate or avoid it; if it doesn't, catch it".

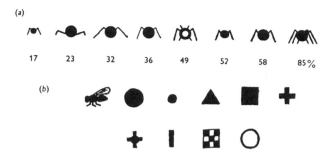

(a)

17   23   32   36   49   52   58   85%

(b)

Fig.11. Stimuli found by DRESS (1952) to evoke courtship (a) and prey-catching behaviour (b) in male jumping spiders (Epiblemum scenicum). The numbers beneath each figure in (a) are the percentage of trials on which courtship was evoked

By way of confirmation of DRESS's conclusions, many spiders have markings on various parts of their bodies which reinforce the pattern made by their legs, and which, one would suppose, should thereby make the task of recognising them as jumping spiders rather easier. Epiblemum (Salticus) itself has a series of diagonal dark and light stripes on the abdomen, a pattern which is common in many other genera. Metaphidippus imperialis (fig.12) is particularly interesting in that it displays on its face a brilliant black and white pattern which mimics rather exactly the leg patterns DRESS found to be effective. Also, during courtship and threat displays between males, the activities of the animal are always such as to increase the number of contours making appropriate angles with the vertical; this may involve raising the front legs diagonally above the head, or raising several legs slightly to create a front-on view in which several parallel sets of legs are visible (see CRANE, 1949, or BRISTOWE, 1958). In each of the above examples the spiders appear to be increasing the extent of their specifically recognisable attributes.

Fig.12. Metaphidippus imperialis (male) showing diagonal stripe pattern on face

b) Role of eye movements   The retinae of the principal eyes can be viewed opthalmoscopially, while they are being presented with stimuli of various kinds, and the movements made by them can be observed and recorded (LAND,1969b). Fig.13 shows the appearance of the retinae viewed in this way while at rest; only the central region of each retina is actually visible by virtue of the fact that the receptors in Layer 4 scatter a certain amount of light, the outline of the rest of the two retinae has been inferred from histological measurements.

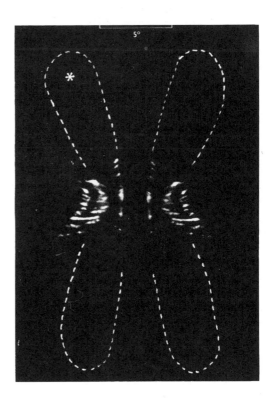

Fig.13. Appearance of the retinae of the principal   eyes of Phidippus johnsoni viewed by ophthalmoscopy while at rest. The visible strictures in the centre of each retina are the receptors of layer 4 (see Figs. 3 and 4), the outline of the rest of the retinae is derived from the anatomy (fig.4). The asterisk (*) corresponds in the body to the ventral region of the right retina. The ophthalmoscope picture corresponds geometrically to the fields of view of the retinae, not to their anatomical geometry in the body. Note that the fields of view of two retinae do not overlap

Using this technique, four types of eye movement are distinguishable (fig.14). These are:
(1) Spontaneous activity. This is simply a descriptive expression to cover all those movements made by the eyes when there was no stimulus in their field of view. Such movements usually consist of side-to-side movements, generally conjugate, i.e. both eyes moving together maintaining the pattern shown in fig.13, but sometimes disjunctive with one eye moving faster than the other or with only one of the eyes moving. Up and down, or diagonal movements also occur, although horizontal move-ments are more common. Sometimes the eyes would take less than a second to cross the 28° field of the opthalmoscope, at other times as long as ten seconds. It is not at all clear what the function of such movements in an unrestained animal might be. (2) Saccades. When a stimulus, e.g. a small dot

is moved across the retina of either the principal or antero-lateral eyes, the retina of the principal eyes move directly towards it, and stop moving when their central regions are looking directly at it. This movement is thus directly analogous to a human saccade which centres the fovea on a target seen by the periphery of the retina. A saccade may lead to tracking or scanning, or alternatively the retinae may return slowly (5 - 10 sec.) to the resting position. (3) Tracking. A moving stimulus onto which the retinae have already fixated will be tracked by the retinae which maintain their fixation on it. This was first discovered by DZIMIRSKI (1959) who saw the retinae of young transparent spiders tracking moving stripes, and making nystagmic movements. (4) Scanning. Following a saccade what frequently happens is that the two retinae begin to scan back and forth from one side to the other of the target, while at the same time rotating, conjugately, about the visual axes (fig.15).

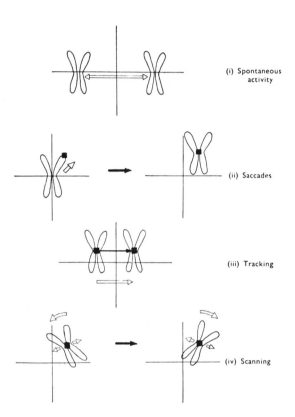

(i) Spontaneous activity

(ii) Saccades

(iii) Tracking

(iv) Scanning

Fig.14. Diagram summarising the four kinds of movements of the retinae of the principal eyes described in the text

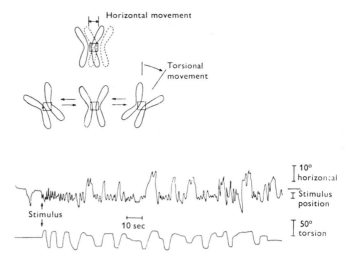

Fig.15. Recording of the movements of the principal eye retinae during an unusually long bout of 'scanning'. The record was made with the aid of a graticule line which could be moved by the observer in such a way as to follow the movments of the edge of one of the retinae. The movements of the graticule line are transducted by potentiometers writing out on a pen recorder. Note that the period of the horizontal scanning movments is very much shorter than that of the torsonial movements (about 1 sec as opposed to 5–8 sec.). During the later stages of scanning the retinae frequently depart from the target to look elsewhere for a time

Scanning has been observed in all four species studied, and takes the same form in each. The period of the horizontal movements is between one and two seconds, and the amplitude varies with the size of the target; targets up to about 10° are scanned from edge to edge, but with targets larger than 10° scanning tends to be concentrated first around one side, then the other. Bouts of scanning generally last a few seconds, but occasionally, as in fig.15, they may last much longer, punctuated by periods when the retinae stray from the target. The torsional movements that accompany horizontal scanning are much slower, with a period of 5–8 seconds and an amplitude of about 50°. The amplitude of torsional movements stays approximately constant throughout the bout, and horizontal and torional movements always begin and terminate at the same time.

Since scanning occurs during the period following detection of a target, one would expect that it is a procedure concerned with the identification of the target. If this is accepted, then it is reasonable to suppose that during scanning the retinae are searching the target for 'legs', the criterion used by the spider for distinguishing mates from prey. Scanning is thus likely to be a search procedure of a rather special kind, and is possibly best understood if the assumption is made that the retina contains rows of receptors organised neurally as line or edge detectors – similar perhaps to 'simple' cells in the cat cortex (HUBEL and WIESEL, 1959). If one then asks, how could such a detector be used to detect the presence and orientation of a line (or 'leg' as here) the answer would be: moved in a direction roughly at right angles to its length in order to generate relative movement of the image and the detector (horizontal scanning) and secondly that it must be rotated through a range of angles to determine the orientation of the contours in the image (torsion). It is actually difficult to conceive of a procedure for accomplishing the task set out in the introduction to this section that would not look like scanning. This hypothesis is summarised in fig.16.

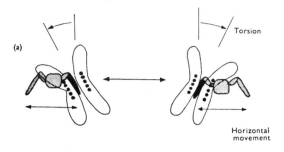

Fig.16. a) Proposed function of scanning eye movements in stimulus identification. Each retina is presumed to contain a row of receptors arranged as an edge or line detector. The torsional component of scanning serves to align the detectors at different predetermined inclinations, while the horizontal component introduces the required relative motion between detector and stimulus image (here shown as a simplified spider). b) Tracing of a photograph of Phidippus johnsoni showing the preponderance of contours making angles of 25 - 30° with the vertical

5. Conclusions

I have attempted to describe in this article some of the requirements of the visual system of an actively predatory animal, and the ways in which these are fulfilled in a particular case - that of jumping spiders. The most interesting general conclusion that emerges from this study is perhaps that in the strategy of prey-capture several kinds of control system are available, and the extent to which each is called into play will depend on the details of each encounter. Thus the spider may face the stimulus with a single large turn, or a combination of large and small turns, all based on the "open" system of the side eyes. After this the principal eyes cause the target to be tracked by an interlocking combination of leg-movements, saccades and tracking eye movements. The situation approaches in complexity the situation in primates where eye, head and body movements contribute to visual locating and following. The second conclusion is a reflection on the extreme economy, both of neurones and space, with which the spider's nervous system functions. The layout of the optical systems, with movement detection and recognition occurring in separate eyes of different focal length, is a unique and elegant way of saving space. The facts that the principal eyes, capable of making the fine distinction between other spiders and flies, have less than 1000 receptors each, and are driven by six muscles receiving only one axon each, are illustrations of the neural parsimony with which arthropods, in contrast to vertebrates, manage their nervous system.

Acknowledgement. I am grateful to the U.S.Public Health Service (Grant EY-00044) for sponsoring much of the research outlined in this review.

References

BRISTOWE, W.S.: The world of Spiders. London: Collins (1958).

CRANE, J.: Comparative biology of salticid spiders at Rancho Grande, Venezuela. Part IV. An analysis of display. Zoologica 34, 159-214 (1949).

DREES, O.: Untersuchungen über die angeborenen Verhaltensweisen bei Springspinnen (Salticidae). Z.Tierpsychol. 9, 169-207 (1952).

DZIMIRSKI, I.: Untersuchungen über Bewegungssehen und Optomotorik bei Springspinnen (Salticidae). Z. Tierpsychol. 16, 385-402 (1959).

EAKIN, R.M., BRANDENBURGER, J.L.: Fine structure of the eyes of jumping spiders. J.Ultrastruct.Res. 37, 618-663 (1971).

EBERLE, H.: Cone length and chromatic aberration in the eye of Lebistes reticulatus. Z. vergl. Physiol. 57, 172-173 (1967).

HOMANN, H.: Beiträge zur Physiologie der Spinnenaugen. Z. vergl. Physiol. 7, 201-268 (1928).

HUBEL, D.H., WIESEL, T.N.: Receptive fields of single neurons in the cat's striate cortex. J. Physiol., Lond. 148, 574-591 (1959).

KAESTNER, A.: Reaktionen der Hüpfspinnen (Salticidae) auf unbewegte farblose und farbige Gesichtsreize. Zool. Beitr. 1, 13-50 (1950).

KIRSCHFELD, K.: Molecular orientation of photopigments in the rhabdomeres. Neurosciences Res. Prog.Bull. 8, 474-475 (1970).

LAND, M.F.: Structure of the principal eyes of jumping spiders in relation to visual optics. J.exp.Biol. 51, 443-470 (1969a).

LAND, M.F.: Movements of the retinae of jumping spiders in response to visual stimuli. J.exp.Biol. 51, 471-493 (1969b).

LAND, M.F.: Orientation by jumping spiders in the absence of visual feedback. J.exp. Biol. 54, 119-139 (1971).

LAND, M.F.: Stepping movements made by jumping spiders during turns mediated by the lateral eyes. J.exp. Biol. (in press) (1972).

MAGNI, F., PAPI, F., SAVELY, H.E., TONGIORGI, P.: Research on the structure and physiology of the eyes of a lycosid spider. II. The role of different pairs of eyes in astronomical navigation. Archs. ital. Biol. 102, 123-136 (1964).

MAGNI, F., PAPI, F., SAVELY, H.E., TONGIORGI, P.: Research on the structure and physiology of the eyes of a lycosid spider. III. Electroretinographic responses to polarised light. Archs. ital. Biol. 103, 146-158 (1965).

MAST, S.O.: Behavior of fireflies (Photinus pyralis) with special reference to the problem of orientation. J. Anim. Behav. 2, 256-272 (1912).

MITTELSTAEDT, H.: Control systems of orientation in insects. Ann.rev.Entomol. 7, 177-198 (1962).

SCHEURING, L.: Die Augen der Arachnoiden, II. Zool. Jb. (Anat). 37, 369-464 (1913-1914).

9. Observations on the Visual Reactions of Collembola

F. Schaller
Department of Zoology, University of Vienna, Austria

Abstract.   Most of the Collembola show simple phototactical reaction only. Shape perception and "Formensehen" could be observed in some species of Sminthuridae (Symphyloena) and in Ar- thropleona in the swamp forest area of the Amazon only. The minimum size of visual angle deter- mined in these observations is $4°$ – $8°$. The ommatidia of the Collembola are provided with an opening angle of $20°$ – $80°$, their axes of field of vision varying from about $16°$ – $60°$, for which reason the field of vision of the adjoining ommatidia frequently overlap. Binocular fields of vi- sion occur only in Symphyleona and in some Arthropleona provided with well developed eyes. The question whether changes of intensity of light between the ommatidia complexes, the single omma- tidia or the rhabdomeres affect shape perception could not be aswered.

The soil–surface dwelling Collembola have atypical complexes of ommatidia consisting of at most 8 eucone ommatidia which are different in size and situated in no special order in a pigment spot on both sides of the head. The visual angle of the single ommatidia is wide and varies from $20°$ – $50°$ up to $80°$. They overlap in different species to a very different extent as the divergency ang- le of the longer axes of the ommatidia are very different, too (e.g. in one case $16°$ – $33°$, in another $34°$ – $61°$). Therefore, a considerably wide field of vision for the complex of each side is resulting (e.g. $130°$ – $150°$ in Sminthurides; PAULUS, 1972). Binocular fields of vision, i.e. overlap of the fields of vision of both complexes, are occurring exceptionally only. In Sminthurides it can reach $45°$ (males of Sminthurides, PAULUS, 1972).

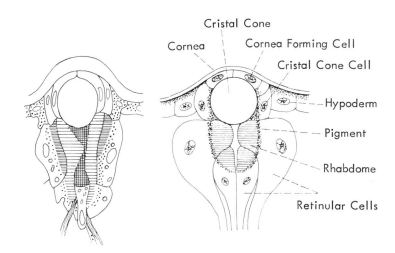

Fig.1.   Ommatidium of Orchesella; right HESSE (1901), left PAULUS (1972)

The visual reactions of Collembola have been investigated more precisely in very few cases. It is known that many species normally exhibit negative phototaxis. Positive phototaxis is known only in species living on the water surface or on herbs and bushes, such as Podura aquatica, Isotomurus palustris, Archisotoma besselsi, Sminthurides aquaticus, Sminthurus viridis. Sminthurides aquaticus reacts to sudden shading ("shade reflex"). Two inhabitants of the water surface (Podura aquatica and Sminthurides aquaticus) show reactions to sharper light–dark edges by running straight along these while swinging their body "clinotactically" to and fro.

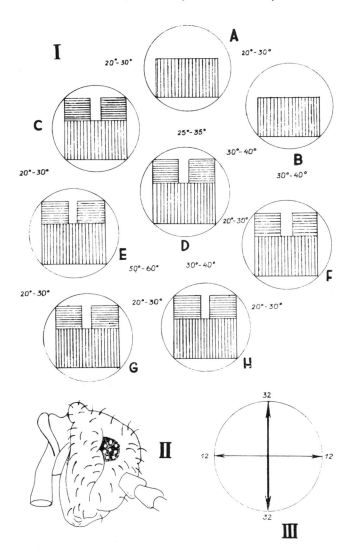

Fig.2. Podura aquatica: position of eye (II) and scheme of orientation of the microvilli of the rhab-domeres (I). Between the ommatidia A - H the diver-gency angles of the ana-tomic ommatidia–axes are stated. III shows that 12 rhabdomeres are straighten-ed out horizontally, 32 vertically (PAULUS, 1972)

Scototactic reactions, i.e. reactions connected with shape perception, were found in Dicyrtomina minuta, a species of forest-dwelling Sminthurides, which will run straight towards treetrunks from a distance of 2 m (MAYER, 1957). By simple behaviour tests with black stripes in a white arena it could be shown that a stripe had to have a minimum breadth of 8 - 10° before Sminthurides would set a course for it. The males of the same species show even more strikung visual reactions in their mating behaviour: they run straight towards their partners or towards any suitable moving visible object of about 1 - 2 mm diameter (e.g. plasticine balls) from a distance of about 2 . The visible objects thus have an angle size of about 4 - 5°.

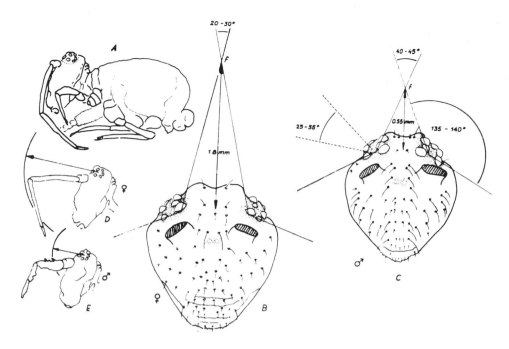

Fig. 3. Sminthurides aquaticus: A female laterally seen; B binocular field of vision of female; C binocular field of vision of male; D and E: position of f in front of the forehead (PAULUS, 1972)

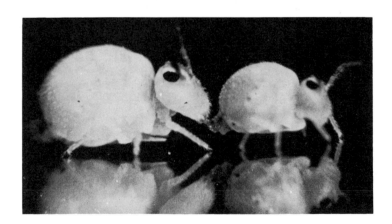

Fig. 4. Bourletiella viridescens – mating behaviour (BERTFELD, 1970)

In the species already mentioned above, Sminthurides aquaticus, a strong sexual dimorphism also with regard to the eyes can be observed. The females have an angle of the binocular field of 20° - 30°. The point of intersection (f) of the two fields of vision is situated 1.8 mm in front of the forehead. The males have a binocular field of vision of 40° - 45°; f is situated in these only 0.55 mm in front of the forehead, namely exactly in the field of traverse of this segment of antenna with which the males grab the antenna of the female while mating. Moreover, these animals are even able to jump directly to blades of grass growing in the water from a distance of 2 - 3 cm (PAULUS, 1972).

Some more precise observations on the visual reactions of Collembola which are flying over inundations have been made. These scototactically shape a course for landmarks (SCHALLER, 1968). This was observed in the arthropleonic genus Lepidocyrtoides living in the Amazon area which are provided with a characteristically strong orthognathy. If they come upon water (where they can run without resting for one hour only), they show extremely remarkable behaviour: (1) They run in close orbits, changing direction frequently. (2) They turn on the spot for about 360° (sometimes repeatedly with change of direction). (3) They swing their body to and fro on the spot up to 180°. (4) They follow a zigzag course. (5) They run straight but swinging their body regularly for 10° - 30° to and fro. (6) They jump, often repeatedly, but straight in one line.

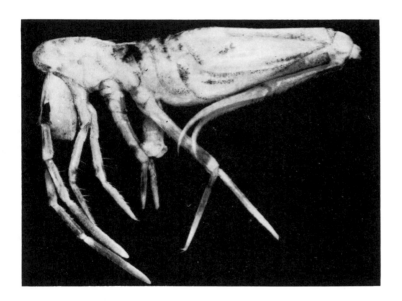

Fig.5. Lepidocyrtoides spec.: position of eye on the orthognathic head, total length: 3 - 4 mm

The aim of all these orientation movements undoubtedly are dark shapes on the horizon of the water surface. Such shapes are reached by them with a high degree of accuracy even under experimental conditions (see fig.6) as long as the breadth of the visual angle is not smaller than 8°.

The behaviour of these animals causes the question whether there takes place a change of "clinotaxis" to "tropotaxis" with increasing angle of the object straightened out for. However, it has to be discussed whether the expression "clinotaxis" is applicable in this case at all. However, there is a clear quantitative relationship between the visibility of the releasing structures and the number of certain orientation movements: the initial and intermediate cycles, turns and rotations are observed most frequently when the dark shape is extremely small and decreasing with its breadth. In an homogeneously white area the animals rotate on spot up to 20 times, changing the direction; at least they jump there, unorientated. In front of well visible stripes they turn and rotate only 3 - 6 times, then approach directly while the lateral movements of the body distinctly decrease. At least they jump directly to the wall. The final aim very often is the ledge of the black shape. Unfortunately, the field of vision of these Lepidocyrtoides could not yet be determined more precisely. However, there is no doubt that the orthognathic position of the head results in quite a wide binocular field of vision.

In connection with the modest observation herewith submitted, the following questions arise: (1) Which are the changes of intensity of light being computed in the shape perception of the optical system of these Collembola? Are these changes between the (a) right and left complex of ommatidia, (b) the ommatidia of one single complex, (c) the rhabdomeres of one ommatidium, or (d) the neighbouring rhabdomeres being located in the same direction? (2) Which of the behaviour described is to be understood as clinotactical, which as tropotactical orientation movement? (3) Which are the degrees of intensity of light perceptible for the various substructures of this primitive visual apparatus? (4) What does the connection of the optical system in the central nervous system of the Collembola look like?

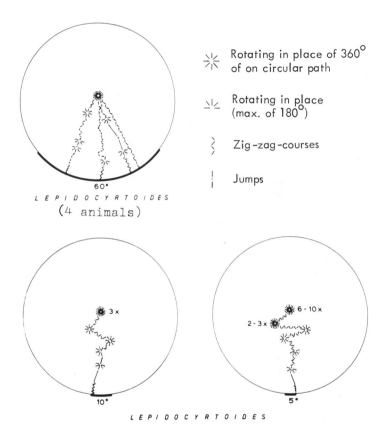

Fig.6. Collembola of the swamp forest area of the Amazonas; courses in a water arena with white edge (diameter 80 cm) towards black stripes

VII. Visual Control of Orientation Patterns

1. Processing of Cues from the Moving Environment in the Drosophila Navigation System

K. G. Goetz
Max-Planck-Institute of Biological Cybernetics, Tübingen, Germany

Abstract.   The properties of the navigation system are derived from the open loop optomotor reactions of the fruitfly Drosophila which is either flying or walking under stationary conditions. The horizontal component of a movement stimulus controls the difference of the propulsive forces of legs and wings on either side and enables the freely moving fly to counteract involuntary deviations from a straight course. The vertical component controls the sum of the propulsive forces of the wings and enables the fly to maintain a given level of flight. Different properties of the navigation system are found in normal and mutant fruitflies. Comparative studies of open loop and closed loop behaviour disprove the separation theorem of KALMUS according to which the fly is always minimizing the rotatory component, and maximizing the translatory component, of a movement stimulus. There is evidence that inversion of the rotatory response can be elicited by the translatory component of the stimulus. Interaction of the stimulus components must be considered in order to determine the closed loop responses of the freely moving insect.

One of the most important sensory cues for the control of locomotion is the displacement of surrounding objects within the visual field of the moving observer. The application of these cues in insect navigation implies that the information on rotatory and translatory displacements is retrieved by an appropriate set of movement detectors in the visual system and conveyed to the appropriate sites of the motor system. The relations between the sensory input and the motor output are accessible either by "closed loop" experiments with free moving insects or by "open loop" experiments with the insects held in fixed positions. The sensory input depends on the motor output in the closed loop experiment, whereas it is disengaged from the motor output in the open loop experiment.

We shall give a comparative account on the open loop optomotor reactions of the fruitfly Drosophila melanogaster, which is either flying or walking under stationary conditions. The resulting catalog of optomotor reactions enables us to comment on a striking discrepancy between open loop and closed loop behaviour. However, there is a way to reconcile the different results. We will see that a commonly accepted principle of movement perception is misleading and has to be replaced by a more rigorous treatment of the optomotor equilibrium.

The fruitfly Drosophila resembles the housefly and the blowfly with respect to the organization of the visual system. The visual field of the faceted eyes comprises about 86% of the surroundings and is imaged onto the array of about 1400 visual elements in the first optic ganglia (BUCHNER and GOETZ, 1972). The resolving power and the acuity of the eyes are determined, respectively, by the density and the overlap of the receptive fields of the "visual elements". The relation of

density and overlap matches the requirements for optimum imaging in bright light (GOETZ, 1965).
Each of the visual elements consists of subsystems of the photoreceptors 1 - 6 and 7 + 8, respectively, which show maximum absorption of the incoming light at different colours and different
inclinations of the plane of polarization (FRANCESCHINI, 1971; WEHNER and SCHUEMPERLI,
1969). The neuroanatomy of the visual elements suggests, in analogy to similar systems in other
flies, the presence of a neuronal device for the improvement of dim-light sensitivity (BRAITEN-
BERG, 1967; KIRSCHFELD, 1967). A protective shielding of the photoreceptors in bright light is
achieved by light-induced migration of pigmented granula in the receptor cells (FRANCESCHINI,
1971).

Movement detection requires the neuronal interaction of at least two visual elements which are
exposed sequentially to an event, i.e. a change of the luminance within their receptive fields.
The nature of the movement detectors in Drosophila is specified by the following observations.
(1) The output of the movement detectors is proportional to the square of the contrast (HENGSTEN
BERG and GOETZ, 1967). (2) The output is determined by the repetition frequency of the Fourier
components of the event function G (t) and not by the angular velocity of the event (GOETZ,
1964). (3) The output is apparently invariant to changes in the phase relations of the Fourier components (ZIMMERMANN, 1972). This is to say that the movement detectors of Drosophila probably
correspond to the Chlorophanus model (REICHARDT, 1970). The process of movement detection is
adequately described by first-order autocorrelation (multiplication and subsequent temporal averaging) of the event functions, G (t) and G (t-$\triangle$t), of any two interactive elements. The property
of phase invariance facilitates the experimental analysis of movement perception. Within the
constraints of linearity it is possible to predict the optomotor reactions to a moving pattern from
the sum of the reactions to each of the Fourier components.

There is experimental evidence that the fruitfly reacts to movement in virtually any part of the
visual field (GOETZ and WENKING, 1972). We therefore expect, at the next level of information
processing, the integration of signals from the movement detectors in different eye regions. To
illustrate the underlying principle we arbitrarily select two distinct zones which contain certain
numbers of visual elements, say n_1 and n_2. The elements are stimulated by the rotatory movement
of a surrounding pattern. The optomotor reactions of the flies indicate the tendency to follow the
rotating stimulus, provided that the spatial period of the pattern is readily resolved. The reaction
is proportional to the total number of stimulated elements, provided that the level of stimulation
is low enough to preserve linearity. What happens if the selected pattern is moving clockwise
within zone No. 1 and counter-clockwise withing zone No. 2? The experimental evidence shows
that the optomotor reaction of the fruitfly is, by magnitude and sign, proportional to the differenc
$n_1 - n_2$ of the numbers of stimulated elements. The reaction disappears whenever n_1 equals n_2
(GOETZ, 1964). We encounter in the present experiment a remarkably simple principle of spatial
integration of visual cues. The fruitfly is apparently unable to focus attention on moving events
in self-selected parts of the visual field.

Fixed flight recordings of torque and thrust have shown the existence of two spatially integrated
movement detecting systems (GOETZ, 1968, 1969). One of the systems responds to the horizontal
components of movement whereas the other is sensitive to the vertical components. The horizontal
system controls the difference of the wing beat amplitudes on either side. Fig. 1 illustrates the
control of the propulsive forces during free flight in a resting environment.

The yaw elicited by a rotatory displacement tends to minimize the visual stimulus. The horizontal
system thus counteracts involuntary deviations from a straight course. The vertical system controls
the sum of the wing beat amplitudes on either side. It acts on the magnitude of the force of flight
which is decomposed into lift and thrust of the flight system. The vertical system counteracts
involuntary deviations from the level of flight. Fig. 1 illustrates the increase of both lift and
thrust by stimulus movements from inferior to superior which occur in a fly that is losing altitude

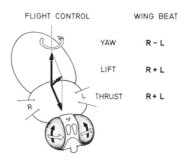

Fig. 1. Rotatory and translatory reactions of the Drosophila flight system are elicited, respectively, by horizontal displacements which control the difference, and by vertical displacements which control the sum, of the wing beat amplitudes on either side

Fig. 2 illustrates the flow of information from the movement-detecting systems to the flight muscles of Drosophila. The equivalent model with the minimum number of neuronal interconnexions has been discussed in an earlier paper. The visual control of the propulsive forces in the housefly Musca appear to be similar and only slightly more complex. The direction of the force of flight in still air as well as a wing pitch and/or the stroke plane are probably not invariant to visual stimulation.

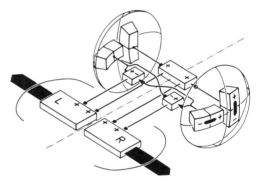

Fig. 2. The sensory systems for horizontal and vertical displacements on either side of the fruitfly communicate with thoracic muscles which determine the beat amplitudes of the ipsilateral and the contralateral wing. The essential properties of the model are derived from fixed flight recordings of torque and thrust

More recently a method has been developed in order to investigate the open loop responses of Drosophila in the walking mode of locomotion (GOETZ and WENKING, 1972). The test fly is allowed to walk for several hours on top of a ball, where it is maintained in stationary position and orientation to the screens of two stimulus projectors on which the continuous movement of a striped pattern is presented. The fly is mounted on a minute sledge which prevents flight and monitors the translatory and rotatory displacements to a servo system. The servo system counteracts the displacements by appropriate rotations of the ball which are further evaluated. The experimental set up is illustrated in Fig. 3. The investigation of several hundred test flies yields an average of 55 forward meters per fly.

The rotatory response to a visual stimulus is given in revolutions per meter pathlength. The experimental results show a close relationship of the movement-detecting systems which control the difference of the propulsive forces in the flying and walking fruitfly. Fig. 4 illustrates the responses on the tread compensator as a function of the frequency of events of a moving periodic pattern. The response has the tendency to minimize the horizontal components of the movement stimulus whereas the vertical components are ineffective. The orientation of the movement detectors is independent of the position in the eye and invariant to the direction and velocity of the movement stimulus.

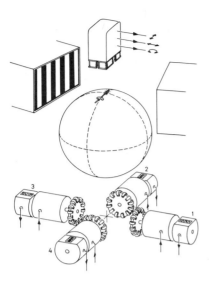

Fig. 3. Exploded view of the compensator with the walking test fly held in stationary position and orientation to the movement stimulus on the screens of the projectors on either side

To investigate the optomotor responses of the limbic system, two arbitrary pairs of legs have been dissected in a number of fruitflies. The average rotatory response of the remaining pairs of legs, in revolution of meter pathlength, was +25 for the fore legs, + 31 for the middle legs and + 32 for the hind legs, respectively. The experiment shows that each of the three pairs of legs is likewise prepared to execute the navigatory commands of the central nervous system.

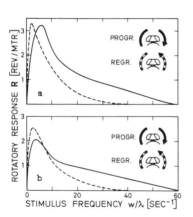

Fig. 4. Rotatory response of Drosophila as a function of the frequency of events in the open loop experiment of Fig. 3. The movement stimulus is characterized by the angular velocity w and by the spatial period λ of the striped pattern. The stimulus is presented to one eye and is either progressing from anterior to posterior or regressing from posterior to anterior. The curves represent the averages of the responses of 50 test flies which are stimulated either in the anterior part (a), or in an arbitrary part (b), of the visual field. The positive sign of the response indicates the tendency to minimize the rotatory displacements of the stimulus

The translatory response depends on the sum of the propulsive forces on either side and is given by the relative change of the walking speed. We remember that the sum of the propulsive forces of the flight system is controlled by the vertical components of the movement stimulus. However lift control is of no use in terrestrial locomotion. It is therefore of interest to know whether or not the walking speed of the fruitfly is modified by the vertical components of the stimulus. Fig. 5 gives a survey on the optomotor responses of normal (+) and mutant (w^a) female fruitflies in the different modes of locomotion. The abbreviations refer to our wild stock 'Berlin' (+), and

to a mutant In (1) sc^8, $sc^8 w^a$ in which the phenotypically predominant eye colour mutation
white-apricot (w^a) is not essential with respect to the present results. We see that the effect of
vertical movements on the walking speed is missing. However, in contrast to the behaviour of
the flight system, the walking speed of the wild type is increased by movements from posterior to
anterior, whereas the walking speed of the mutant is increased by movements from anterior to
posterior and vice versa. The changes of the walking speed are in the order of a few percent.

		WING BEAT	STRIDE
INCREMENT R - L		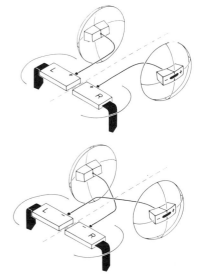	+, w^a
INCREMENT R + L		+, w^a	+
			w^a

Fig. 5. Survey on the optomotor responses of normal (+)
and mutant (w^a) fruitflies in the different modes of locomotion. The direction of the movement stimulus and
the effect on difference and sum of the propulsive
forces on the right (R) and on the left (L) are plotted
on the frontal view of the fly head

Rotatory responses of the walking fruitflies may be elicited by acceleration of the legs on the
outer side of the curve and/or by deceleration of the legs on the inner side of the curve. In
order to simulate both the rotatory and the translatory responses of the wild type and the mutant,
it is sufficient to assume nervous connexions from the movement detectors to the ipsilateral legs
in the wild type (+) and to the contralateral legs in the mutant (w^a). The corresponding models
are shown in Fig. 6.

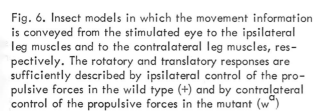

Fig. 6. Insect models in which the movement information
is conveyed from the stimulated eye to the ipsilateral
leg muscles and to the contralateral leg muscles, respectively. The rotatory and translatory responses are
sufficiently described by ipsilateral control of the propulsive forces in the wild type (+) and by contralateral
control of the propulsive forces in the mutant (w^a)

The ipsilateral and the contralateral control system probably coexist in the wild type as well as in the mutant. However, the different contributions to the optomotor responses of these flies suggest the co-operation of two autonomous systems which may be separated by appropriate methods of behavioral genetics.

The closed loop behaviour of the walking fruitfly has been mainly studied under conditions where the fly is free to execute translatory and rotatory movements (HECHT and WALD, 1934; GAVEL, 1939; KALMUS, 1949, 1964; WEHNER, GARTENMANN and JUNGI, 1969; GOETZ, 1970). KALMUS claims that the corresponding displacements of the environment relative to the visual field can be resolved by the visual system into the rotatory and translatory components to which the fly responds by appropriate rotatory and translatory reactions. This is to say that the closed loop behaviour of the fruitfly in a given state of stimulation is simply obtained by superposition of the corresponding rotatory and translatory effects on the propulsive forces, which have been treated separately in the previous paragraphs. However, this is the case neither for the rotatory responses nor for the translatory responses of the freely moving fruitfly. Although the tendency to minimize rotatory displacements is well established in the open loop experiments, it is evident that the flies do not keep on a straight course over extended periods of time. Experiments with the hoverfly Eristalis show that the visual control of locomotion is not interrupted during the turns of the insects (MITTELSTAEDT and v. HOLST, 1953). The deviations from a straight course are therefore ascribed to the commands of higher centers which interfere with the control system, possibly according to the well-established principle of reafference (MITTELSTAEDT, 1971).

Even more striking is the discrepancy between the translatory responses under conditions of open loop and closed loop. The observation of freely walking fruitflies has led to the oversimplifying statement that Drosophila is always maximizing the translatory component of the movement stimulus (KALMUS, 1949, 1964). The survey on the closed loop reaction in Fig. 5 shows that the statement holds for the mutant (w^a) whereas the opposite is true for the wild type (+). To illustrate the complexity of the closed loop behaviour a number of flies is placed in a transparent tube and exposed to the continuous movement of a surrounding pattern (Fig. 7b). The confrontation to the floating environment elicits rotatory responses unless the stimuli on either side are in balance. The optomotor equilibrium is unstable if the fly is heading for the source and stable if the fly is heading for the sink of the floating environment (Fig. 7c). We expect the majority of the flies heading for the sink and therefore accumulating on the right side of the tube. The accumulation may be favoured by the translatory responses of the wild type (+) and diminished by the translatory response of the mutant (w^a). The corresponding experiments are done by means of countercurrent fractionation (GOETZ, 1970). Fig. 7a gives the probability to find the centre of a population of 100 flies at different places of the tube. We see immediately that the results are contrary to the predicted behaviour and incompatible with any other of the existing hypotheses, e.g. after-effects of rotatory movements of the stimulated flies.

The failure to predict the closed loop responses of insects from the open loop responses of the rotatory and translatory stimulus components is evident whenever the fly is exposed to composite stimuli such as in the experiment of Fig. 7. We shall see that the mathematically trivial separation into components is hardly accomplished by the visual system of insects. The arguments against the separation theorem are illustrated in Fig. 8. The curved arrows in the periphery of the left (L) and right (R) eye of the schematic insect denote direction and speed of the moving environment. The optomotor stimulus is either rotatory or translatory or the algebraic sum of the former. The contribution of the visual system on either side to the rotatory response R of the motor system is read from the appropriate frequency response curves such as $R(w/\lambda)$ in Fig. 4. An essential feature of any feasible frequency response curve is the lack of responses at velocities near zero and at velocities beyond flicker fusion. A pair of black arrows indicates the propulsive forces of the motor system. The dashed line refers to the thrust in absence of visual stimulation. The magnitude and sign of the rotatory response is read from the difference of the propulsive forces on either side. The rotatory stimulus produces the expected positive rotatory response. The schematic insect on the left of Fig. 8 decreases the rotatory displacement of the surroundings. The translatory stimulus is unable to contribute rotatory responses. However, the interaction between the rotatory stimulus and the translatory bias produces a negative rotatory response. The schematic insect on the right of Fig. 8 increases the rotatory displacement of

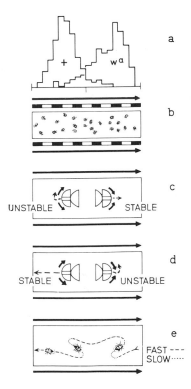

Fig. 7. The wild type (+) and the mutant (w^a) accumulate at the opposite ends of a glass tube when they are exposed to a floating environment (a), (b). The investigation of the optomotor equilibrium shows that the stable position of the fly depends on whether its rotatory activity is comparatively low (c), or comparatively high (d). The complex features of the expected trace have been established in earlier observations

the surroundings. The different curves in Fig. 4 suggest that the response to composite stimuli depends on both the speed and the direction of the translatory bias. Inversion occurs whenever the frequency w/ λ of the translatory bias exceeds the frequency of the rotatory stimulus component by a positive constant.

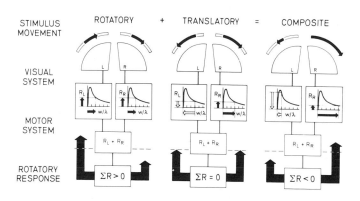

Fig. 8. The effect of the translatory bias on the rotatory responses of a schematic insect. The motor system decreases rotatory displacements and is indifferent to translatory dis-

placements of the environment. However, the superposition of the former displacements elicits negative responses. The rotatory component is actively increased by the motor system of the schematic insect

The predicted influence of the translatory bias on the rotatory responses is established in various experiments (GOETZ, 1972). Table 1 gives the means and standard errors of the responses of 21 stationarily walking fruitflies in comparison to the algebraic sum of the responses to the stimuli on either side, which are derived from the data in Fig. 4a.

Table 1

ROTATORY STIMULUS 3.4 SEC^{-1}	TRANSLATORY BIAS 5.1 SEC^{-1}	COMPOSITE STIMULUS 1.7 SEC^{-1} 8.5 SEC^{-1}	ROTATORY RESPONSE R [REV./MTR.]	
			THEORETICAL	EXPERIMENT.
			+0.59	+0.78 ±.31
			−1.45	−1.13 ±.22

The behaviour of the flies is in agreement with the previous estimates. The rotatory response is still positive in the presence of the progressing bias. However, a regressing bias of the same frequency is already sufficient to produce the expected inversion. The example shows that the rotatory responses are by no means invariant to the translatory bias. The composite stimuli received by a freely moving insect must be considered in order to derive the optomotor behaviour under conditions of closed loop.

The results of the previous paragraphs can be applied immediately to the accumulation problem in Fig. 7. The balance of the stimuli on either side of the fly determines the optomotor equilibrium according to Fig. 7c. However, this is true as long as the fly is in a state of reduced activity where the rotatory fluctuations can be neglected. We expect that the fly is slowly moving toward the sink of the floating environment. The behaviour in a state of increased activity is predominantly determined by the rotatory fluctuations of the fly. The results in Table 1 suggest a stable equilibrium if the fly is heading for the source and an unstable equilibrium if the fly is heading for the sink of the floating environment (Fig. 7d). It is now expected that the fly is quickly moving toward the source of the floating environment. The resulting trace in Fig. 7e is compatible with earlier observations (HECHT and WALD, 1934; GAVEL, 1939; KALMUS, 1949). The apparently complex behaviour of the freely walking flies in the experiment of Fig. 7 can be derived, in a first approximation, from a few fundamental properties of the navigation system in Drosophila. Quod erat demonstrandum.

I would like to thank my friends and co-workers in the institute for their valuable contributions.

References

BRAITENBERG, V.: Patterns of projection in the visual system of the fly. I. Retina-lamina projections. Exp. Brain Res. 3, 271 - 298 (1967).

BUCHNER, E., GOETZ, K.G.: Dunkelanregung des stationären Flugs der Fruchtfliege Drosophila (in press).

FRANCESCHINI, N.: Sur le traitement optique de l'information visuelle dans l'oeil à facettes de la Drosophile; thèse Université de Grenoble, Grenoble (1971).

GAVEL, L.V.: Die "kritische Streifenbreite" als Mass der Sehschärfe bei Drosophila melanogaster. Z. vergl. Physiol. 27, 80 - 135 (1939).

GOETZ, K.G.: Optomotorische Untersuchung des visuellen Systems einiger Augenmutanten der Fruchtfliege Drosophila. Kybernetik 2, 77 - 92 (1964).

GOETZ, K.G.: Die optischen Uebertragungseigenschaften der Komplexaugen von Drosophila. Kybernetik 2, 215 - 221 (1965).

GOETZ, K.G.: Flight control in Drosophila by visual perception of motion. Kybernetik 4, 199 - 208 (1968).

GOETZ, K.G.: Movement discrimination in insects. In: Processing of Optical Data by Organisms and by Machines. Ed. W. REICHARDT. Rendiconti S.I.F. Course XLIII. London and New York. Academic Press. pp. 494 - 509 (1969).

GOETZ, K.G.: Fractionation of Drosophila populations according to optomotor traits. J. exp. Biol. 52, 419 - 436 (1970).

GOETZ, K.G.: The optomotor equilibrium of the Drosophila navigation system (in press).

GOETZ, K.G. and WENKING, H.: Visual control of locomotion in the walking fruitfly Drosophila (in press).

HECHT, S. and WALD, G.: The visual acuity and intensity discrimination of Drosophila (in press).

HENGSTENBERG, R. and GOETZ, K.G.: Der Einfluss des Schirmpigmentgehalts auf die Helligkeits- und Kontrastwahrnehmung bei Drosophila-Augenmutanten. Kybernetik 3, 276 - 285 (1967).

KALMUS, H.: Optomotor responses in Drosophila and Musca. Physiol. comp. ('s Grav.) 1, 127 - 147 (1949).

KALMUS, H.: Animals as mathematicians. Nature 202, 1156 - 1160 (1964).

KIRSCHFELD, K.: Die Projektion der optischen Umwelt auf das Raster der Rhabdomere im Komplexauge von Musca. Exp. Brain Res. 3, 248 - 270 (1967).

MITTELSTAEDT, H. and HOLST, E. v.: Reafferenzprinzip und Optomotorik. Zool. Anz. 151, 253 (1953).

MITTELSTAEDT, H.: Reafferenzprinzip-Apologie und Kritik. Ed. W.D. KEIDEL, K.-H. PLATTIG. Vorträge der Erlanger Physiologentagung 1970, pp. 161 - 171, Springer, Berlin, Heidelberg, New York (1971).

REICHARDT, W.: The insect eye as a model for analysis of uptake, transduction, and processing of optical data in the nervous system; Ed. SCHMITT, F.O. The Neurosciences Vol. 2, pp. 494 - 511. Rockefeller University Press, New York (1970).

WEHNER, R., GARTENMANN, G. and JUNGI, T.: Contrast perception in eye colour mutants of Drosophila melanogaster and Drosophila subobscura. J. Insect. Physiol. 15, 815 - 823 (1969).

WEHNER, R. and SCHUEMPERLI, R.: Das Aktionsspektrum der phototaktischen Spontantendenz bei Drosophila melanogaster. Revue Suisse de Zoologie 76, 1087 - 1095 (1969).

ZIMMERMANN, G.: Personal communication (1972).

2. Behavioral Diagnostics; a Way to Analyze Visual Mutants of Drosophila

M. Heisenberg
Max-Planck-Institute of Biological Cybernetics, Tübingen, Germany

Abstract.   Seven groups of Drosophila mutants are described in which movement detection is disturbed at some neuronal level. In several of the mutants the ability to see at low light intensities is lost, in others contrast transfer is impaired. Some respond only to movement from back to front but not to movement from front to back. With one exception the mutants seem to have electrophysiological defects in the lamina. These preliminary data show that by mutations the fly's visual system can be broken down into its constituents.

One of the promising projects concerning the topic of this symposium is to relate the optomotor behaviour of insects to neural mechanisms and to the nerve patterns in the optic ganglia. In this connection, I would like to introduce a series of neurological mutants of Drosophila in which movement detection is hereditarily disturbed. Studying the visual behaviour of these mutants leads to a description of the brain damage caused by the mutations. The table summarizes some of the properties of the mutants as far as we have investigated them.

We are dealing with 7 complementation groups of which 6 are located on the X chromosome. Most of the mutants were isolated because of their inability to react to specific movement stimuli, others have been known before for certain differences in body pigmentation (tan, ebony). None of the mutants are completely blind. They show normal phototaxis and optomotor responses under certain stimulus conditions. What is more surprising is that with one exception all have abnormal electroretinograms (ERGs). The ERG of Drosophila consists of two parts, one being the extracellular record of the retinula cells, the other an "evoked potential" of the lamina ganglionaris, called lamina potential (HOTTA and BENZER, 1969; PAK, GROSSFIELD and WHITE, 1969; HEISENBERG, 1971). We are concerned here only with those ERG mutants in which the lamina potential is affected. In some of them the whole lamina potential is suppressed more or less strongly, in another one it is decaying during the experiment in still another only the off-effect is reduced.

If the ERG symptoms really indicate a functional impairment of the lamina, behavioral experiments might be able to detect it. In the last few years it has become evident that the fly has two complementary visual input systems, one spezialized for high-contrast transfer (high-acuity system, HAS), consisting at the level of the retina of the retinula cells 7 and 8, which have small rhabdomers and are connected directly to the medulla, the other spezialized for high sensitivity (HSS) receiving visual input via the retinula cells 1 - 6, which have large rhabdomers and transfer their excitation to monopolar cells in the lamina (e.g. KIRSCHFELD, 1969).

Thus one way to test the functioning of the lamina is to compare the low-intensity threshold of the optomotor response using striped patterns with spatial wave-lengths above and below the limit of resolution of the HSS. For instance, in wild type Drosophila the intensity threshold for an 18° stripe pattern (which is resolved by the HSS) is about 50 times below that for a 7.2° stripe pattern (which is resolved only by the HAS). For a mutant defective in lamina functions the threshold intensity for 18° stripes should be nearly as high as that for 7.2° stripes which should be the same as in wild type.

Likewise phototaxis should have a higher threshold intensity in a lamina mutant than in wild type, provided that this reaction is mediated by both visual input systems.

	opm2	tan	ebony	opm18	opm20	opm31	opm37
complementation group	opm2	tan	ebony	opm18	opm20	opm31	opm37
other mutants in this complementation group	opm 17, 48	opm6,24,29, 32,34,42,45,49		opm 47			
ERG defect	LP strongly suppressed	LP suppressed	LP suppressed	LP suppressed	LP labile	none (?) normal ERG	off-effect small
Phototaxis; threshold intensity (X-fold that in WT)	10-25	10-25	200	100		5 – 10	5-10
slope of intensity threshold curve	increased	increased	reduced	reduced		reduced (?)	
Optomotor response; threshold intensity of HSS (18°-stripes) (X-fold that in WT)	0-3	little higher than WT	500	300	higher than in WT	5-10	5-10
threshold intensity of HAS (72°-stripes) (X-fold that in WT)	–	–	10	6		0 same as WT	0 same as WT
visual acuity	reduced $\Delta \varphi = 14°$	reduced	normal	normal		normal	normal
polarisation sensitivity	missing	stronger than normal					

Abbreviations: opm: optomotor mutant; ERG: electroretinogram; LP: lamina potential; WT: wild type; HSS: high sensitivity system; HAS: high acuity system.

Table of Mutants

Some of the mutants in fact meet most of these requirements, others, however, do not. The black mutant ebony, which has its mutation on the 3. chromosome is about 200 times less sensitive in phototaxis than the wild type. In the optomotor response, the threshold intensity for 18° stripes lies only marginally below that for 7.2° stripes; however this common threshold intensity is about 10 times that of the HAS in wild type (i.e. 500 times that of the HSS in wild type). Although this additional shift is not understood, it shows that the HSS does not operate at low intensities and that the HAS does operate at high intensities. Whether the HSS does function at high intensities remains undecided. This will be further discussed below.

Another mutant, opm 18, which is normally pigmented and has its mutation on the 1. chromosome, resembles ebony in all these features, except that the threshold intensity of the HAS is not increased as much as in ebony. We would like to speculate that these symptoms just represent the lack of lamina functions and that the cause for this could be very different in the two mutants. The reduced sensitivity of the HAS in both mutants however remains unexplained.

Two other groups of mutants, opm 2 and tan, in which also the lamina potential is suppressed, show a radically different behaviour. (The mutant tan itself has a slightly lighter than normal body colour, whereas other members of the tan group with similar behavioural symptoms have not.) First of all visual acuity is reduced in these mutants. We measured the average half-width of the visual field of the sampling stations for movement detection ($\Delta \varrho$) in opm 2 following a procedure developed by GOETZ (1964, 1965). It was found to be 14°, which is about twice the value for the HSS in wild type (FRANCESCHINI, unpublished). The spacing of the sampling stations for movement detection ($\Delta \varphi$) seems to be the same as in wild type. This blurring effect must be of neuronal origin, since the optics at least in opm 2 are unimpaired.

Secondly the threshold intensity of the optomotor response using stripe widths well above the limit of resulation in these mutants (51°) is about as low as in wild type. Thus the lamina must function at least partially under these conditions.

On the other hand in phototaxis the threshold is shifted to higher intensities. However, the shift is less than in ebony and opm 18 and in addition the slope of the threshold curve is unusually steep.

These experiments do not tell whether in opm 2 the HAS does function at high intensities, but is suffering from the same blurring defect the HSS does. To check this, we used a distinction of the two systems found by KIRSCHFELD and REICHARDT (1970), who demonstrated that the optomotor response via the HAS depends upon the direction of the E-vector of polarized light, whereas the optomotor response of the HSS does not. If vertical stripes made of polarisation filters alternating in the direction of polarisation between -45° and $+45^{\circ}$ to the vertical are moved around the vertical axis of the fly, an optomotor response is elicited only via the HAS (KIRSCHFELD and REICHARDT, in preparation). Using a 360° stripe pattern, we got a moderate but significant response from wild type, a strong response from ebony, and no significant one from opm 2. It seems that in opm 2 the HAS contributes very little if at all to the optomotor response at high intensity.

Let us assume for a moment that indeed only the HSS is providing visual input to opm 2 and only the HAS to ebony, then we can predict the outcome of a very simple experiment. The double mutant opm 2-ebony should be totally blind. This seems to be the case for the optomotor response. At intensities above the threshold for ebony, and using stripe widths above the limit of resolution for opm 2, the double mutant does not show a normal response. Although this result has to be investigated in more detail, it seems that the above assumptions may be true to a certain degree.

The mutant opm 2 shows still another peculiarity. It reacts only to movement from back to front (regressive) but not or very little to movement from front to back (progressive). This holds also for opm 17, but not for opm 48, nor for tan as preliminary measurements show. The opm-mutants in the tan-group have not been tested in this respect. How these various symptoms of the opm 2 disease are related, we do not know.

A few remarks about opm 31 and 37: In opm 31, which is the only mutant without an obvious ERG defect, the phototaxis threshold intensity is only slightly higher than in wildtype. The optomotor threshold intensity for narrow stripes (7.2°) is not changed and that for broad stripes (18°) is 5 to 10 times above wild type. In which manner the HSS is disturbed here, remains to be investigated.

Again in opm 37, which has a lamina potential with a small off-effect the HSS seems to be damaged without an impairment of the HAS. We assume that the reason for the HSS defect is quite different in opm 31 and opm 37.

The mutant opm 20 has not been studied yet. The description of these mutants will be published in detail elsewhere. The purpose of this preliminary survey is to give an idea which kinds of optomotor mutants we have found, what the means are to investigate them and that with these techniques the fly's visual system can be broken down into meaningful elements. We hope that further optomotor experiments will help to identify the mutational diseases in physiological terms and will tell something about the physiological nature of the "filters", "multipliers" and "summation processes", defined by the optomotor theory (e.g. REICHARDT, 1969).

Acknowledgements. I am indebted to Dr. K.G. GOETZ for his continuous criticism and encouragement and to Mrs. I. SEIBOLD, who isolated the mutants. The help of most people in this institute is gratefully acknowledged.

References

GOETZ, K.G.: Optomotorische Untersuchung des visuellen Systems einiger Augenmutanten der Fruchtfliege Drosophila. Kybernetik 2, 77 - 92 (1964).

GOETZ, K.G.: Die optischen Uebertragungseigenschaften der Komplexaugen von Drosophila. Kybernetik 2, 215 - 221 (1965).

HOTTA, V., and BENZER, S.: Abnormal Electroretinograms in visual mutants of Drosophila. Nature 222, 354 - 356 (1969).

HEISENBERG, M.: Separation of receptor and lamina potentials in the Electroretinogram of normal and mutant Drosophila. J. exp. Biol. 55, 85 - 100 (1971).

KIRSCHFELD, K.: Optics of the compound eye. In: Processing of optical data by organisms and by machines. The International School of Physics, "Enrico Fermi". Academic Press, New York (1969).

KIRSCHFELD, K. and REICHARDT, W.: Optomotorische Versuche an Musca mit linear polarisiertem Licht. Z. Naturforsch. 25b, 228 (1970).

PAK, W.L, GROSSFIELD, J. and WHITE, N.V.: Nonphototactic mutants in a study of vision of Drosophila. Nature 222, 351 - 354 (1969).

REICHARDT, W.: Movement Perception in Insects. In: Processing of Optical Data by Organisms and by machines. The International School of Physics, "Enrico Fermi". Academic Press, New York (1969).

3. Resultant Positioning between Optical and Kinesthetic Orientation in the Spider Agelena labyrinthica Clerck

P. Goerner
Department of Zoology (II), University of Berlin, Germany

Abstract.   The spider Agelena labyrinthica oscillates about a straight line direction during its optically and kinesthetically oriented locomotion upon a horizontal web, on its return to its retreat (fig. 4a, b). If, during the experiment, the optic and kinesthetic reference directions diverge, the spider may then assume a resultant direction. The spider turns at the starting point through the smaller of the two possible angles towards the direction of motion (fig. 5, 6), i.e. the integration of the optic and kinesthetic components must have been performed prior to the start. During the turn towards the direction of motion, the spider already traverses approximately 4 cm of the web. The components of taxis and locomotion within the orientation thus are not separate processes.

1. Introduction

The spider Agelena labyrinthica orients primarily by optical and kinesthetic means, on its return along a horizontal web to its retreat. Orientation by means of the web itself plays a subordinate role (KURTH, in preparation). Other mechanisms have not as yet been found. Shifting the light source prior to the start of the return causes the spider to run along an angle which lies between the direction of the retreat and the new direction of the light. This type of positioning was inter-preted as compromise (resultant) positioning (GOERNER, 1958, 1966; MOLLER, 1970). In these experiments the beginning point and a second point along the return path after the spider had covered approximately ten cm distance were connected by a straight line. With this method a simulation was possible of the resultant positioning which takes place when for example the spider orients initially kinesthetically but then corrects its direction towards the light. Thus, for example, the ant Cataglyphis bicolor orients on its return path according to the sun in areas far from the nest, but according to landmarks near the nest (WEHNER, 1968, 1969; see also this volume).

The experiments described in this paper were intended to determine whether the spider when in conflict between optical and kinesthetical orientation assumes an alternative or a resultant direction.

Fig. 1a, b. Experimental setup, schematic. ca film camera; cy cylinder with the spider's web; L_1, L_2 light sources; op openings for light beam and observation; co cover; R retreat

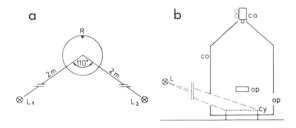

2. Material and Methods

Locally caught spiders Agelena labyrinthica Clerck were used. In the experiments a horizontal web was employed which the spiders had manufactured in a vertical metal cylinder (31.5 cm diameter, 10 cm height).The spider's retreat was always located on the upper end of the cylinder. The cylinder was surrounded by a larger plastic cylinder, which excluded extraneous light. Two holes in the plastic separated in the azimuth by 110 degrees and located 20 degrees above the horizon were used as entrance points for the light. Only one light source was in use at any given time. The illumination strength at the centre of the web was 350 Lux. For details see fig. 1. Film recordings were made vertically, from above, usually with a speed of 48 frames per sec. (occasionally 32 and 64 frames per sec.). Shortly before the experiment the spider was placed on the web and by means of gentle blowing could be driven to its retreat. A fruit fly was then dropped into the middle of the web and caught by the spider. Usually the spider returned immediately to the retreat. Occasionally it had to be induced to return by gentle tapping of the table. Responses of this type were noted in the protocol. In the film-recordings they cannot be distinguished from the normal returns. If the spider did not return to its retreat during the experiment and remained on the web it was then gently driven back. Two control runs were performed before the experiment. In the experiment the light sources were switched while the spider was touching its prey. If it no longer reacted to the vibrations of the fly it was taken from the web and replaced by another. The second control round and the test run were filmed. The film was projected onto a white table top and evaluated frame by frame. The midpoint of the prosoma and the longitudinal axis of the spider were drawn.

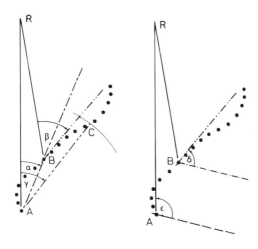

Fig. 2. Angles measured from return runs. See text for details

The following angles were evaluated (fig. 2): (1) The starting angle RAB, in which R is the retreat, A the starting point at which the body axis and the direction of motion coincide for the first time: angle α ; (2) The angle of motion between the axis of the spider in point B and the line BR: angle β. The angle of motion was also measured as β ' against the retreat in relation to point A; (3) The angle of traverse RAC in which C is the crossing point of the path of the spider with a circle around A of 7 cm radius: angle γ . This angle corresponds approximately to the direction which was measured in earlier work with Agelena (GOERNER 1961 etc.). Occasionally abrupt changes of direction of 60 degrees or more occurred during the run. The angle RAC was then measured instead of γ, if the point of the change of direction C' occurred after a distance of less than 7 cm but at least 3.5 cm. If the change of direction occurred before 3.5 cm the data were rejected. Curvilinear runs were not evaluated. The angle between the longitudinal axis of the body and the direction of the retreat taken in the direction of rotation of the spider (ε) and the rotational angle between A and B (δ) were also evaluated. By the further evaluati of the runs it was not taken into consideration that the spider had to maintain different angles with respect to the light source at various positions on the web in order to reach the retreat. This error was maximally 4 degrees. These differences are of no importance in these studies. In order to increase the amount of statistical material available all values of L_2 were added with reversed sign to the values of L_1. Any systematic, light-independent deviations which might occur could thus be compensated for. The dependency of the runs on light, of interest here, remains unaffected by this process. Circular statistics were not used as (1) the scattering of the individual values rarely exceeded 180 degrees, (2) the exact mean value of the direction of a run was not of interest here and (3) the more exact methods for linear statistics for the comparison of two distributions have been worked out (BATSCHELET, 1965).

3. Results

In the control runs the starting angle, angle of motion and traverse angle did not significantly deviate from each other in mean value or in standard deviation (fig. 3a; T-test or F-test; P always >0.05). The differences among α, β and γ may be partially explained by the fact that the spider did not run in a straight line but rather oscillated back and forth in a general direction of motion (fig. 4a, b). Thus the spider maintained approximately the course which it had assumed.

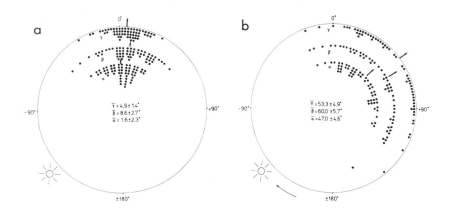

Fig. 3. Distribution of α, β and γ of the individual runs a) in the control and b) in the switch-over experiment. Arrows: mean directions of motion. In the centre of the circles: mean values with standard errors

This direction deviated away from the light by a few degrees from the direction of the retreat. This deviation had already been observed (GOERNER, 1958, 1966) and interpreted (MOLLER, 1970). Curvilinear runs or sudden changes of direction did not occur in the controls. If the lights were switched before the spider started its return path, the spider then assumed a new course which deviated from the direction of the retreat by about 50 degrees (angle of light change 110°, fig. 3b). Once again the mean values and standard deviations for α, β and γ do not vary significantly from one another (P > 0.05); the standard deviations of α, β and γ in the switch-over experiment do vary however from those in the corresponding controls (P < 0.01). This cannot be explained by any appearance of greater changes of direction in the individual runs. The standard deviation of α, β and γ of each individual run around a mean direction of motion in the switch-over experiment (e) is approximately the same as in the control runs (c) ($s\alpha_c = \pm 11.59$; $s\alpha_e = \pm 13.52$; $s\beta_c = \pm 11.78$; $s\beta_e = \pm 13.76$; $s\gamma_c = \pm 5.48$; $s\gamma_e = \pm 6.88$. F-test; P always >0.05).

Curvilinear runs occasionally occurred during the switch-over experiments (in 5% of the runs). The spiders, however, frequently made an abrupt turn (in 67% of the runs). This behaviour which has previously been observed (BARTELS, 1929; GOERNER, 1958 etc.) and studied in detail (MOLLER, 1970) will not be further considered here.

If the longitudinal axis of a spider during the individual phases of motion of a control return are drawn, it then becomes obvious that the spider was already moving forward during the turn towards the direction of motion. This fact applies equally for the control and for the switch-over experiments (see fig. 4). The turning process occurs extremely fast (on the average in 140 msec). During this time the spider travelles a distance of $3,9 \pm 0,1$ cm.

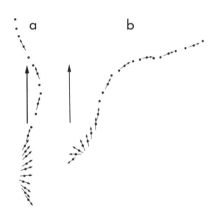

Fig. 4. Return to retreat. a) control run; b) following switch-over from L_2 to L_1. Thin arrows: longitudinal body axis of the spider; heavy arrows: direction towards retreat

In the case of the directional orientation following an external stimulus the animal always turns through the smaller possible angle to the stable direction (principle of the smaller rotational angle; see v. BUDDENBROCK, 1931). If one assumes that this principle is also valid when a kinesthetic component is present, the stable direction for the spider is then most probably the resultant of its run as the spider turns through the smaller angle relative to the direction of

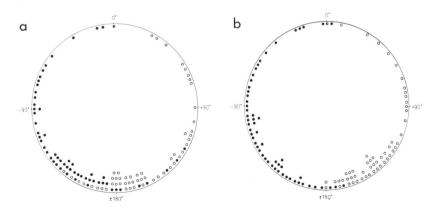

Fig. 5. Direction of turn in the control and in the test runs as measured in relation to the direction a) of the retreat; b) of motion (longitudinal axis at point B; see fig. 2b). Open circles: turns towards left; solid circles: turns towards right

motion (fig. 5a, b). The close correlation between the rotational angle and the direction of the run measured as the angle of the traverse γ is expressed in fig. 6a ($r_1 = 0.959$). For comparison the correlation between the rotational angle δ and the angle towards retreat ε are shown (fig. 6b; $r_2 = 0.387$). r_1 and r_2 are significantly different ($P < 0.001$).

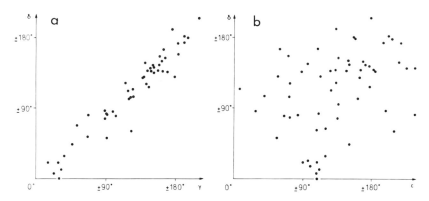

Fig. 6. Switch-over experiment. a) Correlation between angle of traverse γ (abscisse) and rotational angle δ (ordinate); b) Correlation between angle towards retreat ε (abscisse) and rotational angle δ (ordinate)

4. Discussion

Resultant positioning between optical and kinesthetic orientation does not seem to be limited to the spider (LINSENMAIR-ZIEGLER 1970; BURGER, 1972, see also this volume). BURGER demonstrated resultant positioning between optical and kinesthetic orientation in Blatella and Carausius. The insects were drawn off their straight direction of motion by an optical or a gravitational stimulus. If the stimulus was removed the insect would then turn back in the initial direction. The external stimulus thus acted directly on the motion of the insect. With the spider the optical stimulus is already integrated with the kinesthetic component before entrance into the return path. The initial portion of the return path is probably covered without optical feedback as correcting movements are not observed and are in fact rather improbable because of the short duration of the turn (140 msec; see also LAND, 1971 and this volume). It is possible that a comparison between the direction of motion and the light direction occurs after a certain distance (time), in response to which a correction of the course of motion can occur ("abrupt turn").

In A. KUEHN's taxis system (1911) as well as in the orientation system proposed by R. JANDER (1965, 1970), orientation actions, the positioning of the body axis in a stimulus field and purposeful locomotion are - at least formally (JANDER) - distinguished from one another (see also LINDAUER, 1963). In the case of the spider however, positioning reaction and locomotion are not separate processes. The spider already moves more or less straight ahead in the direction of the run while it is simultaneously turning around its longitudinal axis. Further experiments will show to what extent the positioning reaction found in the case of the spider can be applied to other arthropod .

Acknowledgements. My thanks are due to Miss A. MIDDELMANN, Miss A. STUMPNER, Mr. K. DORNFELDT and to my wife for technical assistance, to Dr. M. LICKER for translating the German manuscript and to Mr. A. REISSLAND for critical discussions.

References

BATSCHELET, E.: Statistical methods for the analysis of problems in animal orientation and certain biological rhythms. AIBS-monograph, Washington (1965).

BUDDENBROCK, W. v.: Beiträge zur Lichtkompassorientierung (Menotaxis) der Arthropoden. Z. vergl. Physiol. 15, 597 - 612 (1931).

BURGER, M.-L.: The influence of proprioceptive inputs on the course control in Arthropods (Diplopods and Insects), in press.

GOERNER, P.: Die optische und kinästhetische Orientierung der Trichterspinne Agelena labyrinthica (Cl.). Z. vergl. Physiol. 41, 111 - 153 (1958).

GOERNER, P.: Ueber die Koppelung der optischen und kinästhetischen Orientierung bei den Trichterspinnen Agelena labyrinthica (Clerck) und Agelena gracilens C.L. Koch. Z. vergl. Physiol. 53, 253 - 276 (1966).

JANDER, R.: Die Phylogenie von Orientierungsmechanismen der Arthropoden. Verh. Deutsch. Zool. Ges. Jena, 266 - 306 (1965).

JANDER, R.: Ein Ansatz zur modernen Elementarbeschreibung der Orientierungshandlung. Z. Tierpsychol. 27, 771 - 778 (1970).

KUEHN, A.: Die Orientierung der Tiere im Raum. Gustav Fischer, Jena (1919).

LAND, M.F.: Orientation by jumping spiders in the absence of visual feedback. J. exp. Biol. 54, 119 - 139 (1971).

LINDAUER, M.: Allgemeine Sinnesphysiologie. Orientierung im Raum. Fortschr. Zool. 16, 58 - 140 (1963).

LINSENMAIR-ZIEGLER, C.: Vergleichende Untersuchungen zum photogeotaktischen Winkel-transponieren pterygoter Insekten. Z. vergl. Physiol. 68, 229 - 262 (1970).

MOLLER, P.: Die systematischen Abweichungen bei der optischen Richtungsorientierung der Trichterspinne Agelena labyrinthica. Z. vergl. Physiol. 66, 78 - 106 (1970).

WEHNER, R.: Optische Orientierungsmechanismen im Heimkehr-Verhalten von Cataglyphis bicolor Fab. (Formicidae, Hymenoptera). Rev. Suisse Zool. 75, 1076 - 1085 (1968).

WEHNER, R., MENZEL, R.: Homing in the ant Cataglyphis bicolor. Science 164, 192 - 194 (1969).

4. Idiothetic Course Control and Visual Orientation

M. L. Mittelstaedt-Burger
Max-Planck-Institute of Behavioral Sciences, Seewiesen, Germany

Abstract. The animals used in these experiments are able to maintain course by information about a relation between parts or states of their own body which is correlated to turns ("idiothetic" course control). If one makes the animal deviate from such an idiothetic course by exposing it to a suitably arranged visual stimulus, it shows a compensatory turn when the original situation is reestablished. It is shown that the idiothetic course control operates also in the presence of external orienting cues. Consequences for the interpretation of relevant data are discussed and a new hypothesis about the origin of menotactic light courses is suggested.

Two principles of course control have come into evidence: (1) The organism maintains course by information about a relation between its body and a spatially ordered feature of its environment. That could be done, for instance, by feedback control of the angle between its long axis and a spatially oriented visual stimulus. This type of course control will henceforth be named allothetic. (2) The organism maintains course by information about a relation between parts or states of its own body which is correlated to turns. That could be done by keeping stock of its efferent turning commands or of the proprioceptive inputs resulting from turns. This type of course control without external directing cues will be named idiothetic.

In the first case, the deviation of the animal's path from its original bearing is largely dependent upon the precision of the feedback loop. In the second, however, in addition to that of the running control, the quality of the animal's memory is the limiting factor.

Thus one might expect that the animal will resort to idiothetic orientation only if it is deprived of all external orienting cues. The experiments to be reported here have shown, however, that idiothetic course control operates also in the presence of such cues. This is demonstrated, if one makes the animal deviate from a purely idiothetic course by exposing it to a suitably arranged visual stimulus.

Two experimental set-ups were used. The animals were running free either in a horizontal arena (Ø 150 cm) or on the apex of a rolling sphere. Controls ensured that, during supposedly idiothetic runs, no visual, mechanical, or magnetic orienting cues were actually present. In the arena, the animals (Carausius morosus, Blatella germanica, Schizophyllum sabulosum) started from the center and their path was recorded. In the second set-up, the animal ran on the surface of a large sphere, which, by feedback of the animal's position, rotates such that the animal remains on the apex. This way the animal's path, which is electronically deduced from the sphere's rotation, may become as long as one wants, while the animal stays under readily controllable conditions.

In the first experiment, cockroaches and stick insects were first allowed to run for at least 15 cm under control conditions. They were then exposed to a horizontal beam of light (800 Lux) which suddenly crossed their path under various angles. This usually induced the animal to change course by an angle δ_{ph}. After a run of about 10 cm in the beam, the light was cut off. This again induced the animals to make a turn, usually in the opposite direction, by an angle γ (counter-turn). In figure 1a and b the counter-turn γ is plotted over the phototactic turn δ_{ph} for Carausius and Blatella respectively.

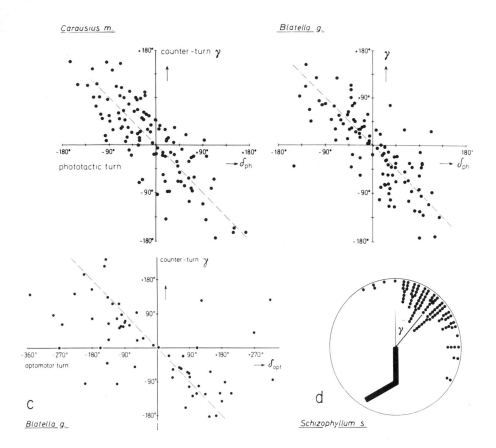

Fig. 1 a, b, c, d.   Counter-turns γ after preceding visually induced changes of course (see text). Distance of the light source 275 cm, light intensity 700 Lux. Points signify runs of, in a, b, and d, 10 animals, in c of 20 animals. c: optomotor deviations. Velocity of the drum 31°/sec; width of stripes 30°. d: counter-turn after a run through a shaded alley-way, bend –60°; arena Ø 30 cm, mean counter-turn γ + 41°, mean length of vector (inverse of variance) 0.89

It can be seen that, on the average, the magnitudes of these angles are equal. Thus the animal, after the entire procedure, has just about the same idiothetic bearing as before. In the second experiment the idiothetic run is periodically interrupted by the sudden appearance of a brightly illuminated striped cylinder which rotates around the animal's high axis. This causes a deviation, which, in contrast to the former experiment, is increasing. After about 10 seconds, the light is switched off. At that moment or, occasionally, after some delay a counter-turn becomes apparent. The optomotor turn (δ_{opt}) is measured as the difference between the tangent to the course at this turning-point and the tangent to the course at the preceding appearance of the drum. The counterturn (γ) is measured as the angle between the first named tangent and the tangent to the course of the subsequent appearance of the drum. The result is shown in fugure 1c. Here the relative magnitude of the counter-turn is slightly smaller than in the preceding case. In the third experiment, millipedes, after an idiothetic run, are led into a shaded path projected on the otherwise brightly illuminated arena. The bent form of the path and a circular histogram of the resulting counter-turns is shown in fig. 1 d.

Taken together, these results show that the animals are able to compensate optically induced deviations in such a way that their original bearing is approximately resumed.

The mechanism which leads to this type of idiothetic orientation has been analysed to some degree in millipedes. There (BURGER, 1971; BURGER and MITTELSTAEDT, in press) proprioceptive inputs measuring the angle between the animal's segments are linearly summed up and stored. The maximum load of the store determines the reference point by which the counter-turning is controlled. The latter gradually unloads the store again. Cockroaches and stick insects behave as millipedes do when induced to deviate from idiothetic courses by means of alleyways or other mechanical influences. Thus, our results in insects may be based on a similar mechanism rather than one which just stores visual information about the initial deviation and quantifies the counter-turn accordingly.

Conflicts.   There is reason to suppose, then, that the idiothetic mechanism is permanently in operation, independent of whether an external cue is present or not. Consequently, conflicts between idiothetic and allothetic tendencies on the course are to be expected. This becomes indeed apparent in the following experiments:After idiothetically running through a bent alleyway. the millipedes were either allowed to counter-turn in diffuse light (controls) or, in addition confronted by a black disk which, in pilot experiments, was shown to be very attractive under those light conditions.

Fig. 2a, b.   Counter-turning tendency in diffuse light with and without black disk (width 16°, hight 14°). Of the initial angle α only the mean vector is shown. Dots represent the points where a circle of radius 10 cm is crossed. a. $\alpha = + 57°$ $s = \overset{+}{-} 22°$, $\gamma = 47°$, $s = \overset{+}{-} 30°$. b. $\alpha = + 44°$, $s = \overset{+}{-} 22°$, $\gamma = + 13°$, $s = \overset{+}{-} 38°$, $n = 30$, 6 animals (Schizophyllum sabulosum)

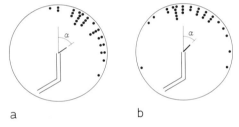

a   b

Fig. 2 shows the initial course α as a mean vector after leaving the alleyway. The dots show the points where the animals crossed a circle of a diameter of 15 cm. Both means show an intermediate between that of the controls and the direction of the black disk.

Whereas in the preceding experiment the allothetic stimulus interfered with the counter-turn, the last series tries to illuminate what happens when an allothetic stimulus is pitted against an otherwise undisturbed idiothetic course. The method is the same as in the very first experiments, yet the evaluation is different as is seen in Fig. 3.

The angle β between the light beam and the tangent to the animal's course at the moment, when the light is switched on, is plotted on the abcissa, whereas the angle β_1 between the beam and the tangent of the course 15 - 70 cm later is plotted on the ordinate. As can be seen in fig. 3a and c the points scatter, albeit considerably, around the line $\beta_1 = \beta_0$. Thus, under those light conditions, the average course did not change. Fig. 3b shows that under increased light intensity the number of turns away from the light becomes greater than that of turns towards the light, that is, the cockroaches show a tendency towards negative phototaxis.

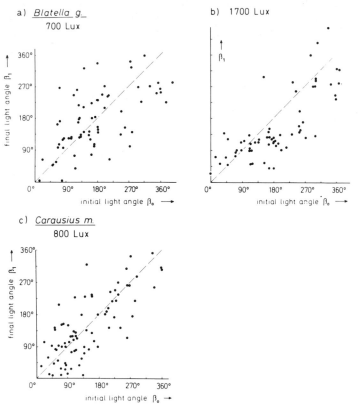

Fig. 3 a, b, c.   Final course β_1 over the initial course β_0 (see text). One dot represents one run. 6 animals

Discussion.   In view of these results, three possibilities of interaction of the two types of course control shall be discussed: (1) The actual course of the animal is the result of an equilibrium bet-ween all idiothetic and allothetic turning tendencies acting at every moment. (2) One tendency is dominant and excludes all others. (3) The two turning tendencies supplement each other (e.g. by alternating) to ensure a more or less straight course under varying environmental conditions.

In fig. 2 the animals ran on an intermediate mean course. This course changes while the animal is approaching the disk. This might indeed be the result of an equilibrium of type A. The results of fig. 3 seem also to be compatible with this idea. One would indeed expect that the relative weig of the light influence decreases with decreasing light intensity. Furthermore, even the distribution of the scatter of the final course over β_0 seems to be in line with what one would expect if the two turning tendencies would conform to the model of v. HOLST (1950) about the interaction of light and gravity in fish. If this hypothesis would turn out to be true, one could make quantitative predictions about those menotactic angles which result from an equilibrium between the tendency to run toward or away from a light source and a previously existing idiothetic tendency. Clearly, such an interaction would make the light course dependent upon the angle between the direction the animal would take under idiothetic control and the direction of the light source (AKRE,1964; LINSENMAIR-ZIEGLER, 1970). That may bias the results of experiments in which the previous history of the animals course is not taken into account.

If the result of the optomotor experiment is also interpreted along these lines, one would expect that, while the cylinder turns with constant speed, the rate of change of the animals course should decrease (because, if not, a constant optomotor stimulus would be confronted with an increasing counter-turning tendency!).

Although, at the present state, the results here reported are compatible with an equilibrium mechanism of type A, it is not at all excluded that further analysis might show that mechanisms of type B or C are realised. In fact, a critical reappraisal of GOERNER's earlier results (1966) seemsto indicate that, in spiders, all three mechanisms might be in operation.

Acknowledgement. The apparatus was constructed by Dr. E. KRAMER, Seewiesen.

References

AKRE, R. D.: Correcting behavior by insects on vertical and horizontal mazes. J. Kansa entomol. Soc. 37, 169 - 186 (1964).
BURGER, M.-L.: Zum Mechanismus der Gegenwendung nach mechanisch aufgezwungener Richtungsänderung bei Schizophyllum sabulosum (Julidae, Diplopoda). Z. vergl.Physiol. 71, 219 - 254 (1971).
BURGER, M.-L., MITTELSTAEDT, H.: in press.
GOERNER, P.: Ueber die Koppelung der optischen und kinästhetischen Orientierung bei der Trichterspinne Agelena labyrinthica (Cl.) und Agelena gracilens (C.L. KOCH). Z. vergl. Physiol. 53, 253 - 276 (1966).
HOLST, E.: Die Arbeitsweise des Statolithenapparates bei Fischen. Z. vergl. Physiol. 32, 60 - 120 (1950).
LINSENMAIR-ZIEGLER, C.: Vergleichende Untersuchungen zum photogeotaktischen Winkel-transponieren pterygoter Insekten. Z. vergl. Physiol. 68, 229 - 262 (1970).

5. The Relation of Astromenotactic and Anemomenotactic Orientation Mechanisms in Desert Ants, Cataglyphis bicolor (Formicidae, Hymenoptera)

P. Duelli
Department of Zoology, University of Zurich, Switzerland

Abstract.   The orientation of the desert ants, Cataglyphis bicolor, has been investigated in a natural habitat in Tunisia, where no efficient landmarks are available.On moonless nights the ants orient anemomenotactically on their return runs to the nest. The overall light intensity in the environment determines which sensory mechanism is used for orientation after displacement to an unknown area. Towards the moon some ants show phototropotactic adjustments up to a light intensity of 1 lux. In the range between 1 to 350 lux only anemomenotactical runs have been recorded, obviously because of the fading contrast between the moon disk and the sky. At a light intensity of more than 350 lux exactly the ants recognize the polarisation pattern of the skylight. The symmetry of this pattern at sunrise leads to a bimodal distribution according to the existence of two presumed home directions. The individual anemomenotactic zero direction of an ant decides which one of the two possibilities to choose. A method to compare the accuracy of different orientation mechanisms is given.

1. Introduction

Cataglyphis bicolor is a desert ant living in semi-arid areas of North Africa and in the western part of Asia. The outside workers are solitary hunters or collectors, never performing mass foraging along scent trails. On foraging runs they often stray from their nests more than 50 meters. Like bees, the single workers can be trained to go to special feeding places.

The experiments dealt with here were performed in a desert area near Maharès in southern Tunisia. As the natural habitat of Cataglyphis bicolor around this place is extremely flat, no efficient horizon-landmarks are available. For the experimenter this is of great advantage allowing him to describe exactly the parameters which are used by the ants for orientation. In the experimental area Cataglyphis bicolor can therefore only use the following four physical stimuli as orientation cues: the azimuth of the sun and the moon, the pattern of the polarized light of the sky and the direction of the wind (WEHNER, 1968, 1970, 1971; WEHNER and MENZEL, 1969; WEHNER and DUELLI, 1971).

2. Methods

As Cataglyphis bicolor hunters are foraging only during daytime the ants had to be trained under daylight conditions. On the other hand testing could take place at any time, day or night. The ants were watched on their way to the feeding place, captured there and rewarded. At the same time data about light intensity, wind direction (W_f), wind velocity and temperature were automatically recorded. The coordinates of that place could be determined by means of a grid of thin white threads, extending 30 by 40 meters over the whole experimental area. The return runs took place in a testing grid of exactly the same size and direction as the training grid has. The mesh width of both the grids was 1.0 meter. The testing place was away far enough from the training grid to make sure that the area was completely unknown to the ants. The return runs taking place either immediately after having captured the ants at their feeding place or up to a hundred hours later were exactly recorded until the animals started to search around at random. The mean direction of each return run was determined graphically (fig. 1).

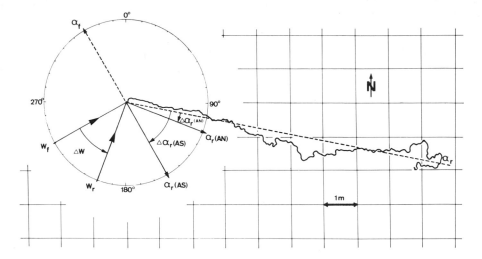

Fig. 1. Example of a protocolled return run (α_r) in the testing grid. The ant has been released in the centre of the ring at total darkness. The zero direction of the wind orientation differs from the permanent zero direction of an orientation towards the polarization pattern or the sun by an angle $\triangle W$. α_f, azimuth of foraging run; α_r, azimuth of return run; α_r (AS), presumed astromenotactic home (zero)direction; α_r (AN), presumed anemomenotactic home (zero)direction; $\triangle \alpha_r$ (AS), astromenotactic error; $\triangle \alpha_r$ (AN), wind error; $\triangle W$, difference between the wind direction during the foraging run (W_f) and the return run (W_r)

Since Cataglyphis bicolor is capable of a time compensated sun orientation, the presumed direction of the return run (zero direction) α_r (AS) of an astromenotactic orientation stays the same as long as the ant's knowledge of the time remains intact. It can be calculated as α_r (AS) = $\alpha_f - 180°$, α_f being the direction of the foraging run. If the ants oriented anemomenotactically, the zero-direction α_r (AN) would be $\alpha_f - 180° - (W_f - W_r)$. Whenever the wind direction during the foraging run (W_f) is the same as the one of the return run (W_r), a discrimination between an anemomenotactic and an astromenotactic orientation is obviously impossible (fig. 2A). To decide statistically whether a number of ants used visual cues or the direction of the wind for orientation, the relation of the two deviations (wind error and astromenotactic error) can be calculated for each return run α_r.

$$\text{astromenotaxis } 1 < \frac{\triangle \alpha_r (AN)}{\triangle \alpha_r (AS)} < 1 \text{ anemomenotaxis}$$

According to statistical methods of circularly distributed data (BATSCHELET, 1965), a mean vector can be calculated for the deviations from the zero direction of different series of ants. The accuracy of orientation mechanisms can be compared by projecting the mean vector on the home direction (zero direction) (fig. 2B). In this way we get the home vector r_o, representing data of the mean error and the dispersion of the series. A diagram (fig. 3) of the r_o vectors of several series enables us to compare the achievement of memory of two orientation performances.

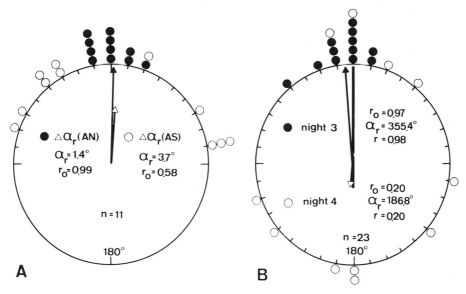

Fig. 2. Circular distributions of home (zero)directions measured after displacement of the ants to an unknown area. $0°$, home direction (presumed return direction). A) Return runs during daytime after the wind direction had changed more than $30°$ since the foraging run. Compound eyes and ocelli were covered with black paint. B) Comparison of the accuracy of anemomenotactic orientated return runs in the third and forth night after the foraging runs. α_r, mean vector of orientation; r, length of the mean vector; r_o, length of mean vector αr projected on the zero direction; n, number of individually trained and tested ants

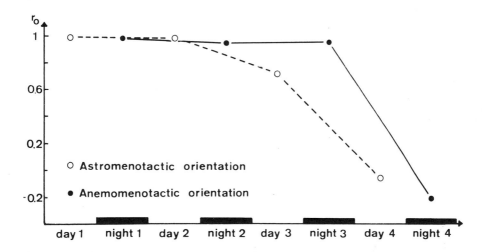

Fig. 3. Comparison of the capacities to memorize the performances of astro- and anemomenotactical orientation. On the fourth night after the foraging runs the ants searched around at random, the runs are no longer orientated. Under daylight conditions already after two days some ants perform unorientated searching runs. r_o, length of mean vector αr projected on the home (zero)direction

The capability of orienting towards the azimuth of the sun declines gradually, whereas the anemomenotaxis collapses only on the fourth night. Considering the complexity of a time-compensated astromenotactic orientation, this finding seems understandable. As MITTELSTAEDT mentioned in the discussion, our results could have been influenced by the effect of our keeping the ants in the dark until they were released in the testing grid. It would be better to expose the animals to normal daylight conditions.

Results. In a summarizing circular diagram we can show in what situation during the course of 24 hours which of the above mentioned orientation mechanisms are performed (fig. 4).

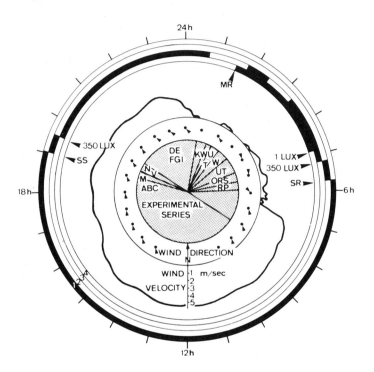

Fig. 4. Circadian diagram of the different orientation performances which can be observed in various combinations during the 24 h of a day. Black ranges of peripheric rings indicate used orientation parameters. 1. Sun azimuth (astromenotaxis). 2. Polarization pattern (astromenotaxis). 3. Wind direction (menotaxis). 4. Moon's azimuth (tropotaxis). MR, moonrise; SR, sunrise; SS, sunset.

The black rings at the periphery indicate the cues which were used for direction-finding at a special moment. The different physical stimuli can be acting isolated or in various combinations. At total darkness (see 24 h) the ants orient anemomenotactically. A menotactic orientation after the direction of the wind has been proved until now only for a few arthropod species (BIRUKOW, 1958; LINSENMAIR, 1968, 1969, 1970). The wind velocity has no effect on the accuracy of the orientation, even if it is not more than 0.4 m/sec. After elimination of the antennae, the ants searched around at random showing no statistically significant preference direction. The same results are obtained when there is no wind (see 1 h). As soon as the moon is visible to the ants, Cataglyphis bicolor shows positive or negative phototropotactic adjustments to the azimuth of the moon instead of the expected homing behaviour. In a situation of simultaneous presence of both moon and wind some ants use the moon, some the wind for direction finding. That alternative decision is influenced neither by the angle height of the moon nor by the wind velocity. Apparently the anemomenotactic zero direction of an ant decides whether it chooses a negative or positive phototropotactic course to the azimuth of the moon. Dusk is the most interesting period of time because it contains several changes of parameters. In the range between 1 and 350 lux only anemomenotactically orientated runs were recorded. Probably the contrast between the moon disk and the sky is too low to serve as an orientation cue. Precisely at an overall light intensity of 350 lux Cataglyphis bicolor is able to perceive the pattern of the polarized skylight. Moreover, runs of antennae-amputated ants confirm this statement. After searching around at random, all the animals almost at the same time start in their individual home directions as soon as light intensity has increased to 350 lux. Due to the symmetrical pattern of the polarized light at the time shortly before sunrise the ants have two zero directions and therefore show a bimodal distribution. In this case the ants' individual anemomenotactic home direction decides which of the two directions is preferred. It is evident, however, that ants with eliminated antennae show the clearest bimodal distribution. If the actual wind direction (W_r) leads to a wind error of more than $\triangle \alpha_r(AN) = 90^\circ$, the 180° counter-direction to the astromenotactic zero direction is chosen.

Immediately after the sun is visible to the ants the distribution becomes unimodal. As bees are able to recognize asymmetries of the polarization pattern of only $2 - 3^\circ$ (LINDAUER, 1957), this unimodality is not necessarily an effect of the ants' notice of the sun but can easily be explained by the rapidly increasing asymmetry of the polarization pattern.

During broad daylight Cataglyphis bicolor orients towards the azimuth of the sun or towards the pattern of polarized skylight. Even on a cloudy day when only 10% of the blue sky is visible the ants orient perfectly. Only in case of an entirely overcast sky or when eyes and ocelli are covered with black paint, Cataglyphis bicolor changes to wind orientation. Here again as in all the alternative decisions between different orientation performances no compromise directions could ever be observed. As we have no exact proof until now of Cataglyphis bicolor using the sun for orientation, only an experiment in which the sun and the polarization pattern artificially compete with each other will solve this interesting problem.

References

BATSCHELET, E.: Statistical Methods for the Analysis of Problems in Animal Orientation and certain Biological Rhythms. American Institute of Biological Sciences. Washington (1965).
BIRUKOW, G.: Zur Funktion der Antennen beim Mistkäfer Geotrupes silvaticus. Z. Tierpsych. 15, 265 - 275 (1958).
LINDAUER, M.: Sonnenorientierung der Bienen unter der Aequatorsonne und zur Nachtzeit. Naturwiss. 44, 1 - 6 (1957).
LINSENMAIR, K.E.: Anemomenotaktische Orientierung bei Skorpionen. Z. vergl. Physiol. 60, 445 - 449 (1968).
LINSENMAIR, K.E.: Anemomenotaktische Orientierung bei Tenebrioniden und Mistkäfern. Z. vergl. Physiol. 64, 154 - 211 (1969).
LINSENMAIR, K.E.: Die Interaktion der paarigen antennalen Sinnesorgane bei der Windorientierung laufender Mist- und Schwarzkäfer. Z. vergl. Physiol. 70, 247 - 277 (1970).
SANTSCHI, F.: Observations et remarques critiques sur le mécanisme de l'orientation chez les fourmis. Rev. Suisse Zool. 19, 303 - 338 (1911).
SANTSCHI, F.: Les différentes orientations chez les fourmis. Rev. Zool. Afric. 11, 111 - 144 (1923).
WEHNER, R.: Optische Orientierungsmechanismen im Heimkehrverhalten von Cataglyphis bicolor. Rev. Suisse Zool. 75, 1076 - 1085 (1968).
WEHNER, R.: Die optische Orientierung nach Schwarz-Weiss-Mustern bei verschiedenen

Grössenklassen von Cataglyphis bicolor. Rev. Suisse Zool. 76, 371 – 381 (1969).

WEHNER, R.: Die Konkurrenz von Sonnenkompass- und Horizontmarken-Orientierung bei der Wüstenameise Cataglyphis bicolor. Verh. d. Dtsch. Zool. Ges. 16, 238 – 242 (1970).

WEHNER, R. and MENZEL, R.: Homing in the ant Cataglyphis bicolor. Science 164, 192 – 194 (1969).

WEHNER, R. and DUELLI, P.: The spatial orientation of desert ants, Cataglyphis bicolor before sunrise and after sunset. Experientia 27, 1364 – 1366 (1971).

6. The Significance of Different Eye Regions for Astromenotactic Orientation in Cataglyphis bicolor (Formicidae, Hymenoptera)

R. Weiler and M. Huber
Department of Zoology, University of Zurich, Switzerland

Abstract.   In freely running ants orientating astromenotactically in their natural environment, different parts of the eye were covered with paint and the middle frontal parts found to be most decisive for astromenotactic orientation.

It has been proved (FRISCH, 1965) that bees cannot orient astromenotactically, when the dorsal parts of both eyes have been covered with paint. This result was confirmed by JENNI (unpubl., see also WEHNER, this volume).

A series of experiments, which were carried out in Southern Tunisia in summer 1970 in which either the dorsal or ventral parts of both eyes of Cataglyphis bicolor were covered, did not show the expected large differences in the orientation ability. Thus, the separation in dorsal and ventral halves proved itself to be insufficient.

In all the experiments described here the covering referred to the mean axes of the rather elliptical eye whose short one is similar to the z-axis, which in the normal head position is approximately horizontal (HERRLING, this volume).

In summer 1971 the following series of experiments were carried out (fig. 1). For research location, training and test conditions see DUELLI (this volume).

Fig. 1.   Group A   Group B   Group C
          horizontal   vertical   quarters covered
          thirds covered   thirds covered

The white parts of the eye are uncovered

1. Methods

An ant which had left the nest in search of a prey was followed and caught after 20 meters foraging run, fed and brought to the experimental site. While the insect was tightly held by the legs under the binocular it bit firmly into the finger holding it, thus keeping the head very still during the following procedure. The antennas were cut away to prevent any influence by the wind (WEHNER and DUELLI, 1971) and an appropriate covering of quick drying acryl-colour was applied with a fine needle.

The ant was released and the return run was noted in the way described by DUELLI in this volume (fig. 2).

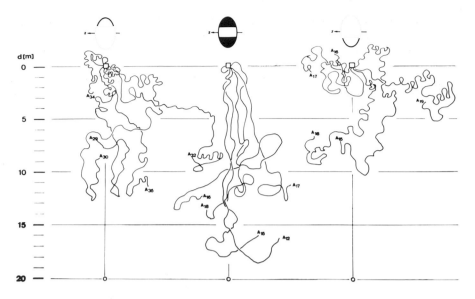

Fig. 2. 15 examples of recorded runs in a "test grid" experiment.

  o hypothetical nest position
  □ releasing point

After the ant was recaught and the covering once again controlled, the head was severed. Photographs were taken of these heads by a photobinocular and measurements were made on the negatives. The further evaluations were only done with run of ants whose both eyes were in order. For a more exact examination of some eyes, stereoscan photographs were taken (fig. 3).

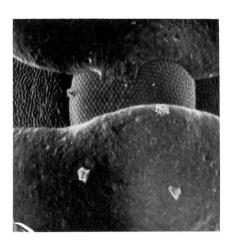 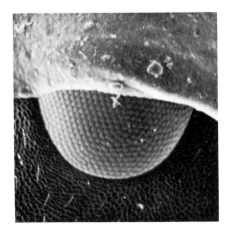

Fig. 3. a) Stereoscan photograph: right eye, middle horizontal third uncovered.
         b) Stereoscan photograph: right eye, ventral horizontal third uncovered

Each series in the group was carried out twice: In the experimental "training grid" the insect was returned to the place of capture, in the experimental "test grid" the releasing point was about 40 meters away in a region unknown to the ant. In fig. 4 the data of two examples were taken to demonstrate the difference between the two experiments. (For calculation of l/d see later.) In both cases the difference between the curves is not significiant. That means that really the value of the visual areas for the astromenotactical orientation was proved and not, for example, for the familiarity of the terrestrial environment.

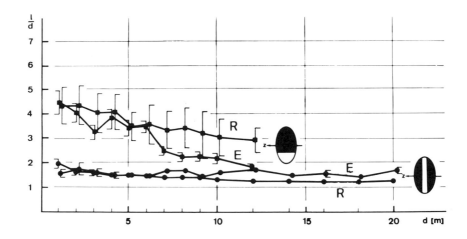

Fig. 4. Two examples to demonstrate the difference between "training grid" (R) and "test grid" (E) experiments. In both cases the difference between the two curves is not significiant

For each single treatment, control series were performed with 20 animals. Fig. 5 shows two of these control series. (For calculation of r see later.) Control A shows the data of ants which were taken 20 meters from the nest and then allowed to return without any treatment. Control B concerns insects whose antenna were cut away and whose eyes following the described method were painted over, but in which covering was removed shortly before the ants were set free again.

2. Results

a) Mean Vector. A transparent paper with concentric circles and 10° angles was placed over the recorded return runs in such a way that the center coincided with the releasing point, the beginning of the run, and the zero direction represented the line between the releasing point and the nest or the hypothetical nest in the "test grid" series. The intersection of a run and a circle gives points that we can represent in a circular distribution. From this data we can calculate a mean vector with length r and angle α to the zero direction (BATSCHELET, 1965) (fig. 6).

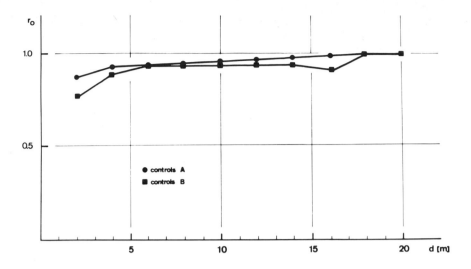

Fig. 5. Control series. For further explanations see text

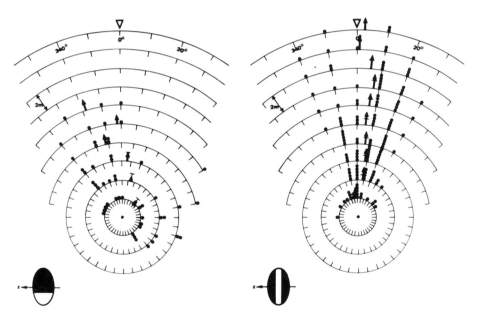

Fig. 6. Circular distributions. ▲ nest entrance. Center of the circles correspond with the releasing point

r can be considered as significiant with p ∠ 0,01 when the appropriate table value for n has been scaled (GREENWOOD and DURAND, 1955). This value is marked by the thin line in each vector r.

We multiply r with the cosine from α and get r_o. The nearer r_o is to 1, the better coincides the angle of orientation with the zero direction and the smaller is the dispersion of the tested population. Figs. 7 to 9 show the r_o data for each group.

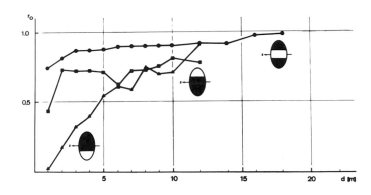

Fig. 7. r_o data. *Horizontal thirds uncovered

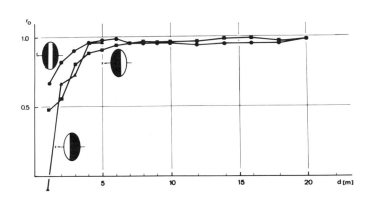

Fig. 8. r_o data. Vertical thirds uncovered

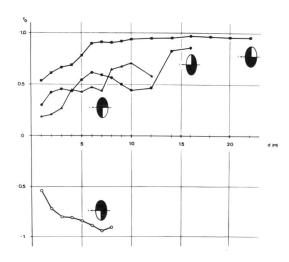

Fig. 9. r_o data. Quarters uncovered

---

*r_o represents the product of r and the cosine of the angle between r and the zero-direction.

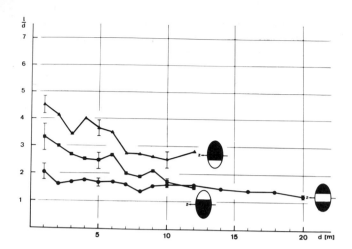

Fig. 10. $\frac{l}{d}$ – data. *Horizontal thirds uncovered

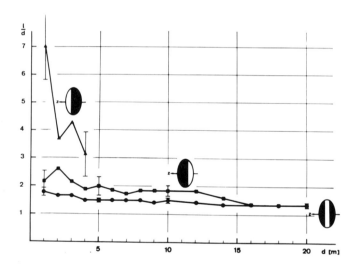

Fig. 11. $\frac{l}{d}$ – data. Vertical thirds uncovered

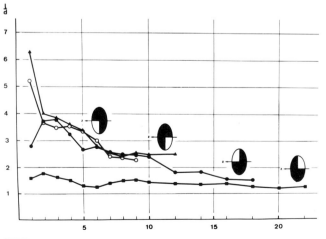

Fig. 12. $\frac{l}{d}$ – data. Quarters uncovered

*l represents the length of the run, when the distance d from the releasing point is reached.

b) Straightness of Runs. The accuracy of direction is not the only criterion which offers information about the orientation. The straightness of the runs also depends upon the orientation ability. Therefore the length of a run was measured from the releasing point to each circle and divided through the radius of the circle presenting us the ratio l/d. The nearer l/d is to 1, the more
direct is the run. Figs. 10 to 12 show the l/d-data for each group.

Fig. 13 shows for each group the reciprocal value of the means calculated from all data l/d for
$1 \leqq d \leqq 10$ m.

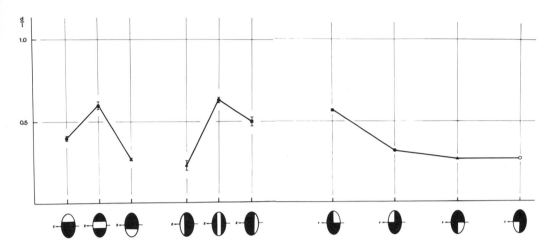

Fig. 13. Reciprocal mean of all data between one and ten meters distance from the releasing point

Each of these two possibilities of evaluation shows that the middle frontal part of the eye of
Cataglyphis bicolor has an important function for astromenotactical orientation.

Histological research of the eye of Cataglyphis bicolor will show whether this difference has any
correspondence in anatomy.

Furthermore, studying runs with high-speed cinematography will give yet more exact information
about the orientation of the eye in space during the runs.

Acknowledgement.   The work was financially supported by a grant awarded to RUDIGER WEHNER
by the Fonds National Suisse de la Recherche Scientifique, No. 3.315.70.

References

BATSCHELET, E. : Statistical methods for the analysis of problems in animal orientation and
        certain biological rhythms. American Institute of Biological Sciences (1965).
FRISCH, K. v.: Tanzsprache und Orientierung der Bienen. Springer Verlag, Berlin (1965).
GREENWOOD, J.A. and DURAND, D.: The distribution of length and components of the sum
        of n random unit vectors. Ann. Math. Stat. 26, 233 - 246 (1955).
WEHNER, R. and DUELLI, P.: The spatial orientation of desert ants, Cataglyphis bicolor, before
        sunrise and after sunset. Experientia 27, 1364 - 1366 (1971).

7. The Visual Orientation of Desert Ants, Cataglyphis bicolor, by Means of Terrestrial Cues

R. Wehner and I. Flatt
Department of Zoology, University of Zurich, Switzerland

Abstract.   When in desert areas Cataglyphis ants are displaced from the nest entrance to any point more than 2 m apart, nearly all ants search around at random. Therefore a method is described to train the ants to a feeding place by simultaneously extinguishing all astromenotactical information about the azimuth of that place. Then the storage and use of information about terrestrial cues can be measured unambiguously. During a set of one foraging and one return run the ants receive information about a "route", extending from the nest entrance to the feeding place and characterized by a specific constellation of tiny terrestrial cues. By offering artificial cues (black, vertically arranged two-dimensional screens) the significance of this route for reorientation can be markedly enhanced. A method is developed to quantify this effect of artificial horizon landmarks. In this way pattern recognition problems can be studied.

1.   Introduction

If a hunter ant of Cataglyphis bicolor is displaced in various directions and distances from the nest entrance, a very poor knowledge of the desert area surrounding the nest entrance can be proved. It is surprising to see how an ant returning from a foraging run by use of astromenotactic orientation (WEHNER, in press) may reach the nest entrance by a straight course ($l/d = 1.1$, mean value; l = length of the trace, d = distance from feeding place to nest entrance, i.e. bee-line) and by a mean speed of the return runs of 15 m/min, even about distances of more than 150 m, whereas the ant does not succeed in finding the nest entrance when it is only 2 m away from it, but directly displaced to that point. In two different places of southern Tunisia (Maharès, Gabès) only ratios of 0.09 and 0.07 of the ants reached the nest entrance by means of $l/d < 1.1$ (shaded areas in fig.1), after having been displaced for 2 m and 5 m in various directions. When displaced for 3 m (as in all succeeding experiments), only a ratio of 0.57 of the ants reached the nest entrance in less than 5 min (n = 30). The mean time of 2.18 $\pm$ /.28 min, needed by these ants for the distance of 3 m, is 10 times the one of an astromenotactically orientating ant. In all these cases, of course, no foraging runs preceded the displacement so that no astromenotactic zero direction could have been built up.

In this study we would like to present experimental evidence for the acquisition of information about terrestrial cues in Cataglyphis bicolor, after the astromenotactic compass orientation has been extinguished.

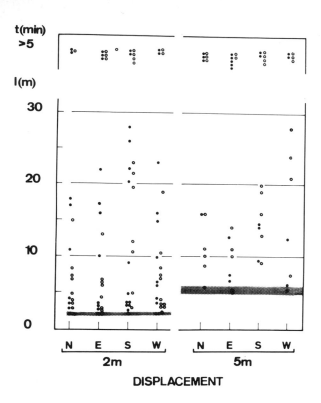

Fig.1. Length (l) of the traces of ants having returned to the nest entrance after displacement for 2 m and 5 m in various directions from the nest entrance. The shaded areas represent the l-interval due to ants returning from foraging. The numbers of ants, which did not arrive at the nest entrance within less than 5 min, are plotted above

2. Methods

a) Training procedure. The experiments were performed from July 17 to September 12, 1970, in a desert area near Maharès (35.53° N, 10.49° E). An individually marked ant was further trained to a feeding place, 3 m apart from the nest entrance (fig.2: foraging run, S.4). After reward it immediately returned to the nest (return run). When the ant again started to leave the nest entrance – a few minutes or even a few seconds later – it was rewarded just at the nest entrance. By this procedure the stored information about the azimuth of the previous feeding place (3 m apart) is completely extinguished, demonstrated by the fact that the ant can no longer orient astrometanotactically from any point of the surrounding area.

b) Test program. After reward at the nest entrance the ant is displaced to the previous feeding point (fig.2: displacement) in the training grid (grid A) or to any point in a test grid (grid B) completely unknown to the ants (for methods see DUELLI, this volume). Training and test procedure are either performed in the unchanged natural environment (S.4) or by additional use of artificial horizon landmarks consisting of an array of three black, vertically arranged square–shaped screens seen under visual angles of 28°x28° from one screen to the other. These artificial horizon landmarks are offered either during both foraging and return runs(S.1), or during the foraging run (S.2) or return run (S.3) alone.

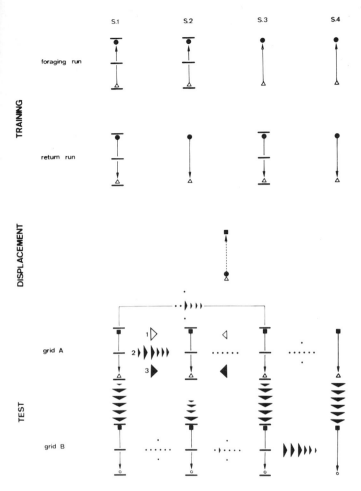

Fig. 2. Training procedure and test program. △ nest entrance; ●, feeding place; ■ , releasing point after displacement; o, hypothetical nest entrance point in grid B; grid A , training area; grid B, test area. The black bars mean artificial cues, the arrows (signatures 1, 2 and 3) indicate statistically significant differenc es between different series. For further explanation see text (methods)

c) Evaluation of data. The return runs in the test situations (grids A and B) are recorded by means of Cartesian coordinates according to WEHNER (in press), DUELLI (this volume), and fig. 3 (left part). The length (l) of the traces are then measured and subdivided into intervals of 1 m. Plotting the distances (d) from the nest entrance after every meter of trace length, each trace is transformed to a l/d function, $d = f(l)$, as seen in the right part of fig. 3. The zero point of that function represents the length of the trace performed by an individual ant having reached the nest entrance. From these particular curves a mean length–distance function $d = f(l)$ can be calculated.

d) Statistical approach. In figs. 4–6 the mean values and standard errors of mean values are plotted in intervals of $\triangle l = 1$ m. In order to prove statistical differences between the l/d functions of various test situations, t-tests are performed for the values of $2 \leqq l \leqq 7$ m. The statistical significance of these differences ($p < 0.001, 0.01, 0.05$) is represented by the strength of the black arrows shown in fig. 2 (test situation, signature 2). Signatures 1 and 3 in fig. 2 are due to temporal parameters: the former characterizing differences in absolute returning times (t_r), the latter corresponding differences in relative returning times (t_r / t_f ; t_f = time interval of the proceeding foraging run). The arrows always point in the direction of less accurate orientation performances. Where a small circular point is given instead of an arrow, no significant differences could be proved ($p > 0.05$).

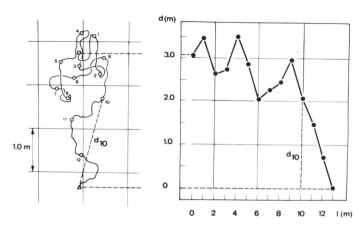

Fig. 3. Evaluation of data.
Left part: record of a single
return run. Trace length intervals of 1m are indicated.
Right part: length-distance
function for the trace graphed
on the left-hand side. 1,
length of the trace; d, distance
from the nest entrance (△)

Results

a) The signifiance of terrestrial cues.   When an ant is trained to a point 3m apart from the nest
entrance, it normally reorientates by means of the astromenotactic mechanism. That mechanism,
however, is switched off in our experime nts by previous reward at the nest entrance. Reorientation
is now only possible by using terrestrial cues learnt during training (foraging).

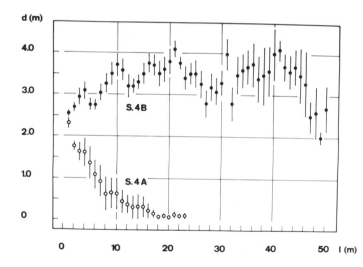

Fig. 4. Length-distance functions (mean values and mean errors of mean values) of series S.
4A and S, 4B (compare fig. 2)

The total extinction of any astromenotactic compass information is proved by fig. 4 (S. 4B), where
the ants are released in the test grid B lacking all information about terrestrial cues in the training grid A. Even after courses of 50m in length, the ants do not reach the corresponding point of
the nest entrance (which lies in grid A, not B), but search around at random. The same test performed in grid A instead of B (fig. 4, S. 4A) delivers strong evidence for the significance of terrestrial cues learnt along that route during the ant's previous foraging and return run.

If, additionally, horizon landmarks are offered along that route (see methods), the ants even in grid
B gradually orient towards the hypothetical nest entrance point (fig. 5, S. 1B). But as the comparison between S. 1B and S. 4A shows (fig. 5, lower part), the minute terrestrial cues in grid A represent stronger information about the visual environment than the array of large-sized black

screens. In all series, the differences between the l/d functions of S. 4B and the series considered can be taken for quantitatively measuring the significance of the cues in study.

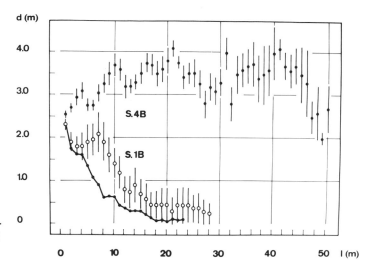

Fig. 5. Length–distance functions of series S. 4B, S. 1B and S. 4A (solid line)

b) The time phase of storaging visual information. The strongest information about the artificial cues is built up, when the screens are offered at both foraging and return runs (fig. 6, S. 1.A). With the presentation of the screens restricted to the foraging phase, the l/d functions of that series (S. 2A) and S. 4 are completely identical. Cues learnt on the return run, however, improve the reorientation ability in the test situation (S. 3.A), as one may deduce from the l/d functions (fig. 6) as well as from the reorientation times (fig. 2).

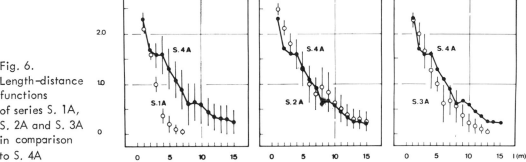

Fig. 6. Length–distance functions of series S. 1A, S. 2A and S. 3A in comparison to S. 4A

While the astromenotactic course used for the return run is calculated by the 180° reversal of the immediately proceeding foraging course (asimuth of the feeding place) and no compass angles have been learnt during previous return runs (WEHNER, in press), nearly the opposite seems to be true for the storage of information about terrestrial cues. Landmarks used in the return situation are predominantly imprinted during preceding return runs. Furthermore, the information about these cues is not directly extinguished by succeeding foraging runs at different feeding places, as it holds for astromenotactic angles.

c) Competition between terrestrial and celestial cues. When during the return run an astromenotactic angle competes with the direction inicated by terrestrial cues, the ants exclusively perform
alternate decisions instead of compromise directions (WEHNER, 1970). In fig. 7 one of hundreds
of examples show the strongly maintained astromenotactic orientation, when the ant as well as the
horizon landmarks leading to the nest entrance are displaced. In 27 succeeding runs that particular
ant started in a mean direction of $104°$ (instead of $98°$, theoretically predicted). The ends of the
straight courses are marked by the so-called searching points determined according to the method
described in BURKHALTER (this volume). After displacement (in the open loop condition) an ant always calculates a compass direction by means of the compass angle of the preceding foraging run,
not by that of the preceding return run. Therefore no different astromenotactic zero directions can
be learnt for both types of runs. With regard to horizon landmarks the next question has to be, whether or not this capacity is due to the storage of information about terrestrial cues.

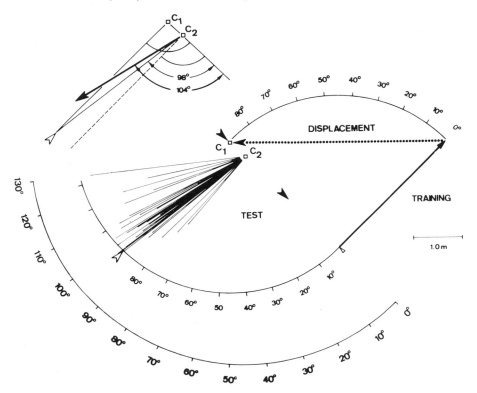

Fig. 7. Competition between an astromenotactic angle and horizon landmarks positioned
in the direction to the nest entrance. The example presented here is based on measurements
of a single ant. ▲ , horizon landmarks displaced from the training direction; Ⱥ , hypothetical nest entrance point expected by a merely astromenotactically orientating ant; △ ,
nest entrance; C1, releasing point (centre of the inner circle); C2, mean point of the beginning of the ants straight courses (centre of the outer circle). The mean direction of the
27 return runs after displacement is shown in the inset figure. For further explanation see
text

4. Discussion

a) As previously shown (WEHNER, in press) and now reconfirmed, the astromenotactically determined azimuth of the return run is always calculated by the $180°$ reversal of the azimuth of the im-

mediately preceding foraging run. Learning of a compass direction on the return run (tested after displacement, i.e. under open loop conditions) could never be proved (compare fig. 7). In bees, correspondingly, the azimuth of the foraging flight and not that of the return flight is transposed

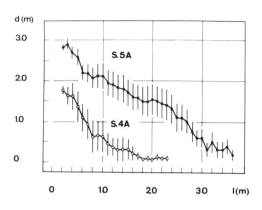

Fig. 8. Length–distance functions of S. 4A and of ants displaced from the nest entrance with no preceding foraging run (S. 5A). In the latter case presenting the same experimental arrangement as in fig. 1, the ants search around at random, until they reach the nest area by chance

to the gravity field in the waggle dance. Only under special training conditions the return flight, too, becomes significant to a certain degree (OTTO, 1959).

As the return course is determined by 180° reversal of the corresponding foraging course, the information about the azimuth of a special feeding place can be completely extinguished by subsequent reward at the nest entrance. Even after one of these treatments, no astromenotactic reference direction can be proved when the ant is displaced to any unknown point in its environment.

b) On the contrary, horizon landmarks used during the return run are predominantly, but not exclusively learnt during the preceding return run. In the digger wasp, Bembix rostrata, learning of terrestrial cues around the nest entrance always occurs in the "pre-learning flights", whereas the "pre-entering flights", i.e. the approach flights to the nest entrance, are shown to be unimportant for imprinting information about landmarks (IERSEL and ASSAM, 1964).The question, whether cues learnt on the foraging run can be used by 180° reversal for reorientation to the nest entrance, has been studied for wasps (K. WEISS, 1954) and ants, Formica chaufussi (B. WEISS and SCHNEIRLA, 1967). Both experimental arrangements, however, are unsuitable to test this problem unambiguously so that the results need not be discussed in that context.

c) An astromenotactic course is precisely learnt by one foraging run. Even information about terrestrial cues along a "route" of 3m are stored during one set of foraging and return runs in such a way that reorientation by the use of landmarks alone is significantly improved with respect to reorientation preceding the training run (fig. 8). For comparison, spectral colours are at least as quickly learnt as in bees (MENZEL, 1967; WEHNER and TOGGWEILER, 1972), and astromenotactic and anemomenotactic directions are able to be stored for 3 or 4 days, when in the meantime the ants are kept in complete darkness (DUELLI, this volume).

d) As ants, displaced to any place only 2 to 5m away from the nest entrance, search around at random and as there is a significantly more precise reorientation after one foraging and return run only by means of terrestrial cues, information of the "route" must be imprinted in the ant's central nervous system. More detailed experiments on the extensions and structures of those routes are described by BURKHALTER in this volume.

e) Alternative decisions instead of compromise directions are found in all competition situations studied in the natural environment: between astromenotactic directions and directions defined by horizon landmarks (fig. 7; WEHNER, in press)as well as between astromenotactic angles (WEHNER and DUELLI, 1971; DUELLI, this volume). As GOERNER points out in this volume. the spider Agelena labyrinthica assumes a resultant direction when optic and kinesthetic reference directions diverge (compare MOLLER, 1970). He also presents experimental evidence for the assumption that in the spider's central nervous system the optic and the kinesthetic components are integrated before the spider's course is calculated. The experiments on Agelena labyrinthica differ from those on Cataglyphis bicolor in so far as in the latter case one studies the competition of two reference directions due to external stimuli, whereas in the former situation the competition exists between allothetic (optic) and idiothetic (= kinesthetic) turning tendencies (see MITTELSTAEDT-BURGER, this volume).

Acknowledgement. The experimental work was financially supported by grant No. 3.315.70 of the Fonds National Suisse de la Recherche Scientifique.

References

IERSEL, J. J. A., ASSEM, J.: Aspects of orientation on the digger wasp Bembix rostrata. Anim. Behav. Suppl. 1, 145 - 162 (1964).

MENZEL, R.: Untersuchungen zum Erlernen von Spektralfarben durch die Honigbiene. Apis mellifera. Z. vergl. Physiol. 56, 22 - 62 (1967).

MOLLER, P.: Die systematischen Abweichungen bei der optischen Richtungsorientierung der Trichterspinne Agelena labyrinthica, Z. vergl. Physiol. 66, 78 - 106 (1970).

OTTO, F.: Die Bedeutung des Rückflugs für die Richtungs- und Entfernungsangabe der Biene. Z. vergl. Physiol. 42, 303 - 333 (1959).

WEHNER, R.: Die Konkurrenz von Sonnenkompass- und Horizontmarken-Orientierung bei der Wüstenameise Cataglyphis bicolor (Hymenoptera, Formicidae). Verh. dtsch. zool. Ges. 64, 238 - 242 (1970).

WEHNER, R.: Visual orientation performance of desert ants, Cataglyphis bicolor, towards astromenotactic directions and horizon landmarks. Pro. AIBS Symp. Animal Orientation and Navgation. Wallops Station, Virginia; in press.

WEHNER, R., DUELLI, P.: The spatial orientation of desert ants, Cataglyphis bicolor, before sunrise and after sunset. Experientia 27, 1364 - 1366 (1971).

WEHNER, R., TOGGWEILER, F.: Verhaltensphysiologischer Nachweis des Farbensehens bei Cataglyphis bicolor (Formicidae, Hymenoptera). Z. vergl. Physiol. 77, 239 - 255 (1972).

WEISS, B.A., SCHNEIRLA, T.C.: Inter-situational transfer in the ant Formica schaufussi as tested in a two phase single choice point maze. Behaviour 28, 269 - 279 (1967).

WEISS, K.: Ueber den Hin- und Rückweg bei der Labyrinth-Orientierung von Bienen und Wespen. Z. vergl. Physiol. 36, 531 - 542 (1954).

8. Distance Measuring as Influenced by Terrestrial Cues in Cataglyphis bicolor (Formicidae, Hymenoptera)

A. Burkhalter
Department of Zoology, University of Zurich, Switzerland

Abstract. By means of backward, forward and lateral displacements the influence of terrestrial cues on the distance measuring mechanism was investigated in ants, Cataglyphis bicolor, performing their return runs. In an unknown area, where visual input from horizon landmarks is excluded, the ants run in a straight course (r) 110% of the distance d_o between nest entrance and feeding place. This ratio r/d_o is systematically altered when the releasing points are situated along special lines in the known area near the nest entrance. By this a "route" can be calculated, along which the ant can find its way by means of horizon landmarks. In the desert regions of southern Tunisia that route extends up to a distance of 15 m from the nest entrance and for 5 m perpendicular to this axis.

1. Introduction

We can illustrate the investigated problem and the questions of our interest by means of the following experiment: If we displace a hunter ant of Cataglyphis bicolor in an unknown area, the ant for a certain distance runs back in the sun compass direction, suddenly interrupting its direct course and starting to perform searching runs. The fact that the animals compensate their foraging run on the backway in an unknown area, allows us to conclude that distance measurement is completely independent of the knowledge of familiar terrestrial cues. In that work the point of interest was the mechanism responsible for distance measurement as influenced by exogenous stimuli. That analysis may present evidence for the internal and external parameters of the distance measuring system. The results of WEHNER and FLATT (this volume) have shown the existence of a route-like specific knowledge of the area surrounding the nest entrance. Following these results, we were interested in the dimensions of the route, valid under natural conditions. Therefore, hunters were captured after foraging runs of 20 and 40 m length and displaced either into the test grid or in the training grid with the releasing point shifted from the training point in backward, forward or sideways directions.

2. Method

The measurement of the length of the direct sun compass return run needs a reliable definition of the searching point. Based on the records of each run, the ratio l/d was calculated in 2 m long steps according to the contribution of WEILER and HUBER in this volume. The real searching point was defined to be that point on the curve (fig. 1), where the curves ascends by at least $45°$. The comparison of the return run distances resulting according to these calculations and of those, measured subjectively in the actual records shows no significant differences. In addition to the correlation mentioned above, we have used another method for data processing. As the dependent variable we did not use the quotient l/d, but d_m/d_o. The criterion for the searching point was again a $45°$ ascent. Inaccuracies on the first 4 m of the run were not considered. This second method did not bring forth any differences in the localisation of the searching point, if the results are compared with the method described previously. The accuracy of the sun compass direction was proved with the aid of statistics of circular distributed data (BATSCHELET, 1965). We only used angle deviations of the courses between the starting points and the searching points.

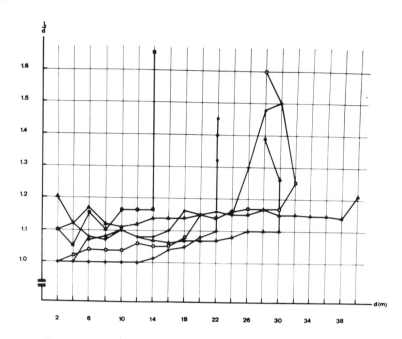

Fig. 1. Ratios of l/d for varying distances from the start and the releasing point. From these graphs the searching points are calculated according to a more than 45° deflection of the mainly horizontal lines. For abbreviations see chapter 2

Mean vectors and standard deviations were calculated. Abbreviations used in the following contribution: d_o = distance between nest entrance and feeding place; r = length of the ant's straight return course, measured from the releasing point (after displacement) to the searching point; b_d = distance of backward displacement; f_d = distance of foreward displacement; l_t = distance of lateral displacement; d_t = length of the course performed by the ant in the training grid as measured after reward at the feeding place, but before the ant is displaced in the test grid.

3. Results

The hunters directly displaced from the feeding point into the test grid slightly overcompensated their foraging run on their return run, with the quotient r/d_o equal to 1.1 and standard deviations equal to ± 0.24, standard errors of the mean values equal to ± 0.06. In the further experiments this percentage was considered to be put forward by the distance measuring mechanism, irrespective (or in the absence of) terrestrial cues.

a) The length of the route. The animals were displaced backwards from their feeding places in distances of 5, 10, 15, 20 and 25 m, 20 m apart from the nest entrance. Starting point, feeding and nest entrance point were situated on a straight line. Fig. 2 shows in its lower part the percentage of the successful return runs. The differences between the test grid data of r/d_o and measurements in this series decrease with increasing backward displacements. The number of ants finding their way back to the nest entrance decreases correspondingly. In the cases of 15 m, 20 m and 25 m displacements of the standard deviations are remarkable. They indicate the great variability of the samples. Therefore the ants need more information to accomplish the run.

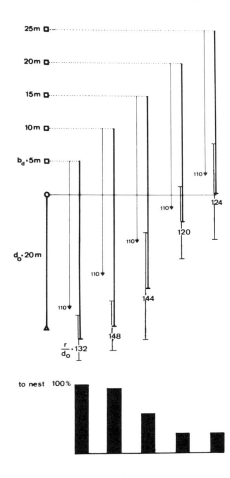

Fig. 2. Results of the backward displacement experiments. For explanations and abbreviations see text

The diagram proves that the achieved information is taken from the nest surroundings along a distance of nearly 15 m.

b) The width of the route. In order to determine the width of the route the ants were displaced in lateral directions from the straight return course and (for reasons not discussed here in detail) simultaneously backwards for b_d = 8.3 m. Fig. 3 shows the decrease of the length compensation with increasing l_d. The series with l_d equal to 10 m, 15 m and 20 m do not differ from the results in the test grid. It has to be pointed out that 70% of the ants in the series with l_d equal to 5 m returned absolutely correct in the mean sun compass direction; the remaining ones were slightly drifted towards the nest entrance in the last third of their return runs.

c) Displacements in the direction towards the nest entrance (forward displacement). The ants were displaced for 5, 10, 15 and 20 m along their foraging courses. As expected, each animal returns to the nest entrance. Under certain circumstances the series f_d equal to 20 m shows the same result. The standard deviations increase from the series of 5 m to that one of 15 m (0.087; 0.134; 0.177). If one supposes distance measuring to be performed by an internal store with its load determined by the foraging run, and influenced by terrestrial cues, then both parameters must compete in the forward displacement experiments. That competition results in a decrease of the r/d_o ratio.

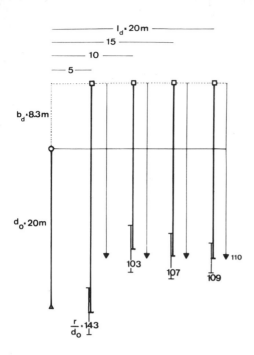

Fig. 3. Results of the lateral displacement experiments. For explanations and abbreviations see text

d) To demonstrate this competition between the internal distance information storage and the terrestrial cues of a known area more precisely, we simulated backward displacements in the well-known area between the nest entrance and the feeding place (fig. 5). Animals started at the feeding place to pass their return runs partially for 5 m, 10 m, 15 m and 20 m in the known area. Then they were captured once again and brought back to the feeding place. In each experimental situation, Cataglyphis bicolor finds the way back to the nest, with exception of the series with d_t equal to 20 m. In that case the distribution was bimodal. 20% started to search around immediately at the feeding place, 10% returned to the nest entrance.

4. Conclusions

All preceding papers concerning visual orientation of Cataglyphis bicolor deal with direction finding by celestial as well as terrestrial cues. Additionally, in that contribution distance measuring by necessity cooperating with astromenotactical orientation is investigated. Displacement of the ants in a test grid lacking all familiar horizon landmarks presents clear evidence for an internal central nervous store computing informations on the length of the foraging run. As these runs, recorded in our experiments up to a distance of 20 m and 40 m, respectively, are nearly straight ($l/d = 1.1$; l = length of the performed run, d = distance from the nest entrance), at the present status we cannot decide whether l or d is computed in the store.

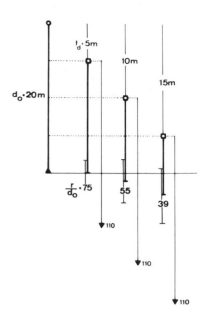

Fig. 4. Results of the forward displacement experiments. For explanations and abbreviations see text

In the unknown test grid the ants perform a straight return run, the length r of which is 1.1 with
respect to the expected value of d_o = 20 m resp. 40 m. By releasing the ants after reward in
the training grid, i.e. in the nest entrance surround, different ratios r/d_o are obtained according
to the different types and distances of displacement. Differences between the ratio r/d_o = 1.1
and the ratios obtained in the particular series can be only due to the terrestrial cues in the
known area of the nest entrance surroundings. By this a method is found to investigate the topology of that known area. According to WEHNER and FLATT (this volume), a "route" of a known
constellation of terrestrial cues exists along the first part of the foraging pass. The cues along
that route are highly efficient near the nest entrance, so that a gradient can be calculated from
the experimental results in the direction of the course (figs. 2, 4 and 5) and in the one perpendicular to the ant's course (fig. 3). When a ratio of r/d_o = 1.1 is reached farther than 15 m from
the nest entrance or more than 5 m laterally to the foraging route, the ants stop their straight
courses, i.e. reach their searching points, and hence do not further enlarge the r/d_o -ratio.
Therefore the known landmarks studied by WEHNER and FLATT (this volume) are totally situated
within that route.

As this presentation only deals with the influence of terrestrial cues on the distance measuring
mechanism, no remarks will be made on the latter mechanism itself.

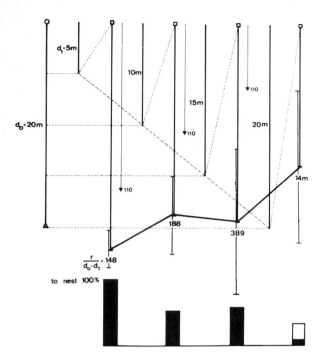

Fig. 5. Results of backward displacement experiments with ants having performed varying parts d_t of their return runs in the known area. The ants were always released at the previous feeding place. For explanations and abbreviations see text

Acknowledgement.   The work was financially supported by grant No. 3.315.70 offered to RUDIGER WEHNER by the Fonds National Suisse de la Recherche Scientifique.

References

BATSCHELET, E.: Statistical methods for the analysis of problems in animal orientation and certain biological rhythms. American Institute of Biological Sciences. Washington (1965).

VIII. Storage of Visual Information

1. The Time – Dependent Storing of Optical Information in the Honeybee

J. Erber
Department of Zoology, Institute of Technology, Darmstadt, Germany

Abstract.   In several series of experiments the learning behaviour of honeybees and its dependence on the parameters "quantity of reward" and "duration of reward" is analysed. The bees were trained to learn colour marks of artificial foodsources. The experiments show that over a wide range of different quantities of reward the learning behaviour is not influenced. In contrast to these findings, temporal parameters of the program of reward have a specific influence on the learning efficiency in honeybees.

In experiments with different programs of reward it can be shown that interruptions of sucking lead to an improvement of the certainty of reproduction of the signal learned. The temporal consolidation of the learned signals was analysed by registering the learning behaviour for a duration of 16 minutes after a single reward. The effect of repeated testing on the choice behaviour and the motivation is investigated and its significance discussed for very short rewards. Information-theoretical considerations show a convergence between the dynamic and stochastical description of the learning system.

In an analysis of the learning behaviour of honeybees MENZEL (1968) showed that a colour as a learning signal is stored in a short-term or long-term form. With rewards lasting no longer than 5 seconds, the learning signal is always stored in the short-term form; with rewards lasting longer than 10 seconds, it is always stored in the long-term form. These results are the basis for further investigations analysing the learning capacity of bees after short rewards. Parts of these experiments will be discussed in this paper.

1. Methods

Freely flying bees were trained with the experimental arrangement shown in fig. 1. By means of lamps and three mirrors, interference or broad-banded glass filters (SCHOTT, Mainz), mounted on a turn-table, were lighted from below. Centered above the turn-table is the training plate with three holes, which are covered by convex ground discs.

During the training phase the bees are rewarded with a sucrose solution out of a capillary on the ground disc in the centre of the training plate. The disc is illuminated with the learning colour. During the test situation the two other ground discs are illuminated with the learning colour and an alternative colour; the bees are able to choose between the two alternatives. The "approach" towards one of the two discs is registered as a positive decision for one of the colours. Very often an "approach" ends in landing and "sitting" on the illuminated ground discs. By means of a motor with four gears and a syringe filled with sucrose solution, different quantities and programs of reward can be offered to the bees in the training situation.

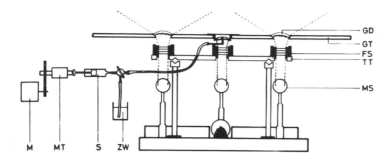

Fig. 1. Training arrangement for freely flying bees. MS = mirrors, FS = Filter systems, GD = convex ground glass discs, TT = turn-table supporting the filter systems, GT = training plate, top grained gray, diameter 1.50m, M=synchronousmotor, MT = microtransmission, S = syringe, ZW = storage vessel for sucrose solution

2. Different influx velocities of the sucrose solution

To decide whether the duration of reward is the decisive parameter for the short-term storing of learning signals in the bee, a series of experiments with different quantities of reward but equal duration of reward was performed. The learning curves did not show significant differences between the quantities 25 μl, 9 μl and 2.5 μl of sucrose solution. Though the reward of 2.5 μl is about the same quantity that the bee sucks, when it is rewarded for 2 seconds, the learning signal is stored in a long-term form. Thus we have to draw the conclusion that not the quantity but the duration of reward is the decisive factor for the short-term storing of learning signals in the bee (MENZEL and ERBER, 1972). The analysis of these experiments indicated that short interruptions of sucking lead to better learning results, therefore another series with different programs of reward was performed.

3. Different programs of reward

In contrast to the experiments described above bees were rewarded in the following experiments only once according to three different programs of reward. The bees were tested at different times after the reward, thus one is able to analyse the dynamics of storing learning signals. The learning colour was spectral light, 444nm, the alternative colour in the test was 532nm. Fig. 2 shows the percentage of right choices at different times after the reward. These were the three programs of reward: (1) a control group, rewarded 30 seconds without interruption (1 x 30: 0), (2) a group rewarded six times for a period of 2 seconds with 5 interruptions of sucking lasting at least 3 seconds each (6 x 2 : 3), (3) a group with 10 short one second long rewards and 9 interruptions of at least 3 seconds each between the rewards (10 x 1 : 3). Each curve gives the medium value of the percentage of right choices from at least 27 bees.

To avoid an experimental extinction of the signal learned, the bees were tested for only 2 minutes, these "primary decisions" were computed in five temporal groups. If a bee returned after this first test, the decisions then made,"secondary decisions", were evaluated separately.

The curves show that the certainty of reproduction of the learned signal augments during the time after the reward, a consolidation of the stored information is apparent. The comparison of the control group with the two other groups reveals an improvement of learning efficiently through interruptions of sucking during the reward. If sucking is interrupted very often, e.g. with the 10 x 1 : 3

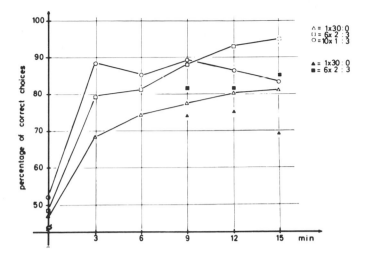

Fig. 2. Time-dependent consolidation curves for three different programs of reward. 0 minutes denote the spontaneous choice, the filled symbols mark the percentage of right choices for "secondary decisions"

reward, the percentage of right choices rapidly reaches a high level, a further consolidation does not take place. Though there are clear differences between the three curves, significant differences can not be found with the chi-square test. Up till now no statistical test which compares two curves with one another has been described. A mathematical analysis of the three consolidation curves proved that they can be described by the same e-function (ISERMANN, 1971), the time constant of consolidation for all curves is the same, interruptions of sucking lead to an increase of the amplification factor of the time-dependent consolidation function. In addition fig. 2 shows the percentage of right choices for some points of the 6 x 2 : 3 and the 1 x 30 : 0 curves, computed from "secondary decisions". Though statistical significant difference between these points and the corresponding of the original curves can not be shown, it is yet obvious that the repeated testing of the bees leads to a decrease of the certainty of choice.

4. Short rewards

Fig. 3 shows the time-dependent certainty of reproduction after a short reward lasting 2 seconds. This curve is computed from the medium decision scores of 29 bees. The percentage of right choices leaps to a very high level shortly after the reward. The percentage then decreases to a lower level which is identical with that of long-lasting rewards without interruptions of sucking. Contrary to the findings of MENZEL (1968), the curve does not decrease to the level of spontaneous choice. This discrepancy may be due to the difference in the chosen testing time. MENZEL (1968) tested the bees indefinitely (see MENZEL, 1968, figs. 8 and 9), whereas in the experiments described above, bees were tested only for 2 minutes; the influence of an experimental extinction should therefore be minimal.

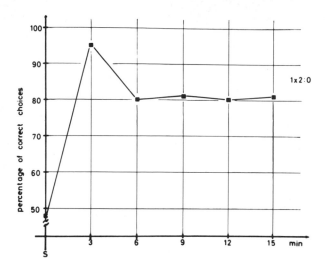

Fig. 3. Time-dependent consolidation curve for a short reward of 2 seconds, S denotes the point of spontaneous choice

5. Analysis of the motivation of choice

The frequency of decisions during the test may be regarded as a measure of the specific motivation of choice after a certain reward. Fig. 4 shows the frequency of approach to the ground discs during the tests. There is a statistical significant difference of the motivation of choice between the short reward and the other programs of reward (t-test, $p < 1\%$). The analysis of the frequency of decision shows significant differences between "primary" and "secondary" decisions, which signifies that not only the percentage of right choices but also the motivation of choice decreases with tests lasting longer than two minutes (t-test, $p < 1\%$).

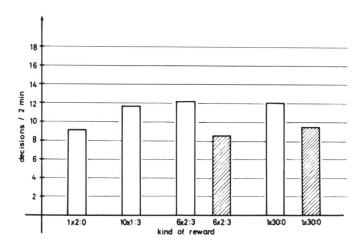

Fig. 4. Motivation in dependence of the kind of reward and the kind of testing. Hatched columns mark the frequency of "secondary decisions"

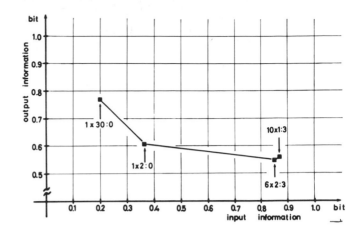

Fig. 5. Relation between input and output information in bit. Information is computed by formula : $H(s) = -\sum_{i=1}^{n} p_i \, ld \, p_i$, input information is calculated by evaluating $H(s)$ for the probability of the occurrence of the beginnings of sucking during the reward. Output information is calculated by evaluating $H(s)$ for the mean probability of right choices averaged over 16 minutes of testing

6. Considerations involving information theory

An improvement in the percentage of right choices with different programs of reward can be interpreted as an increase of the input information of the learning system (SCHULTZE, 1969). Repeated interruptions of sucking and therefore the repeated beginning of sucking enables the bee to store more information about the characteristics of the food source than without interruptions. The percentage of right choices or the output information will therefore be better in the following tests. If the beginning of sucking and the interruptions of sucking after a short sucking period are calculated as equivalent in the information processing, one is able to compute a relative measure of the input information. Fig. 5 shows the reaction between input and output information of the learning system for the different kinds of reward. An increase in the number of interruptions of sucking from 5 to 9 only leads to a small increase of the input information, therefore the output information for these two programs of reward is about the same. The described estimation of the input information illustrates that with very short rewards the input information is higher than with a long-lasting reward without interruptions; this difference manifests itself in the output information.

7. Discussion

The experiments discussed in this paper indicate that over a wide range the parameter "quantity of reward" does not influence the efficiency of learning. This means that even unproductive food sources are learned by the bee with a certainty that does not differ from that of very good food sources. The influence of temporal parameters of the reward is far more relevant for the learning system of the bee, which has its own temporal dynamics. Under natural conditions the bees suck the nectar of a certain flower only for a few seconds; after one foraging flight the bee will be able to recognize identical food sources with a high grade of certainty. Short interruptions of sucking, occurring permanently under natural conditions, even lead to an increase of the certainty of reproduction. With very short durations of sucking the percentage of right choices leaps to a high level shortly after the reward. This "differential reaction" enables the bee to quickly find with a high probability a food source with identical characteristics, when the yield of nectar of the flower visited suddenly decreases. The certainty of choice with short rewards is subjected to an experimental extinction, after a few negative experiences the percentage of right choices decreases rapidly. This effect can be measured by testing the bees for more than two minutes. These interpretations, based on the dynamics of consolidation, is a first step towards a dynamical and stochastical description of the learning system of the honeybee.

References

ISERMANN, R.:Experimentelle Analyse der Dynamik von Regelsystemen, BI Hochschultaschenbücher, Mannheim (1971).

MENZEL, R.: Das Gedächtnis der Honigbiene für Spektralfarben . 1. Kurzzeitiges und langzeitiges Behalten, Z. vergl. Physiol. 60, 82 - 102 (1968).

MENZEL, R., ERBER, J.: The influence of the quantity of reward on the learning performance in honeybees. Behaviour 41, 27 - 42, (1972).

SCHULTZE, E.: Einführung in die mathematischen Grundlagen der Informationstheorie. Springer Verlag, Berlin (1969).

2. Learning Experiments on the Use of Side – Specific Information in the Olfactory and Visual System in the Honey Bee (Apis mellifica)

T. Masuhr and R. Menzel
Department of Zoology, Institute of Technology, Darmstadt, Germany

Abstract.   Provided that the honey bee is engaged in feeding–search behaviour, the functional difference of the olfactory and visual system in the perception of signal patterns at close range was hypothetically formulated and tested by means of a training with scent and colour stimuli.

In order to record the effect of a reversal of the pattern, which was based on the stimulation of only one of the paired sense organs, the side of the receptor organs was changed between training and test. The fact that in regard to the olfactory system, the shifting of the scent stimulus gives rise to a loss of the release function of this stimulus for the learned behaviour, suggests a signal pattern as the true learned releaser.

The optic stimulus remains a cue function even after shifting. In the visual system the stimulus quality per se is the releaser. The difference between the orientation–specific and the learning–specific storing related to both sense systems is discussed. In addition, the stimulus–response associations are interpreted combined with the possibility of   intracerebral transfer.

1. Introduction

MARTIN (1964) proved that the position of the sense organs, namely the right and the left antenna, is the decisive instance for the orientation of the honey bee in accentuating the scent as "coming from the right" and "coming from the left". Moreover, he showed in 1965 that, when the olfactory system is engaged, a spatial pattern of scent signals can be stored and used for the discrimination of other patterns of the same signals.

In virtue of these indications it appears efficient to test whether the bee which has learned to produce a distinct behaviour on a one–sided applied stimulus, can reproduce it if the entrance of this stimulus is shifted to the contralateral side.

If a signal pattern had been learned, based on a different excitation of the sides of the paired sense organs, then the shifting of the stimulus, relative to the bee, should only have a consequence in the direction of inefficiency of the stimulus for the release of the learned behaviour. If the stimulus quality per se had acquired the character of a signal, this should not occur.

It is advisable to tackle this complex of questions at the two levels of the olfactory and of the visual system.

The function of a short–distance orientation system is ascribed to the olfactory sense, whereas the optic sense accounts for the long–distance orientation. In addition to this functional difference, the visual system is predestinated by a light–compass orientation to supply other bees in social communication with objective data on the position of the food source.

The following hypothesis seems to be justified: Regarding the olfactory system as the basis, the bee learned a signal pattern, serving itself as the fixed point in recording the inherent spatial relations.

In contrast to that, it is not within the power of the visual system to store such a signal pattern, only recognizable as relative constant at a short distance, over a long period of time and to make use of it for the identification of the food source. In this connexion we have to imagine that the visual system is stimulated in the course of flight by identical objects in the manner of a frequently changing right-left pattern and has the task to determine the spatial relations of the food source to a superior fixed point.

This functional difference – on one side the learning of a signal pattern, on the other side the learning of the signal quality per se – should express itself in a different effect of the shifting for the release of a learned behaviour.

2. Scent-training with flying bees. Experiments on the use of side-specific information in the olfactory system

a) Methods.   The test bees were rewarded in the centre of a table on an artificial food source, which was standing on a wire netting over an open petri plate. A filter paper soaked with the scent oil, fennel, had been placed in the plate. For the initial spontaneous choice and the crucial tests two identical models of the food places, one with the training scent, fennel, and the other with the alternative scent, rosemary, were offered to the bees for decision on the right and on the left side of the table. The decisions, calculated as the numbers of settings, were registered during each test lasting four minutes, in the course of which the models were exchanged with each other every 60 seconds.

In order to enable the shifting of the scent stimulus from one antenna in the learning situation to the other in the decisive test situation, the ten bees of the experimental group were treated in the following manner. For the learning situation their "test antennae" were spread over with a cosmetic (Relax Beauty Mask, Ayer), which doesn't have any scent of its own, the dehydratation process is completed. The tarsi of the first two legs on the treated side were amputated to prevent the bees from slipping off the mask.

With these conditions for the learning situation a spontaneous choice and the first crucial test were carried out after the fifth learning act. The bees after having been caught in the beginning of the seventh learning act were prepared for the decisive test. The cosmetic film on the "test antenna" was drawn off as a thin membrane and the "learning antenna" was cut off near the base. After two hours at the occasion of the first return to the food place a further crucial test, the decisive test, took place. After another five learning acts, this time under the condition of the test situation, the bees were additionally tested with the purpose of recording the effect of learning solely with the previously treated antenna.

Along with the experimental group, four control groups were set up. The bees of the control groups A, B, C learned corresponding to the schedule above, but from the beginning with only one antenna, which was treated with the cosmetic for half-an-hour to three hours respectively to the group A and B. That was to test the function of the antenna after treatment. The animals of the control group D learned till the first crucial test the five learning acts utilizing both antennae. After which one antenna was cut off for a subsequent test to prove the effect of the loss of one antenna on the certainty of the decisions already known.

b) Results.   The bees of the experimental group showed a nearly equal distribution of their decisions for the training and the alternative scent in the spontaneous choice (I). After five learning acts in the first crucial test (II) the bees chose the training scent in 91% of their decisions. In the decisive test situation with only one antenna, previously isolated from the scent, the crucial test (III) resulted in an equal distribution of the choices. It is possible to interpret this result by saying that through the olfactory system the signal pattern – on the right, fennel, on the left, no scent, – was associated with success in the feeding search. Not at all the fennel scent per se was learned as a signal, then that being the case the bees should not have failed in the test situation, when the stimulus pattern – on the right, no scent, on the left, fennel, – was experimentally produced. This interpretation is supported by the results of the control group D. These bees learned to choose the training scent in 94% of their decisions (II),

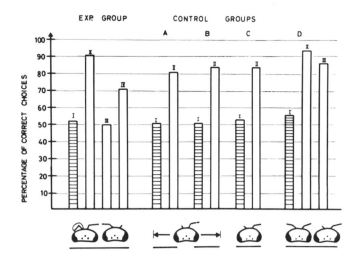

Fig. 1. The percentage of correct choices for the training scent, fennel, is registered with regard to each spontaneous choice (I) and each crucial test over the experimental group and all control groups. Under the abscissa the conditions of learning are skeletonized

given the possibility to learn with both antennae, however they also chose this scent up to 87%, if the original signal pattern - on the right, fennel, on the left, fennel, - was only reduced to the pattern - on the right, no scent, on the left, fennel, - by cutting off one antenna between test II and test III.

If the bees of the experimental group continued to learn the training scent in five learning acts with the test antenna only, an increase of the correct choices to the percentage of 71 can be demonstrated (IV). This result compared with that of the control groups A and B, which also learned with only one treated antenna, suggests a process of learning anew the spatial relations in the signal pattern, see also MENZEL (1969) and KOLTERMANN (1969).

3. The establishment of a conditioned reflex released by a light stimulus. Experiments on the use of side-specific information in the visual system

a) Methods.   The basis of the learning was the elaboration of a classical conditioned reflex released by the applicance of light on one complex eye only, a method, which KUWABARA (1957) was the first to demonstrate. The raise of the proboscis from the experimentally determined position for more than one mm is registered as a positive reaction. Analogous   to the method of VARESCHI (1971) the bee was kept in a narrow glass tube and its head was fixed with wax-collophonium (see fig. 2 A). For the training only one complex eye of the bee was stimulated by a ray of light as "US", bundled by a converging lens. The focal point was placed in the centre of the head. The angle of incidence comes to 32°. The limitation of the visual field of the contralateral eye prevents it from being stimulated by the ray of light (see fig. 2 B, C).

A blue filter (BG 9, Schott) and a light stop were inserted in the light beam. This part of the experimental arrangement was built up twice, on the right and on the left side of the bee, to stimulate both eyes without being forced to alter the position of the bee. The light intensity of both lamps was identical.

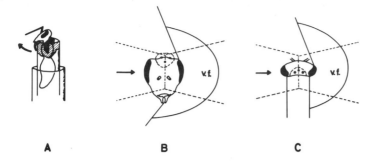

Fig. 2. A Scheme of the position of the bee in a glass tube. The arrow indicates the positive reaction of the proboscis. B frontal, C dorsal view of the head with a description of
the light ray and of the visual field (v.f.), drawn from BAUMGAERTNER (1928)

Pilot tests had revealed that the success in learning depends solely on the number of the conditioning trials within a large scope of different learning rates. It had turned out that the bees,
which had received over 40 linkages of light stimulus (CS) and sugar water (US) in the first day
of the experiment, reached their maximum level of learning in the second day. Because the
"tube bees" did not show a strong increase of mortality until the third day, we adhered to a two-
days scheme in my chief experiments.

In detail the following timing was observed: During the first day of the experiment the bees received 50 conditioning trials on the learning eye in two groups 25 each and in every case at the
regular interval of three minutes. The CS was presented for a period of three seconds. At the end
of the first second, during which the behaviour was recorded, a drop of sugar water was given as
US. For sucking up the food two seconds were available to the bees. In the second day of the
experiment the learning eye received ten additional conditioning trials to consider the increment
of learning after the night period. After a brief interval of three seconds the conditioning of the
proboscis reaction began, now using the test eye. In this case the bees also received 50 conditioning trials in the same manner.

Two series of experiments were performed. In the first series, containing 13 bees, the ocelli were
accessible to the incidence of the light, in the second series, containing five bees, the ocelli
were isolated against light with a layer of a mixture of soot- shellac. The last series was set up
to avoid light-dependent effects of the ocelli on the optic ganglia, as described by JANDER and
BARRY (1967).

b) Results.   The records of the positive and negative reactions of the bees with regard to each
light stimulus were treated at first in such a manner, that a mean probability was calculated for
the appearance of a positive reaction by averaging these effects of the conditioning for each
period of ten trials covering all bees in every series of experiments. These data are equivalent
to the levels of learning.

Because the levels of learning in both series were not statistically different from one another
(Chi-square test, p >0.01), it seems permissible to combine the data to a single curve, see fig.3.

This curve, in which the effect of shifting of the light stimulus from one eye to the other is represented, shows a statistically secured increase in the amount of learning between the conditioning periods 1 - 10 and 51 - 60 (Chi-square test, p < 0.01). When the test eye is used for the
application of the CS, the probability of the appearance of the positive reaction in the first period
of 10 conditioning trials is almost on the same level as when the learning eye is used during the last
period of the conditioning trials.

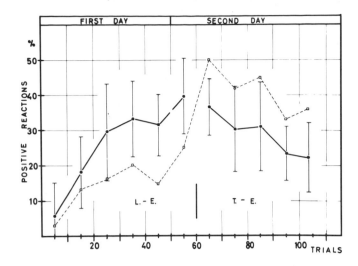

Fig. 3. The curve ——— · ——— represents the percentage of the positive reactions, averaged over each period of ten conditioning trials and over all bees of the two series of the experiment. The abbreviations L.-E. = learning eye and T.-E. = test eye. The statements "first day" and "second day" stand for this curve. The range of each level of learning over a whole training period is percentally expressed and inserted in form of standard deviation in this curve.

The curve ----- . ----- showed the analogous data, characterizing the results of bees which have learned with one eye only over the whole 110 conditioning trials

According to pilot experiments, the following decrease of the levels of learning have to be ascribed to satiety of the bees conditional on the numerous feedings. But by no means is this specific for the shifting of the light stimulus, as the perforated curve, which represents the corresponding data for the bees learning with one eye only, shows a similar course in its qualitative aspects.

To be sure that the relative high level of learning recorded during the first 10 trials on the test eye cannot be explained by a learning process over this whole period, therefore the percentages of the probability of the appearance of the positive reaction were evaluated separately for each trial, see fig. 4.

It can be demonstrated that the mean of the level of learning for the first 10 conditioning trials doesn't result from averaging data of a continually ascending curve.

These results suggest that the difference in stimulation, originating from a difference of the application side of the same stimulus, cannot be used by the visual system for building up two different signal ·patterns. In the visual system the stimulus quality per se is the signal to release an adequate behaviour.

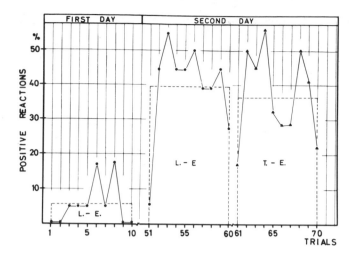

Fig. 4. Ordinate: Percentage of the probability of the appearance of the positive reaction in each trial. Abscissa: The number of the conditioning trials. L.-E. = learning eye, T.-E. = test eye

4. Discussion

Provided that the honey bee is engaged in the feeding-search behaviour, a difference can be demonstrated between the olfactory and the visual system with regard to the use of a side-specific application of stimuli.

For the olfactory system the following formal approach seems to be valid: The short-time store, necessary for orientation, with its side-specific storing of stimulus differences, measured on the basis of the position of the receptor organs, will be directly linked with the store of the learning system. That should not be the case when the visual system is involved. Then at this store only the messages arrive, which, with regard to the results, don't have the contents of these spatial relations. A model is imaginable, which is based on direct feeding of the learning store by the optic sense organs, whereby the data are reduced on the signal quality per se by the special kind of feeding. The origin of the data, coming from the right or from the left receptor organ, is extinguished.

This solution of the problem is based on the assumption of a uniform store in the learning system, which receives not only data from the olfactory system but also from the visual system, and which is able to store the data according to their origin from the different regions of the head, so far as it gets them in the right order.

HORRIDGE has hypothetically formulated that the calyces of the corpora pedunculata should have the ability to store the directions of stimuli relative to the head as in "hemispheric charts".

The problem is also related to the transfer of a learned achievement from one brain hemisphere to the other. With regard to the results it is necessary to abandon the conception that the anatomic fact of transferring paths which, in accordance with VOWLES (1955) exist between the two sense systems, and last not least between the corpora pedunculata, is already a sufficient condition for extinguishing the special information about the side of stimulation.

In contrast with the visual system, which is responsible for the long-distance orientation, only the olfactory system has the possibility to store spatial information , acquired by difference measurement of the stimulation on both receptor organs, in the learning system, presumably the corpora pedunculata.

References

BAUMGARTNER, H.: Der Formensinn und die Sehschärfe der Biene. Z. vergl. Physiologie 7, 56 - 143 (1928).

JANDER, R. and BARRY, Ch.K.: Die phototaktische Gegenkoppelung von Stirnocellen und Facettenaugen in der Phototaxis der Heuschrecken und Grillen. Z. vergl. Physiologie 57, 432 - 458 (1968).

BULLOCK, T.H. and HORRIDGE, G.A.: The structure and function in the nervous system of invertebrates. Vol. II, 1239 - 1264 (1965).

KOLTERMANN, R.: Lern- und Vergessensprozesse bei der Honigbiene - aufgezeigt anhand von Duftdressuren. Z. vergl. Physiologie 63, 310 - 334 (1969).

KUWABARA, M.: Bildung des bedingten Reflexes von Pavlovs Typus bei der Honigbiene. Jour. Fac. Sci. Hokkaido Univ. Ser. VI, Zool. 13, 458 - 464 (1957).

MARTIN, H.: Zur Nahorientierung der Biene im Duftfeld. Zugleich ein Nachweis für die Osmotropo-taxis bei Insekten. Z. vergl. Physiologie 48, 481 - 533 (1964).

MARTIN, H.: Leistungen des topochemischen Sinnes bei der Honigbiene. Z. vergl. Physiologie 50, 254 - 292 (1965).

MENZEL, R.: Das Gedächtnis der Honigbiene für Spektralfarben. II Umlernen und Mehrfach-lernen. Z. vergl. Physiologie 63, 290 - 309 (1969).

VARESCHI, E.: Duftunterscheidungen bei der Honigbiene. Einzelzellableitungen und Verhaltens-reaktionen. Z. vergl. Physiologie 75, 143 - 173 (1971).

VOWLES, D.M.: The Structure and Connexions of the Corpora Pedunculata in bees and ants. Quart. Jour. of Micr. Sci. 96, part. 2, 239 - 255 (1955).

IX. Methods of Quantifying Behavioral Data

1. The Relationship between Difference in Stimuli and Choice Frequency in Training Experiments with the Honeybee

O. v. Helversen
Department of Zoology, University of Freiburg/Br., Germany

Abstract. When honeybees are trained to discriminate between colours, the choice frequency between training and test stimulus can be measured. This measurable relation between stimuli and choice frequency can be split into two parts assuming an intermediate variable, the "perceived stimulus difference". If the perceived stimulus difference is defined as a metric function, the relationship between "perceived stimulus difference" and choice frequency can be derived from experimental measurements by means of a mathematical evaluation. The characteristic curve of this relationship is sigmoidal.

An ethologist measuring actions or action frequencies of his experimental animal is often not interested in these actions themselves but in "inner conditions" upon which the observed actions depend. He uses the observable actions as a measure for the inner condition. Under favourable circumstances there is a one-to-one relationship between the inner condition and measurable action. However, this function will very seldom be linear. The determination of the characteristic curve (or "calibration curve") is a fundamental problem of behavioural physiology, but for various reasons it is quite difficult to approach. In the first place, an unequivocal definition of the "inner condition" concerned may not be possible; and secondly, in most cases no information regarding the relations involved can be obtained from the actual measurements.

I shall now attempt to demonstrate one possibility for determining this relationship for the following very special and particularly simple case: the frequency of right choices between a test and a training stimulus. Such a relationship is in some respects analogous to the calibration of a measuring instrument.

As K. v. FRISCH has shown in his famous experiments, a honeybee can be trained to discriminate between two colours when rewarded with a drop of sugar water in connection with one of the colours. The velocity with which the bee learns can be described by a learning curve (MENZEL, 1967) which relates the probability of right choices to the number of learning acts. Learning curves of the bee are characterized by a more or less steep slope, which then reaches a saturation plateau. This saturation level of the learning curve can be used as a measure of the similarity (or difference) of two stimuli as perceived by the bee. With help of this relation (height of saturation level) a great number of investigations have been conducted, especially experiments involving colour and form perception of the bee. I wish to consider the following problem: How is the "difference between two stimuli" (an exact definition will be presented later) translated within the bee to the observable choice frequency?

1. Qualitative formulation of the problem

The following two examples shall be used to clarify this question. Figs. 1 and 2 show the results of typical training experiments with honeybees. The bees were trained to obtain sucrose solution in connection with a black stripe (at an angle of 45° to the horizontal) – fig. 1 – or in connection with a spectral colour of 407 nm wavelength – fig. 2. The frequency with which the bees chose the training stimulus at saturation of the learning curve was measured when the bees were required to discriminate between the training stimulus and similar test stimuli.

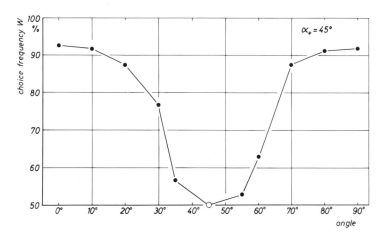

Fig. 1. Honeybees were trained to discriminate between black stripes at varying angles to the horizontal on a vertical screen. Choice frequency was measured when the test stimulus (varying angle on the abscissa) was at an angle greater or smaller than that of the training stimulus (45° angle). When the bees cannot discriminate between the stimuli, a choice frequency of 50 : 50 is obtained. From WEHNER (1966)

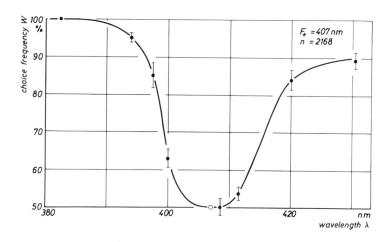

Fig. 2. Discrimination between monochromatic colours of varying wavelength. Training stimulus was violet (407 nm)

From these curves one can recognize two properties characteristic for response curves in training experiments of this type: (a) For small differences between test and training stimulus the choice frequency rises gradually, but with increasing difference it reaches a saturation level. This saturation may be 100% (full discrimination) or lower (for instance, 90%) depending upon the training situation (colours, forms etc.) and the training method. (b) The second property is the U–shape of the curve in the neighbourhood of the point where the training stimulus is equal to the test stimulus. The tangent near this point is almost horizontal, or at least, less steep than in the ascending branches of the curve.

In these experiments the choice frequency expresses something like the "stimulus difference as perceived by the bee" (Reizverschiedenheit). But the perceived stimulus difference should not be limited by the saturation of the choice frequency. Therefore it seems plausible to split the observable relation: stimuli S_1, $S_2 \longrightarrow$ choice frequency W into two hypothetical parts S_1, $S_2 \rightarrow V (S_1, S_2) \rightarrow W (S_1, S_2)$, where $V (S_1, S_2)$ means the "perceived stimulus difference" between S_1 and S_2. The stimuli are converted within the sense organ to neuronal excitation patterns (this relation will be discussed later), and they are then computed to something like "perceived stimulus difference", which further leads to the observable actions, measured as choice frequency W. (See below fig. 4).

Considering the relation between V and W we can expect (in agreement with results as shown in figs. 1, 2) that the relation is monotonous, i.e. increasingly different stimuli are discriminated with increasing (at least, not decreasing) accuracy. But the relation cannot be assumed to be linear, because the two properties mentioned above contradict linearity: (a) the saturation level, and (b) the U–shaped course of the curves. In sufficiently close proximity to the training stimulus we would expect that the stimulus difference is related linearly to the physical (or geometrical) parameter. This would result in a V–shaped curve. If the U–shape depends upon an accidental distortion of the relationship between the geometrical or physical parameter and the stimulus difference (the first relation of our hypothesis), two such curves for slightly different training stimulus values could not both exhibit a U–shape. But since the U–shape is a general phenomenon, the non–linearity seems to be a property of the second translation (perceived stimulus difference V into choice frequency W).

These considerations permit us to establish a rough hypothesis: The distortion leading to U–shape and saturation of the choice frequency curves is achieved between V and W (close to the output). The relation between perceived stimulus difference and choice frequency has a sigmoidal course and levels off in a saturation plateau, as pictured in fig. 3.

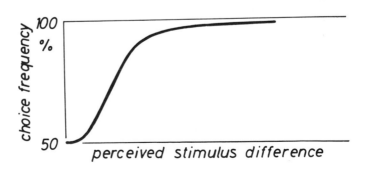

Fig. 3. Hypothetical relation between perceived stimulus difference and choice frequency

How can we proceed to confirm this hypothesis and to determine the curve quantitatively?

Earlier approaches to the problem involved the proposal of models of the choice action (JANDER, 1968). I wish to suggest another way: attempting to construct the curve from measured data only.

How can we precisely formulate the stimulus distance $V(S_1, S_2)$? If we have no unequivocal definition of V, any one-to-one correspondence between V and W can be assumed without contradicting the measured result (and without being of further value either). How can we define V?

Let us ask what we expect. I would anticipate the following: Take three stimuli A, B and C with B resting between A and C along the shortest line of connection between A and C (in the sensory space of the bee); then the distance between A and C will be the sum of the distances A, B and B, C. (We are familiar with this distance function from Euclidean geometry). This may seem to be rather trivial, but it represents a serious restriction.

We now shall use this relation to define the "perceived stimulus difference" $V(S_1, S_2)$ as a distance function.

2. Mathematical formulation of the problem

Let us express our question in more mathematical terms. The stimuli S_1, S_2, ..., defined physically or geometrically, can be regarded as points of a space Q, the "stimulus space". When we measure the ability to discriminate between two stimuli S_1 and S_2, the choice frequency will be a real number between 50% and 100%. This interval can be easily transformed to $\langle 0.1 \rangle$. We obtain a relation between each pair of stimuli and the measured value W: $W = W(S_1, S_2)$; $Q \times Q \longrightarrow \langle 0,1 \rangle$. Such a function is a "distance (or metric) function" in Q, if the following holds

(1) symmetry: $W(S_1, S_2) = W(S_2, S_1)$
(2) $W(S_1, S_2) \geq O$; $W(S_1, S_2) = O$ only if $S_1 = S_2$
(3) $W(S_1, S_2) + W(S_2, S_3) \geq W(S_1, S_3)$ for every S_1, S_2, S_3 and
$\quad W(S_1, S_2) + W(S_2, S_3) = W(S_1, S_3)$ if S_1, S_2, S_3 lie along a straight line.

Assuming that in the expression $W(S_1, S_2)$ the first of the stimuli stands for the training stimulus, the second for the test stimulus, the condition (1) (the "symmetry" condition) means that the result of the experiment must be independent from the choice of S_1 or S_2 as training stimulus. This symmetry has been shown by DAUMER (1956) to be valid for colours.

Condition (2) is always valid if the physically defined stimuli are assigned to equivalence classes, as in the sense organ. If there is no one-to-one correspondence between the chosen set of stimuli and the sensory or neuronal excitation patterns, we have to regard the space of excitation patterns, not the stimulus space. This has to be discussed for each case.

Condition (3) is the so-called triangular inequation in an especially rigorous form, such that equality will be attained if all three stimuli lie along a "straight line".

This does not hold true for the choice frequency W, as can be shown considering three stimuli all discriminated by the bee at the 100% level, or in view of stimuli near the discrimination threshold as well.

But this very condition (3) is implied by our definition for the stimulus difference $V(S_1, S_2)$. Symmetry and condition (2) must be valid as well.

Therefore our main question can be formulated as follows: Regarding the experimentally measurable function $W(S_1, S_2)$, is there a strong distance function $V(S_1, S_2)$ and a monotonous function $y = h(x)$, such that $W(S_1, S_2)$ can be expressed as $W = h(V)$?

The question is illustrated and summarized in fig. 4.

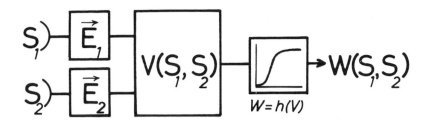

Fig. 4. Information flow from stimuli S_1, S_2 to choice frequency W. The stimuli S_1, S_2 are defined physically (or geometrically). They are transformed by the sense organ into patterns of excitation (vectors $\vec{E}_1$, $\vec{E}_2$). According to the hypothesis, the excitation patterns then lead to the perceived stimulus difference V, which, in turn, is transformed by a monotonous (probably sigmoidal) function h into choice frequency W. $V(S_1, S_2)$ is defined as above, where the main point is additivity for certain sets of stimuli

Is this concept reasonable? – This is a question for the bee.

In attempting to answer the question we can treat the following point right away. Even with relatively weak prerequisites (satisfied in the case of colour vision), it can be shown that at least one rigorous distance function always exists in the stimulus space (to be exact, in the space of equivalence classes of stimuli), and likewise that there are shortest lines of connection between every two points. (The formal proof is not presented here. The idea is to go by "threshold steps" along all connecting lines between two stimulus points. The connecting lines, along which a minimal number of steps is necessary, can be defined as "straight lines", and the minimal number itself as V. Then V is a rigorous metric function).

This reduces our question: (1) Does a monotonous function $y = h(x)$ exist, which satisfies the equation $W = h(V)$? (2) How can we ascertain this function h on the basis of experimental measurements?

I will begin with the second question and solve it for one particular simple case by stating a procedure for construction. Whether such a function exists and the range of validity will be discussed later.

3. Construction of V and h for the "bee-purple" from measured data

In his well-known publication on the colour vision of the honeybee DAUMER (1956) displayed the following set of curves (fig. 5).

The regularity of these curves is striking. Obviously two trends are superimposed: All the curves become steeper from right to left, and they all have a similar sigmoidal shape independent of their steepness. Let us discuss the first trend: The increase of steepness from right to left means that the more yellow contained in a training colour the less ultraviolet must be added in order to achieve discrimination by the bee. The steepness of the curves can be expressed by measuring the discrimination thresholds, i.e. the differences $\Delta\alpha$ of the mixture coefficient necessary for discrimination at the 80% level. In fig. 6 these values of $\Delta\alpha = f_{80\%}(\alpha)$ – interpolated from fig. 5 – are plotted against the mixture coefficient α.

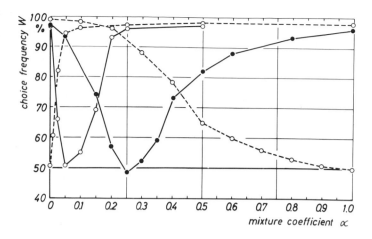

Fig. 5. Colour discrimination in the "bee-purple". All colours were mixed from yellow (588 nm) and ultraviolet (360 nm) in various proportions, described by the mixture coefficient α ($\alpha = 0$ means pure yellow, $\alpha = 1$ pure ultraviolet). The training stimuli (corresponding to each of the four curves) were: $\alpha = 0$, $\alpha = 0.05$, $\alpha = 0.25$ and $\alpha = 1$. From DAUMER (1956)

Now let us assume that the colours obtained by mixing yellow and ultraviolet in various proportions correspond with a one-to-one relation to changes in only one parameter perceived by the bee, i.e. that they are along the "shortest line of connection" between yellow and UV. This assumption is plausible. (Further explanation, see fig. 10 and text there. In human chroma-ticity diagrams as well, the purple colours lie along a straight line, and what we perceive are colours between violet and red, thus varying in only one parameter.) The equation V (A, B) + V (B, C) = V (A, C) then holds true for points A, B, C along this line. This means, that the sti-mulus distances between the different bee-purple colours, characterized by their mixture co-efficients α, are additive.

As a result of this additivity, a single function R (α) is sufficient to describe all possible sti-mulus distances: Take R (α) = V (0, α), where V (0, α) is the stimulus distance between a purple colour and the point $\bar{\alpha} = 0$ (pure yellow). Then the stimulus distance of any pair of purple colours, given by their mixture coefficients α_1 and α_2, is described by $|R(\alpha_1) - R(\alpha_2)|$.

What is the relationship between R (α) and $f_{80\%}(\alpha)$? First, it is easy to derive $f_{80\%}(\alpha)$ from R (α). Because of the one-to-one correspondence between V and W, for the 80% level of choice frequency there is a value k of stimulus distance. Therefore, for any given colour, characterized by a mixture coefficient α_1, a second colour α_2 can be found, such that $|R(\alpha_1) - R(\alpha_2)| = k$. When $\Delta\alpha = |\alpha_1 - \alpha_2|$ the value of the function $f_{80\%}(\alpha) = \Delta\alpha$ is determined.

On the other hand, it is possible to derive R (α) from $f_{80\%}(\alpha)$. I will not present the formal proof here but only mention, that it is a more general form of the considerations which led FECH-NER to his interpretations of WEBER's findings in psychometry. I did not apply a differential equation, but rather a recursion formula. The result is the curve shown in fig. 7. The points have been constructed from $f_{80\%}(\alpha)$ of fig. 6. R (α) can be obtained from the experimental data, except for an appropriate factor (arbitrarily assuming that R (1) = 1) and an additive constant which, however, disappears in all arguments, because only differences of R are used. The curve was not interpolated but derived from theoretical considerations which are discussed later. For the moment we can assume it is won by interpolation.

What good is the function R(α)? It allows us to find directly the desired function h, the "cali-bration curve". We can plot the measured values of choice frequency W against the values of V, given by $|R(\alpha_1) - R(\alpha_2)|$, for each pair of α_1, α_2. Fig. 8 shows the result. The points des-cribe a clearly distinguishable curve, corresponding to our hypothesis from fig. 3.

The construction of R (α) corresponds to the derivation of V. It can be carried through from DAUMER's set of curves under two conditions, namely additivity of V and monotony of h. Only one point of each training curve from its steep portion (here at a choice frequency of 80%) is used to arrive at R (α). This describes the "first trend" (see above) of the set of graphs. The plot of W against V enables us to grasp the "second trend". Thus the information contained in this set of curves is split according to our concept (fig. 4) into a first part $S_1, S_2 \rightarrow V (S_1, S_2)$ and a second part $V (S_1, S_2) \rightarrow W (S_1, S_2)$, and this without further assumptions about the form of both functions.

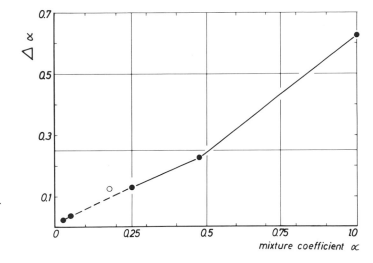

Fig. 6. The differences $\Delta\alpha$ = $|\alpha_1 - \alpha_2|$ of the mix-ture coefficients discrimi-nated with a choice frequen-cy of 80%. Interpolated from DAUMER's set of curves (fig. 5)

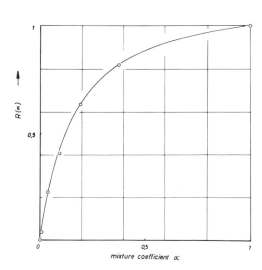

Fig. 7. The function R (α), constructed from the $\Delta\alpha$ curve of fig. 6. For further expla-nation see text

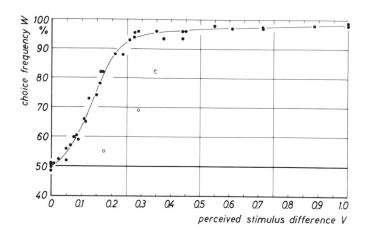

Fig. 8. The "calibration curve" W = h (V) for DAU-MER's measurements of discrimination between colours of the "bee-purple"

This curve is dependent upon the training method of DAUMER (and upon the stimuli used, namely colours). Two measured points do not fit the curve, but I believe this should not bother us since DAUMER did not plan such a quantitative evaluation of his experiments. Furthermore we can explain this deviation rather simply, assuming that some of the bees had not reached the optimal level of learning.

We have thus constructed the "calibration curve" for one special example. How can we prove the concept W = h (V) more generally?

4. Is the concept valid for the whole space of colour perception?

First of all, I have evaluated a further series of DAUMER's experiments involving colour mixtures of blue and ultraviolet. Employing the same procedure, we arrive at a curve (fig. 9) very similar to the first constructed for the bee-purple. (Deviation of points from the derived curve is not greater than the experimentally determined deviation).

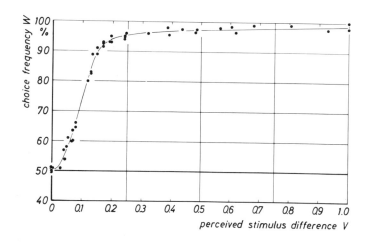

Fig. 9. The "calibration curve" W = h (V) for DAU-MER's measurements with colour mixtures between blue (440 nm) and UV (350 nm)

This second series is also based on experiments with colour mixtures. Does the concept of the "calibration curve" W = h (V) hold for all colours, in particular for monochromatic colours?

In order to examine this question we must deal with the honeybee's space of colour perception, i.e. how the physically defined stimuli are transformed into patterns of sensory excitation.

In a trichromatic colour system all physically possible spectral compositions are mapped into a three-dimensional space of equivalence classes. Processing of colour data in the CNS commences with the excitation values of three types of receptors characterized by their spectral sensitivity curves. (In the case of the bee, the receptors are those for UV, blue and yellow-green.) These three-dimensional excitation patterns $\vec{E} = (E_1, E_2, E_3)$ comprise all the information the system can possibly hold about a colour.

The sensitivity curves of the three receptor types are therefore decisive for the establishment of equivalence classes and for comparison of two colours as well.

The ratios of the excitations of the receptor types define the chromaticity (hue and saturation) of a colour. Since there are three types of receptors, the ratios considered as functions of wavelength are rather complicated. Therefore the monochromatic colours lie along a strongly curved line in the bee's (or man's) space of colour perception. This curve definitely does not represent the shortest line between two points. Thus the equation V (A, B) + V (B, C) = V (A, C) is not valid for the points of this curve, and therefore the previously used procedure does not hold for wavelength as a parameter.

Is it still possible to find curves for which the additivity of stimulus distances is valid, in spite of the relatively complicated sensitivity functions which characterize the space of colour perception? We need curves for which the sensitivity functions of the three receptor types are as simple as possible, ideally straight lines, and we arrive at just such curves when we regard the points of those colours obtained by mixing two given colours A and B in various proportions. When for the colour A the three receptor types absorb with values A_1, A_2, A_3, and for the colour B with values B_1, B_2, B_3, then for mixed colours they absorb with values which correspond linearly to the mixture coefficient (fig. 10). The curves are particularly simple when colours from both ends of the spectrum are mixed, assuming, more simplified, that only one receptor type absorbs at each end of the spectrum.

This, of course, explains why DAUMER's set of curves for the bee-purple is particularly easy to analyse.

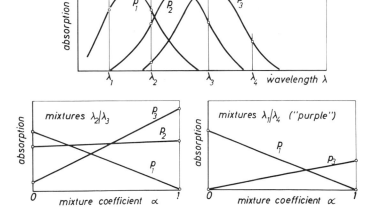

Fig. 10. Diagrammatic absorption curves p_i of the receptor types of a trichromatic system as functions of wavelength λ and mixture coefficient α. The case of purple colours is particularly simple

For testing the validity of the relation $W = h(V)$ for monochromatic colours, the basic idea is the following: If one could find a mathematical model for the stimulus distance $V(S_1, S_2)$, then one could test this model for the simple case of purple colours against the measured values of V. If it proves to be suitable, the model can be applied for the whole space of colour perception. I will now proceed in this manner.

Information about wavelength (or the mixture coefficient for a bee-purple colour) is not contained in the receptor's excitation values themselves but in their ratios. For the example of bee-purple this means that the expression $\frac{P_3}{P_1}$ is decisive - or $\frac{P_3}{P_1 + P_3}$, to avoid difficulties when $p(\alpha) \longrightarrow 0$. A plausible hypothesis would be to assume that

$$V(\alpha_1, \alpha_2) = \left| \frac{P_3(\alpha_1)}{P_1(\alpha_1) + P_3(\alpha_1)} - \frac{P_3(\alpha_2)}{P_1(\alpha_2) + P_3(\alpha_2)} \right|$$

where $P_i(\alpha_k)$ is the sensitivity value of the receptor type i for the mixture coefficient α_k. In actual fact this assumption corresponds quite well with experimental results. The function $R(\alpha) = \frac{P_3(\alpha)}{P_1(\alpha) + P_3(\alpha)} = \frac{k_1 \cdot \alpha}{k_1 \cdot \alpha + (1-\alpha) \cdot k_2}$ (see fig. 10) is solely dependent upon the parameter k_1/k_2. For the ratio $k_1 : k_2 = 7.5:1$ we observe excellent agreement with the function $R(\alpha)$ constructed from experimental data. The ratio 7.5 : 1 means that for the interference filter used by DAUMER, the UV receptor was about 7.5 times as sensitive as the yellow-green receptor. The agreement is so good (fig. 7) that it is hard to imagine that something physiological is not involved.

Thus the model $V(\vec{A}, \vec{B}) = \left| \frac{A_1}{A_1 + A_2} - \frac{B_1}{B_1 + B_2} \right|$, where $\vec{A} = \begin{pmatrix} A_1 \\ A_2 \end{pmatrix}$ and $\vec{B} = \begin{pmatrix} B_1 \\ B_2 \end{pmatrix}$ are the excitation patterns of the receptor types, thus $A_i = P_i(\alpha)$ for a purple colour and $A_i = I \cdot p_i(\lambda)$ for a monochromatic colour of wavelength λ and intensity I, is applicable for this special case. How can this model be extended from a two-dimensional to a three-dimensional case?

The formula stated for V is the special two-dimensional case of the following function for n-dimensional space

$$V(\vec{A}, \vec{B}) = \sum_{i=1}^{n} \left| \frac{A_i}{\sum_{\nu=1}^{n} A_\nu} - \frac{B_i}{\sum_{\nu=1}^{n} B_\nu} \right|$$

if one disregards a constant factor.

For a three-dimensional case, this distance function has the following obvious significance: It expresses the distances between the straight lines which pass through the origin by determining their points of intersection with the plane $x_1 + x_2 + x_3 = 1$ and applying the norm $\| \vec{x} \| = \sum_{i=1}^{n} |x_i|$ to these intersection points. (To be exact, we are not dealing with a distance function of three-dimensional vector space, but of the respective two-dimensional projective space). In the space of colour perception the straight lines through zero contain in first approximation (assuming linear characteristic curves of intensity for the receptors) all colours of the same chromaticity but different brightness. As DAUMER has demonstrated, the bees take little notice of the brightness of colours, for the range of intensity used here, when they have to differentiate between two colours.

Thus we have attained a mathematical model for $V(S_1, S_2)$. We have confirmed this model by comparison with the values V constructed from measurements in the simple case of the "bee-purple". We can now use this model to check for our hypothesis regarding the "calibration curve" $W = h(V)$ in the whole space of colour perception; however we are limited by not exactly knowing the sensitivity functions of the three receptor types of the bee.

In order to determine the wavelength discrimination curve of the honeybee, I measured the choice frequency of the bee for more than 50 pairs of monochromatic colours (v. HELVERSEN, 1972). With the assistance of a computer program I calculated the excitation vectors $\vec{E} = (E_1, E_2, E_3)$, using approximated receptor sensitivity curves, and then applied the mathematical model of stimulus distance stated above. I thus obtained values for the stimulus distance between these monochromatic colours and compared them with the values for the measured choice frequency. Fig. 11 shows the result.

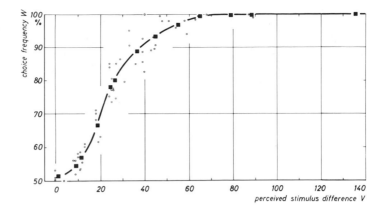

Fig. 11. The "calibration curve" W = h (V) obtained from discrimination measurements between monochromatic colours using a mathematical model of V. (o: measured values of choice frequency, ■ : mean values). Further explanation see text

Regarding the scattering, it should be considered that the experimental inaccuracy is insignificant compared with errors which arise through insufficient knowledge of the sensitivity functions and through the simplifications of the model. Nevertheless, we again obtain a function similar to the one postulated in the beginning.

I believe that the hypothesis W = h (V) is valid for the entire space of colour perception with a universal function h, but that the proposed mathematical model for the stimulus distance is still not complete.

5. Discussion

The function W = h (V) can thus be determined quantitatively (figs. 8, 9, 11). It reaches a saturation level at 100%, using colours and my training method (not described in detail here, v. HELVERSEN, 1972). Using DAUMER's training method saturation is reached at about 95 to 97%. The course is sigmoidal; the point of inflexion is at about 65 to 70%.

With this "calibration curve" we are now able to compare previous solutions. In experiments dealing with the optokinetic reactions of the curculionid beetle Chlorophanus viridis, HASSEN-STEIN (1958) assumed a probit transformation for the translation of turning tendency into frequency of right and left choices on the y-maze globe. JANDER (1968) proposed likewise a probit transformation for problems similar to the one discussed here. In addition, the probit transformation was preceeded by a "mingled-decision transformation" in order to be able to explain saturation values less than 100%. JANDER developed this hypothesis from a model of choice action.

My results show, that the "calibration curve" W = h (V) for the choice behaviour of bees discussed here cannot be a probit transformation because of its sigmoidal course.

It is uncertain how this sigmoidal course is to be interpreted. One could imagine that it is actually produced by a "smeared threshold". Because of statistical variation the threshold is sometimes reached, sometimes not, and we thus obtain the smooth rise. Experimental inaccuracy, of course, also contributes to the fact that no sharp angle can be measured. The question whether the threshold is a characteristic of the differentiating system or of the learning system must remain open.

Regardless of how the curve may be interpreted, I hope to have shown that the "calibration curve" can be determined solely from experimental data and independently of this interpretation. Once one has obtained such a curve, problems can be approached such as the perception of combined trigger mechanisms according to SEITZ's stimulus summation rule (see JANDER, 1968), or a point-for-point construction of the honeybee's space of colour perception, or perhaps even the dimension and metric of perceptual space in pattern recognition, i.e. the number of independently measured form parameters in the honeybee.

References

DAUMER, K.: Reizmetrische Untersuchung des Farbensehens der Bienen. Z. vergl. Physiol. 38, 413 - 478 (1956).
HASSENSTEIN, B.: Die Stärke von optokinetischen Reaktionen auf verschiedene Mustergeschwindigkeiten. Z. f. Naturforsch. 13b, 1 - 6 (1958).
v. HELVERSEN, O.: Zur Empfindlichkeit und Unterschiedsempfindlichkeit der Honigbiene für Licht verschiedener Wellenlänge. (in preparation).
JANDER, R.: Ueber die Ethometrie von Schlüsselreizen, die Theorie der telotaktischen Wahl- handlung und das Potenzprinzip der terminalen Cumulation bei Arthropoden. Z. vergl. Physiol. 59, 319 - 356 (1968).
MENZEL, R.: Untersuchungen zum Erlernen von Spektralfarben durch die Honigbiene. Z. vergl. Physiol. 56, 22 - 62 (1967).
WEHNER, R., LINDAUER, M.: Zur Physiologie des Formensehens bei der Honigbiene. I. Winkel- unterscheidung an vertikal orientierten Streifenmustern. Z. vergl. Physiol. 52, 290 - 324 (1966).

Springer-Verlag
Berlin
Heidelberg
New York

London · München · Paris
Sydney · Tokyo · Wien

Handbook
of Sensory Physiology

Editorial Board: H. Autrum, R. Jung,
W. R. Loewenstein, D. M. MacKay,
H. L. Teuber

Vol. 7, Part 1:

Photochemistry of Vision

Edited by H. J. A. Dartnall
296 figures. XII, 810 pages. 1972

Vol. 7, Part 2:

Physiology
of Photoreceptor Organs

Edited by M. G. F. Fuortes
342 figures
Approx. 800 pages. 1972

Vol. 7, Part 3:

Central Processing
of Visual Information

Edited by R. Jung

Part A: Integrative Functions and
Comparative Data
Approx. 210 figures
Approx. 700 pages. 1973

Part B: Visual Centers in the Brain
215 figures. Approx. 700 pages. 1973

Vol. 7, Part 4:

Visual Psychophysics

Edited by D. Jameson, and
L. M. Hurvich
297 figures. X, 812 pages. 1972

Sir John C. Eccles
The Physiology of
Synapses

101 figures. XII, 316 pages. 1964

Information Processing
in the Nervous System

Proceedings of a Symposium held at
the State University of New York at
Buffalo 21st-24th October, 1968
Edited by K. N. Leibovic
69 figures. XVII, 373 pages. 1969

This symposium volume presents a series
of papers with the objective of correlating
neuronal 'machinery' with psychophysiol-
ogical phenomena. The theoretical and 'experimental' contributions are thus an attempt
to reflect the biological unity, which is
often dismembered conceptually into
anatomy, physiology and psychology.

K. Motokawa
Physiology of Color
and Pattern Vision

137 figures, 26 tables
283 pages. 1970

Published by Igaku Shoin Ltd., Tokyo.
Distribution rights for all countries
excluding East Asia: Springer-Verlag

The author has specialized in sensory
physiology, and his work in this field has
attracted international acclaim. This book
is arranged as a monograph of his contributions to research in this field over
more than 30 years.

H. Hensel
Allgemeine
Sinnesphysiologie

Hautsinne Geschmack, Geruch,
184 Abb. XII, 345 Seiten. 1966
(Lehrbuch der Physiologie in Einzel-
darstellungen).